Rethinking Rights and Regulations

Published in association with the Telecommunications Policy Research Conference

The Internet Upheaval: Raising Questions, Seeking Answers in Communications Policy
edited by Ingo Vogelsang and Benjamin M. Compaine, 2000

Communications Policy in Transition: The Internet and Beyond
edited by Benjamin M. Compaine and Shane Greenstein, 2001

Communications Policy and Information Technology: Promises, Problems, Prospects
edited by Lorrie Faith Cranor and Shane Greenstein, 2002

Rethinking Rights and Regulations: Institutional Responses to New Communication Technologies
edited by Lorrie Faith Cranor and Steven S. Wildman, 2003

Rethinking Rights and Regulations

Institutional Responses to New Communication Technologies

Edited by
Lorrie Faith Cranor and Steven S. Wildman

The MIT Press
Cambridge, Massachusetts
London, England

This book was set in Sabon by The MIT Press and was printed and bound in the United States of America.

Library of Congress Cataloging-in-Publication Data

Research Conference on Information, Communication, and Internet Policy (30th : 2002 : Alexandria, Va.)
Rethinking rights and regulations : institutional responses to new communication technologies / edited by Lorrie Faith Cranor and Steven S. Wildman.
p. cm. — (Telecommunications policy research conference series)
Includes bibliographical references and index.
ISBN 0-262-03314-3 (hc. : alk. paper)
1. Information technology—Social aspects—Congresses. 2. Information technology—Law and legislation—Congresses. 3. Information technology—Management—Congresses. 4. Internet—Government policy—Congresses. 5. Telecommunication policy—Congresses. 6. Information society—Congresses. I. Cranor, Lorrie Faith. II. Wildman, Steven S. III. Title. IV. Series.

HC79.155R47 2002
384—dc21

2003044504

10 9 8 7 6 5 4 3 2 1

Contents

Contributors

Stefan Bechtold University of Tuebingen Law School, Germany

Yochai Benkler New York University School of Law

Colin J. Bennett Professor of Political Science, University of Victoria

Sandra Braman, University of Wisconsin-Milwaukee

Dan L. Burk Oppenheimer, Wolff & Donnelly Professor, University of Minnesota Law School

Robert Cannon Senior Counsel for Internet Issues, Federal Communications Commission

Barbara A. Cherry Deputy Chief, Office of Plans and Policy, Federal Communications Commission

Lorrie Faith Cranor Principal Technical Staff Member, AT&T Labs Research

Avi Goldfarb Assistant Professor of Marketing, Rotman School of Management, University of Toronto

David J. Farber Alfred Fitler Moore Professor, School of Engineering and Applied Science, University of Pennsylvania

Gerald R. Faulhaber Professor, Wharton School, University of Pennsylvania

Chris Forman Assistant Professor, Graduate School of Industrial Administration, Carnegie Mellon University

Shane M. Greenstein The Elinor and Wendall Hobbs Professor, Kellogg School of Management, Northwestern University

Dan Hunter Robert F. Irwin IV Term Assistant Professor of Legal Studies, The Wharton School, University of Pennsylvania

Dennis Karjala Jack E. Brown Professor of Law, Arizona State University

Raul L. Katz Vice President, Booz-Allen & Hamilton, Inc.

Mark A. Lemley Boalt Hall School of Law, University of California at Berkeley

Stephanie Lynch University of Alabama

J. Scott Marcus Federal Communications Commission

Lee W. McKnight Syracuse University Center for Science & Technology

David J. Phillips Assistant Professor of Radio/Television/Film, University of Texas at Austin

Priscilla M. Regan Associate Professor of Public and International Affairs, George Mason University

Burkhard N. Schrage Tufts University, Fletcher School of Law & Diplomac

Paul M. Vaaler Tufts University, Fletcher School of Law & Diplomacy

Steven S. Wildman James H. Quello Professor of Telecommunication Studies, and Director of the James H. and Mary B. Quello Center for Telecommunication Management & Law, Michigan State University

Foreword

It is a time of reflection. We have done the boom. We have done the bust. What remains? Whatever became of the Information Revolution? Were the details of the revolution sold at a liquidation auction or shredded just prior to the investigation? And what has become of the Telecommunications Act of 1996? Are rumors of its death greatly exaggerated?

In 1998 *Wired Magazine* declared that the revolution was over. Four years later and we see that the declaration of victory mistook a preliminary skirmish for the battle. As reflected in the papers at the 2002 Telecommunications Policy Research Conference, now comes the hard questions. Intellectual property papers at the conference questioned the role and ownership of information in the new age. Eli Noam questioned the role of academia on the path from boom to bust. Gene Crick gave a keynote on the plight of community networks in underserved neighborhoods. Dan Hunter reflected on the transition of Internet culture from open network to closed property where exclusion is normative. And the never-ending conundrum of broadband deployment was thoroughly revisited. We are in a period of transition from the open innovative end-to-end Internet of old to an uncertain future where some prophets foretell a network controlled by a few incumbent massive corporations with a different vision of the revolution.

TPRC is a showcase of the best academic research examining communications and information policy. It is an oasis, replacing shallow and tired rhetoric with deep analysis and hard data. The work presented becomes woven into the policy dialogue for years to come. I should know; I am one of those senior government officials whose thinking is annually altered by the excellent work of TPRC.

TPRC is also an annual Herculean sprint of a dedicated program committee divining each new conference and a devoted board of directors with the mission of perpetuating and overseeing the 35-year TPRC tradition. Thanks to the Program Committee: Ed Balkovich, Rand Corporation; Andrew Blau, Flannery Works; Paula Bruening, Center for Democracy and Technology; Barbara Cherry, Federal Communications Commission; Robert Frieden, Pennsylvania State University; Anne Hoag, Pennsylvania State University; Mark Lemley, University of California—Berkeley; Elliot Maxwell, Sam Paltridge, OECD; Bill Rogerson, Northwestern University; and Sharon Strover, University of Texas. A special thanks to Anne Hoag who played a vital role in the success of this year's conference.

Thanks to the board of directors: Benjamin M. Compaine (chair), MIT Program on Internet & Telecoms Convergence; Robert Blau (treasurer), BellSouth D.C.; Marjorie Blumenthal, Computer Science & Telecom Board; Michael Nelson (vice chair), IBM Corporation; W. Russell Neuman, University of Michigan; Adam Clayton Powell III, Powell Communications; Lawrence E. Strickling, and George Vradenburg, AOL Time Warner. A special thanks to the leadership of Ben Compaine, whose efforts were tireless.

There must also be a particular thanks both to the program committee and to the board for giving me this opportunity, tolerating my antics, and rewarding my efforts by volunteering me to chair the conference for yet another year. Additional thanks go to Danya International, Inc., for their tireless effort in administering the conference.

Finally, thanks to the TPRC community. It is an unusual and unique group. Participating in TPRC is like presenting to a community of friendly vultures, always ready to pounce. It is peer review at its best. For the young, TPRC is a resource of mentors with a wealth of knowledge. TPRC is an ongoing discourse in a democratic ideal of beating on ideas until translucent to light.

Long live the dialogue!

Robert Cannon
Chair, TPRC Program Committee 2002

Introduction

There is strong tendency, especially in the popular press, to think and write of new technologies as exogenous forces shaping society. The reality is that the interplay between society and technology is a complex improvisational dance in which each partner responds dynamically to the moves (or anticipated moves) of the other. New technologies are introduced in response to opportunities perceived in preexisting social, economic, and technological relationships. But these opportunities may be enhanced or diminished as affected interests respond; and the perceived opportunities themselves reflect institutional embodiments of societal attitudes toward technology. Because technologies are embedded in social systems and are understood in this context,[1] responses to new technologies may be as varied and complex as the social systems that incorporate (or reject) them. And just as social systems develop over long periods of time, so may societal responses to technologies evolve in stages over time as the long-term implications of new technologies become more evident, and as the social context of technology changes with society itself.

The papers presented in this volume, selected from those presented at the 30th Research Conference on Communication, Information and Internet Policy (TPRC 2002), reflect on and at the same time are reflections of this process of institutional adjustment. Collectively, they address three broad themes associated with institutional adaptations to new technologies: changes in regulatory institutions, changes in the legal rights associated with the use of new technologies and the goods and services they make possible, and attempts to better understand new technologies and their implications for various societal interests.

Evolving Conceptualizations of the New Communications Arena

The first, and perhaps most difficult, challenge to developing policy responses to a new technology is the construction of a conceptual framework to guide policy development. Unfortunately, there is no recipe for framework development. Rather, a new understanding emerges in fits and starts (or misstarts), building on insights drawn from eclectic sources and efforts to resolve the contradictions between predictions based on early conceptualizations and experience that arise inevitably over time. The more revolutionary a technology, the more arduous and protracted must be the process of figuring out just what it is and what it means in the long term.[2] The five chapters in Section I reflect various aspects of the process by which conceptual frameworks are developed, while at the same time making their own contributions to a conceptual framework to guide the development of Internet policy.

Analogies and metaphors based on technologies and institutions with which we are already familiar are natural starting points for developing conceptual frameworks appropriate to new technologies and the new economic and social structures they spawn. While convenient as a starting point and often productive of valuable insight, explaining the new in terms of the old carries with it the danger of obscuring critical differences between old and new technologies that should be reflected in policy responses to new technologies. In chapter 1 Dan Hunter describes how metaphors operate within our cognitive system, and then examines the use of the physical metaphor "cyberspace as place" in understanding online communication environments. He argues that reliance on this metaphor exercises a strong and unrecognized influence on the regulatory regimes of cyberspace and leads to undesirable private control of the previously commons-like Internet. In his chapter 2 companion piece to the Hunter chapter, Mark Lemley focuses on logical steps that he believes courts are missing as they move from metaphor to decision, arguing first that the "space as place" metaphor has significant deficiencies and then suggesting ways that courts might take into account differences between cyberspace and the off-line world to which it is compared.

Implementation of technology policies depends critically on legal definitions of technologies and applications that determine the range of situations to which laws, regulations, and judicial decisions may be applied.

Consistency in policy thus depends critically on definitional consistency. Definitional consistency might also be used as an index of the extent to which a consensus understanding of a technology and its context has emerged. From this perspective, Robert Cannon's examination of legal attempts to define the Internet in chapter 3 suggests that our understanding of the Internet may still be in an early formative stage. While the term Internet appears frequently in legislation and legal decisions, he shows that there is little consistency and often disagreement in the definitions employed. He asks whether attempts to define the Internet for legal purposes may be fundamentally misguided and offers a foundational exploration of what a good understanding of the Internet might look like.

Sometimes the attempt to understand a new technology reveals hitherto unappreciated features of older technologies. The ongoing controversy over institutions created to govern the Internet Domain Name System highlights the critical role that domain name allocation plays in Internet governance. In chapter 4, Stephan Bechtold defines a namespace as "the collection of all names in a particular system" and argues that namespaces are both ubiquitous and an overlooked facet of governance, whether in real space or in cyberspace. Based on his study of a variety of namespaces, including the Internet Domain Name System, IP addresses, ENUM, peer-to-peer systems and instant messaging, he concludes that technical control over a namespace creates levers for political influence and regulatory control. He develops a taxonomic structure to guide debates about the legal and policy implications of various namespaces and to inform choices among options that arise in their design.

A critical aspect of the understanding of any new technology is who uses it and why. In chapter 5, Forman, Goldfarb and Greenstein present the findings of the first census of the dispersion of Internet technology to commercial establishments in the United States. Defining "participation" as the use of the Internet as a requirement for doing business and "enhancement" as the adoption of Internet technology to enhance computing processes for competitive advantage, they focus on answering questions about economy-wide outcomes, such as variation in participation and enhancement rates among industries, and any differences in participation and enhancement that may be apparent in comparisons of cities and states. They conclude that dispersion patterns for participation differ from those for enhancement and that there are geographic differences in dispersion patterns as well. While participation has approached saturation rates in most industries, enhancement rates are lower and vary with long standing differences in computer usage among industries. They find, however, that the geographic distribution of industries across geographic regions explains much of the regional variation in dispersion. They also find that commercial Internet use is more widely dispersed than previous studies have shown.

The Evolution of Legal Rights

Legal rights set limits on the permissible range of actions for political and economic actors—in some cases by reserving rights to engage in specified actions to certain types of actors, in other cases by forbidding certain actors from engaging in specified activities. Economic rights are commonly evaluated in terms of their effectiveness in creating conditions conducive to the effective exploitation of economic opportunities, including those presented by new technologies. Political rights may be viewed as playing an analogous role with respect to promoting effective political governance; but independent of concerns with the quality of governance, they may also guarantee vital liberties, the rights to which may be viewed as self-evident and inalienable, as in the U.S. Declaration of Independence.

Technological change poses challenges to systems of legal rights that are both philosophical and economic. The challenge to economic rights comes from new and altered sets of economic opportunities, especially when the use of new technologies threatens established interests. We see this, for example, with peer-to-peer file sharing technologies, which make the transfer of recorded audio and video files easier and less expensive for users, but at the same time may diminish the ability of authors and artists to collect payment for the value users derive from their creative endeavors, not to mention the profits realized in conventional distribution channels. New opportunities and threats may call for new or redefined rights to achieve that balance of contending interests that would effectively promote the larger social good. Thus, economic rights should evolve with technology. Each of the chapters in this section illustrates how technological change may force a fundamental rethinking of rights.

The clash of contending interests created by new technologies is perhaps nowhere more evident than in the current debate over proposals that would change dramatically the way that uses of the electromagnetic spectrum for communication are governed. Owing to the long-term trend of declining cost and size of microprocessors accompanied by dramatically increased computational power, many now argue that more efficient wireless services could be built by reallocating processing power from network centers to the devices at their peripheries. This would allow peripheral devices to both coordinate with and communicate directly with neighboring devices. A potential benefit of such coordination would be much more efficient utilization of spectrum, possibly by allowing services currently relegated to separate spectrum bands to jointly use commonly available spectrum. Such an allocation scheme is incompatible with the current dominant model of assigning specific units of spectrum to designated licensees for their exclusive use and would depend on the development of a new set of rights to govern spectrum use. Chapters 6 and 7 take us to the heart of this fascinating debate over spectrum rights and regulation.

In chapter 6 Yochai Benkler describes the technological developments at the heart of the current wireless policy debate and explains why these developments may favor a system of open wireless networks. However, implementation of such a system would require a dramatic revision of the current system of spectrum property rights. He develops an informal model for evaluating the comparative efficiencies of the two property rights regimes, and concludes with a set of proposals for a multi-year market test of the two approaches that would help us select an appropriate mix of the two.

While Benkler focuses on the tradeoffs involved in selecting one type of spectrum management regime over another, Gerald Faulhaber and David Farber approach the debate over appropriate property right regimes for spectrum from a very different perspective in chapter 7, asking whether property rights in spectrum might be designed to permit the simultaneous co-existence of different spectrum management regimes. They argue that the positioning of different spectrum management regimes as alternatives creates a false dichotomy, when in fact a system of rights might be designed that would allow the simultaneous existence of markets in spectrum with well-defined rights of exclusivity while allowing for the rapid introduction of new radio technologies under somewhat restricted open access conditions. They describe two such systems. One would give individuals the right to buy and sell designated spectrum with a noninterference right while granting other users a right to use the same spectrum subject to the restriction that they not interfere with users of the first type. The second system would grant exclusive rights to designated frequencies while allowing those holding the exclusive rights to permit shared uses of their frequencies subject to payment of sharing fees. They describe institutional arrangements that might arise to coordinate uses in each system.

The distinction in intellectual property law between patent and copyright has become increasingly unclear in the context of computer programs, which encompass both intangible creative expressions that have traditionally been accorded copyright protection and tangible productivity innovations traditionally protected under patent. In chapter 8 Dennis Karjala proposes that functionality be employed as a principle that may be used to define a boundary between patent and copyright that will promote effective application of these two bodies of law.

Sandra Braman and Stephany Lynch explore the intersection of commercial rights and constitutionally protected political rights in their analysis of ISP terms of service in chapter 9. They show that ISP terms of service and acceptable use policies include numerous rules that forbid constitutionally protected content and behaviors and, in many cases, grant ISPs licenses to all content uploaded or posted. They explore both economic and legal responses to what they argue is a tendency of ISPs to continually add new restrictions on users.

In chapter 10 Dan Burke presents a good example of how policy responses to new technologies must often evolve over time. He examines the anticirumvention provisions of the Digital Millennium Copyright Act—dubbed "paracopyright"—and argues that they are ripe for abuse. He explains that these provisions effectively grant copyright holders sweeping new ability to impose terms of access on content users: consumers who access content without accepting the content owner's terms would violate the owner's paracopyright even if the material accessed is not itself copyrighted or copyrightable. Additionally, where a particular use would be permissible under copyright law, content owners may be able to exclude the use as a condition of access. Content owners may use paracopyright to require purchase or use of related products; for example, DVD access controls require that the disc be played on approved hardware, effectively dictating the consumer's purchase of playback equipment. At some point, Burk argues, such leveraging of access control seems certain to overstep the bounds militated by sound policy or intended by Congress. Thus the statute has critical flaws that have become increasingly evident over time. To address these deficiencies he proposes that improper leveraging of paracopyright should be curtailed by application of the misuse doctrine, guided by the standards established in its previous applications to patent and copyright law.

Regulatory Innovation and Responses to Technological Change

Rights influence behavior by setting bounds on what is permissible. Regulation involves substantially more direct involvement of government in the control of behavior of private entities, possibly relying on fairly direct oversight by an expert agency with some discretion over interpretation of legislative mandates and the responsibility to implement what are often vague prescriptions or statements of intent in enabling legislation. Regulatory authorities must therefore constantly adjust their plans and reinterpret enabling statutes as technological change alters the industries over which they have oversight responsibility.

Barbara Cherry's study of liability regimes for common carriers examines an intersection of rights and regulation, and thus is an appropriate transition from section II's focus on legal rights to the predominant focus on regulation of section III. Cherry describes in chapter 11 how the liability regime for telecommunications carriers is shifting from one based on an absolute limit on liability in tariffs to a form of common law strict liability. The FCC contributed to this process when it instituted mandatory detariffing for interexchange carriers in 1996 and announced that with detariffing carriers would no longer be able to unilaterally limit their liability for damages. She argues that this claim was based on flawed reliance on earlier judicial decisions incorrectly viewed as supporting the position that filed rates were the legal basis for limited liability contracts, and furthermore, that the FCC failed to consider the implications of eliminating the strict limits on liability that had prevailed for more than a century for other public policy goals, such as universal service and broadband deployment. The 1996 detariffing order was itself in part a response to changed industry conditions attributable in no small part to technology driven changes in underlying industry economics. Cherry's study demonstrates the need to think in larger system terms when modifying or replacing existing regulatory frameworks.

Until fairly recently, the emergency response systems in the United States and Canada relied largely on wireline connections. Growing penetration of wireless telephone services has created an opportunity to increase the coverage of the emergency response system through the integration of wireless carriers, and efforts are underway in both countries to incorporate wireless systems into the existing emergency response framework, in what is called "wireless enhanced 9-1-1," or E–9-1-1. In chapter 12, Priscilla Regan, Colin Bennett and David Phillips compare implementation efforts in three regions in the United States and Canada, and conclude that several trends are driving technologies and organizational configurations in directions that might conflict with traditional goals of privacy protection.

Transfer from government to private ownership (or the reverse) is probably the most dramatic form of institutional change affecting communications industries. The last two decades have witnessed a strong world-wide trend toward the privatization of state-owned telecommunications carriers. How this has been accomplished has varied substantially among countries. In chapter 13 Lee McKnight, Paul Vaaler, Burkhard Schrage, and Raul Katz write about how they took advantage of this natural experiment to test two competing models of the effect of residual state ownership interests on the performance of privatized firms. What they term the mainstream model predicts shareholder returns increase as residual state ownership interests decrease and more time elapses from the commencement of privatization. The predictions of the alternative model are just the opposite, with shareholder returns increasing with the degree of state ownership and decreasing with time from privatization. The authors' event studies provide only weak support for the mainstream hypothesis, limited mainly to privatized carriers from industrialized countries, but fairly strong support for the alternative hypothesis, particularly for carriers privatized in emerging market economies. This supports the argument for the alternative hypothesis that in some countries a nontrivial state interest in the performance of privatized carriers may be necessary to ensure against subsequent state appropriation. These findings suggest that the larger institutional environment in which privatization (or other institutional reforms) occur has a substantial influence on the outcomes.

Communication industries have traditionally been defined by fairly tight pairings of services and technologies, such as television with broadcasting and telephony with wireline copper networks. For the most part, the regulatory frameworks that evolved in different countries over time took these pairings for granted and contained rules that differed substantially among industries. With convergence, the old tight pairings of technologies with services are breaking down. This raises problems of sustainability for affected industries and inconsistency in the effects of regulations applied in similar circumstances in different industries. In the final chapter, J. Scott Marcus examines regulatory responses to convergence in the E.U. and draws lessons from that experience for the development of policy in the United States.

Conclusion

The impact of new technologies might be measured in a variety of ways, but one particularly revealing index of impact is the degree to which the societies into which they are introduced are forced to make adjustments. Incremental improvements on old technologies and new technologies that are fairly straightforward substitutes for (or complements to) old technologies can be handled with fairly modest alterations to existing institutions. Revolutionary technologies call for more fundamental institutional innovations. Collectively, the research presented in this volume suggests that major institutional changes may be necessary to best respond to advances in communication technology. However, the response process may be slow, as policy makers are still coming to terms with how to conceptualize the new technology. In addition, major institutional changes will have far reaching effects and many of those potentially affected may oppose such changes. Nonetheless, if the new technologies currently challenging policy makers are to have a truly revolutionary impact, ultimately we can expect to see fundamental institutional changes in response.

Notes

1. See, e.g., Carolyn Marvin (1988), *When Old Technologies Were New: Thinking About Electric Communication in the Late Nineteenth Century* (New York: Oxford University Press).

2. For example, more that 150 years after its invention, scholars were still discussing the economic implications of the telegraph. See, e.g., Yates, J (1986), "The Telegraph's Effect on Nineteenth Century Markets and Firms," *Business and Economic History,* Second Series, 15:149–63.

I

Evolving Conceptualizations of the New Communications Arena

1

Cyberspace as Place

Dan Hunter

In the early days of computer networks it seemed a slightly far-fetched metaphor to describe . . . sites as "places," since bandwidth was narrow. . . . As bandwidth burgeons and computing muscle continues to grow, cyberspace places will present themselves in increasingly multisensory and engaging ways. . . . We will not just look *at* them; we will feel present *in* them.[1]

It is not too portentous to say that we stand at the fork between two possible futures of intellectual endeavor. Down one road lies a future of completely propertized and privatized ownership of intellectual activity. Down the other road is a future where the interests of society at large are fostered, which at times leads to private ownership of intellectual activity, and other times demands that some public intellectual space be kept in commons for all.[2]

This observation has been made by others within the spheres of intellectual property rights such as copyright and patent. However, the focus on intellectual property interests masks the area where the trend to property in ideas has its most pernicious effect: on the Internet. Cyberspace was once thought to be the modern equivalent of the Western Frontier. It was a place, albeit an abstract place, where land was free for the taking, where explorers could roam, and communities could form with their own rules. It was an endless expanse of space: open, free, replete with possibility. No longer. As with the Western Frontier, settlers have entered this new land, charted the territory, fenced off their own little claim, and erected "No Trespassing" signs. Cyberspace is being sub-divided. Suburbs and SUVs cannot be far off.

Since cyberspace appears to be like a place this trend seems preordained: the progression of property interests over the last five hundred years shows that places tend to be enclosed, and privately exploited.

Nonetheless, it is a surprising trend because legal commentators have convinced us that cyberspace is not a place at all. However, these commentators have confused the descriptive question of whether we think of cyberspace as a place, with the normative question of whether we should regulate cyberspace as a regime independent of national laws. Whatever the answer to the normative question, there is significant evidence that, purely as a descriptive observation, we do think of cyberspace as a place. Cognitive science investigations into how people think provide ample evidence of this.

Thinking of cyberspace as a place has led judges, legislators and legal scholars to apply our physical assumptions about property in this new, abstract space. Owners of Internet resources start to think of their systems as their own little claim in cyberspace, which must be protected against the typical encroachments we find in the physical property world. This has lead to a series of cases and statutes that enshrine the idea of these property interests in cyberspace.

The effect of this is to take the hitherto commons-like character of cyberspace, and splinter it into millions of tiny landholdings. Privatization is not, of itself, a problem: private interests are the dominant forms of resource allocation in our world. However, modern property theorists have demonstrated the dangers of splintering interests: the undesirable consequence is to create "anticommons property." Anticommons property emerges where multiple people hold rights of exclusion to a property such that no one has an effective right of use. As a result, a "tragedy of the anticommons" occurs, where property is locked into sub-optimal and wasteful uses, because the preclusion rights-holders block the best use of the resource.

This chapter suggests that thinking of cyberspace as a place, and the consequent legal propertization of cyberspace, is leading us to a tragedy of the digital anticommons. Recent laws and decisions are creating millions of splintered rights in cyberspace, and these rights are destroying the commons-like character of the Internet. If we continue down the fork we currently are traveling we risk creating a digital anticommons that has the potential to limit many of the innovations we have created to date. Historians will look back on our time and wonder—when we have seen what the Internet could be—how we could have sat and watched as the tragedy of the digital anticommons occurred.

Space forbids a full development of the necessary conditions proving my argument. Specifically, I don't present any argument rebutting the legal assumption that cyberspace is not a place, and Part I merely sketches how metaphor works in structuring our understanding of regulatory regimes. Part II explores the evidence that the specific structuring metaphor of CYBERSPACE AS PLACE[3] operates on us all, in both lay discussion of online interactions, and in legal analysis. Part III leads on to the very disturbing implications of this metaphor. If we accept the CYBERSPACE AS PLACE metaphor, then it is a very short step to assume that there is some kind of property online, and that this property should be privately owned, parceled out, and exploited. Part IV demonstrates how private interests are reducing the public ownership of, and public access to, ideas and information in the online world. As we all come to stake out our little claim in cyberspace, the commons that is cyberspace is being destroyed.

I. Thinking in Metaphors

Metaphors are more evocative, and conjure up more associations, than their purely literal counterparts. But is this all they do? Are they merely rhetorical "flourishes," which may leaven language, but which are not vital to the way that we express or even conceive ideas? For a very long time, this was the dominant view in philosophy, and linguistics. At its most charitable, this view suggested that metaphor was sometimes useful, but never essential, to the way we express ideas. The view was also held that metaphor was not necessary for us actually to *think* our ideas.

In the last few decades, linguists, philosophers and cognitive scientists have suggested that metaphor is more central to language and thought than the prior conception would have it. Within philosophical and cognitive science circles, metaphor has been rehabilitated, and become the subject of great interest. Metaphor studies have burgeoned, and now form a major category of study in the philosophy of language, in linguistics and in cognitive science.

One of the most important theories has been that of Lakoff and his colleagues. Their influential theory has a number of distinguishing features, but it is most notable for its assertion that our everyday concepts are structured and molded by a series of cognitive metaphors that all human beings share. Their cognitive metaphors are not linguistic metaphors like Shakespeare's "Juliet is the sun" or Justice Cardozo's "trackless ocean of law." Instead, they look to general conceptual metaphors that organize our thinking. These metaphors break down into various types, most of which stem from our physical experiences as humans in the world. These conceptual cognitive metaphors are reflected in linguistic utterances; that is, what we would normally think of as a "metaphor."

Lakoff suggests that we can excavate our underlying conceptual metaphorical structures, by a close examination of our use of language: "Since metaphorical expressions in our language are tied to metaphorical concepts in a systematic way, we can use metaphorical linguistic expressions to study the nature of metaphorical concepts and to gain an understanding of the metaphorical nature of our activities."[4]

An example may make the approach clearer. According to Lakoff, we all hold a communal, cognitive conceptual metaphor that ARGUMENT IS WAR. We see this metaphorical structure reflected in language thus:

- Your claims are *indefensible.*
- He attacked every weak point in my argument.
- His criticisms were *right on target.*
- He *shot down* all of my arguments.

Lakoff is not saying that we actually view arguments *exactly* as war. Rather, he suggests that features from the source domain (WAR) are mapped onto the target (ARGUMENT). Thus, we cognitively structure our perception of arguments in terms that adopt elements of war; and this structuring is reflected in our use of language.

The next part of this chapter is devoted to the proposition that we see the same thing happening in cyberspace regulation. We all hold a cognitive metaphor of CYBERSPACE AS PLACE and this leads to a series of metaphorical inferences: cyberspace is like the physical world, it can be "zoned," trespassed upon, interfered with, and divided up into a series of small landholdings that are just like real world property holdings.

II. Cyberspace as (Legal) Place

The Lakoffian view of metaphor explains why the language that we use to describe our experience of the Web is a reflection of the underlying conceptual metaphor of CYBERSPACE AS PLACE. Adopting the metaphor means that we understand the medium of internet communication as having certain spatial characteristics from our real world experience. Telling in this regard are the references that have been variously used to describe the "consensual hallucination" of life on the Internet. Consider the way in which everyone talks about events, transactions, and systems that exist or occur online. At its most fundamental, consider the term *web*, an allusion to the "web-like" connections between computers. Then there is the *Net*, referring to the network of connections, and also to the net-like character of the material caught in the network. We *surf* this *web, moving* from one *site* to the next, *entering* or *visiting* the site, or *homepages*. We *hang out in chatrooms* communicating with our *online* buddies. We *roam around* multi-user *dungeons and domains* (MUDs and MOOs). Software programs called *robots*, or *agents*, or *spiders*, are allowed to *crawl* over websites, except where they barred by terms and conditions of *entry* or *access*, or by the robot *exclusion* standard. We *navigate* the *web*, using computer programs with names like *Navigator* and *Explorer*. We use Uniform Resource *Locators* and *domain* names, to find our way.

Information is sent to us using hypertext *transport* protocol (http) or simple mail *transport* protocol. We use email *addresses* to send messages to others, and the machines themselves use IP *addresses*. We log *into* or log *onto* our Internet Service Provider. Malignant wrongdoers *access* our accounts, hack *into* the system, using *backdoors, trapdoors* or stolen *keys*, and engage in computer *trespasses*.

The point here is not to document exhaustively all of the evidence that we hold a conception that cyberspace is a place. Rather, this is merely a short demonstration that the language we use in discussing cyberspace is shot through with physical references and implications. All of these examples take the physical characteristics of the real world and map them onto the abstract virtual world.

This process, of mapping the real onto the virtual, is pervasive in legal academic discourse, in judicial pronouncements, and in legislative enactment. Lakoff explains how we cannot help but see abstract concepts in physical terms such as this, and so we should expect that lawyers would adopt the CYBERSPACE AS PLACE metaphor. However, we need to recognize what the adoption of the metaphor entails. As Lakoff and Black have shown, the metaphor carries with it the implication complex that constrains the sorts of outcomes we can expect. I will later argue that adoption and use of the CYBERSPACE AS PLACE metaphor means that we begin to see cyberspace as land that may be fenced off and privatized, but now we need better evidence that the metaphor is present in law.

Crime Online

Laws against computer crime represent one of the earliest areas where the conception of the network as a place emerged. In fact the most obvious example, the Computer Fraud and Abuse Act of 1986 (CFAA), relied on the spatial characteristics of computers and networks before the Internet was in the minds of legislators. The offenses in the Act all proscribe access to computers and networks. For example, §1030(5)(A)(ii) provides that whoever "intentionally accesses a protected computer without authorization, and as a result of such conduct, recklessly causes damage" above a specified value, is guilty of an offense. Other sections deal with the special cases of unauthorized access of government computer systems or information from government agencies, unauthorized access to obtain credit card information or credit reports, amongst a slew of other crimes stemming from the trespass. However all offenses share the basic requirements of unauthorized access of the computer—to deal with outsiders hacking into a system—or accessing a computer beyond the limits of authorization—to deal with insiders who have access to the system, but who move beyond the scope of their authorization. The mere access of the system without any damage will, generally, be sufficient to invoke the Act.

These types of offenses are generically labeled "computer trespasses," and it is not hard to understand why. The computer is "accessed" without permission, implying the illegal entry into the system. Not for nothing do we see movie depictions of hackers announcing triumphantly "I'm in" when they access the system. They may use a "backdoor" to enter the system, or just bypass (i.e., "pass by") the security. Authorized users type a word that allows them passage into the system, hence the name "password." Though we might bemoan the inexactness of the use of the expression "accessing" a computer, it is an obvious example of linguistic usage reflecting a deeper physical metaphor of the COMPUTER AS PLACE.

It is not surprising then that, when the CFAA was applied to internet-connected computers, the initial COMPUTER AS PLACE metaphor would be supplanted by the CYBERSPACE AS PLACE metaphor. This latter metaphor has emerged as the dominant way to understand computer trespasses over networks. The network provides the means of moving around, and the servers, hosts, or websites within the network are the access points. So, for example, an Internet dating service was entitled under the CFAA to a temporary restraining order, prohibiting a former programmer from accessing the dating service's website via the Internet, and diverting its clients and users to a pornography site. The idea of CYBERSPACE AS PLACE can be seen clearly. The Net is conceived in familiar terms, just like the public roads that lead to private properties, on which the defendant trespasses.

The CYBERSPACE AS PLACE metaphor has lead to the application of the CFAA to problematic uses of publicly accessible websites, in circumstances that are a long way from the original criminal scope of the CFAA. In a series of cases involving unsolicited bulk email (the "spam cases") and companies downloading information from competitor's websites (the "website cases"), courts have used the CYBERSPACE AS PLACE metaphor, and extended the scope of the CFAA beyond recognition.

In the spam cases, providers of email services prevailed over spammers who harvested email addresses from them. In each case, the courts had no difficulty in applying laws directed at criminal action to the civil subject matter before it. The most interesting issue was how the spammer's access was forbidden, since computer trespass requires "unauthorized access" of the computer system. One might imagine that email, even unsolicited bulk email, could not amount to an unauthorized access. Yet the courts found that the spammer did not have permission to undertake this type of "access" of the email system, since the terms of use of the providers' systems specifically forbade access for this purpose.

Spam is such an egregious intrusion of online space, and such a pernicious evil, that we should hardly be surprised or troubled that courts would extend criminal liability to this type of activity. It is hard to feel sorry for spammers. We might hope, however, that courts would limit the CFAA to these sorts of outrageous activities. This is not so, however. In the website cases, courts have been exceedingly swift in applying criminal sanctions to activities that can only be characterized as competitive, but hardly criminal, practices. In *Register.com v. Verio*,[5] the practice involved Verio's "scraping" of WHOIS data from Register.com's website. Register.com, in keeping with any organization that is granted the right to register domain names, is obliged to publish the WHOIS records of the domain names it registers. A WHOIS record details the name, address, and contact information of the domain name holder, as well as certain technical data about the domain name. Register.com published the WHOIS records of its registrants on its publicly accessible website. Verio, a competitor in the domain name registration and website hosting business, collected these records from Register.com, for domain names that had been recently registered. Verio then used this information to contact the domain name registrants, offering them various website hosting services.

At no point did Verio hack into Register.com's database: the information was all gathered from the publicly available website. The sum total of Verio's "access" of Register.com's computer system was a series of well-formed http requests to Register.com's webserver, which Register.com made available to the web at-large. Moreover, the data that Register.com sent in response to these http requests was information that, under the terms of its agreement with ICANN, it was obliged to make public. These facts notwithstanding, the court had no qualms holding Verio guilty of a computer trespass under Title 18. Verio had accessed the computer system of Register.com, and the access was clearly unauthorized, since Register.com maintains a "Terms of Use" policy for their website. These terms, tucked away on a separate page from the default page, forbade anyone from using an automated search mechanism to poll the Register.com WHOIS database. The court's analysis was straightforward: "Because Register.com objects to Verio's use of search robots they represent an unauthorized access to the WHOIS database." That is, the simple objection to access provides the basis for the criminal liability. This is true even though Verio was only making a request on a publicly accessible database. Verio exceeded the terms of use since all bots were forbidden from making requests on the site. The court therefore concluded that Verio was therefore criminally liable under CFAA §1030(a).

"Terms of Access" are the most obvious way that a competitor may render a rival's access to its website to be "unauthorized." However, in *EF v. Explorica*,[6] the court considered whether a confidentiality agreement could provide the same limitation on access. The data in this case were pricing, routing and associated information about vacation packages offered by EF, a specialist in cultural and student tours. The information was publicly available on EF's website. Explorica was set up to compete in the student cultural and language vacation market, and Explorica commissioned a programmer to write a routine that "scraped" all of the information from EF's publicly accessible site. Explorica had knowledge of EF's fare structure, gathered from the ex-employee. It was therefore a simple task to write an automated query agent that—as with the agent in *Verio*—sent a series of requests to EF's webserver and collected the responses in a database maintained by Explorica. Armed with this data, Explorica could offer similar or identical tours to EF, and, having collected EF's pricing structure, could systematically undercut EF. EF sought a preliminary injunction to stop Explorica's actions.

There was no question that Explorica had accessed anything other than information that was publicly available. The District Court concluded however that placing material on the web did not affect the analysis of whether the access was unauthorized, concluding that EF's copyright notice "should have dispelled any notion a reasonable person may have had that the 'presumption of open access' applied to information on EF's website." Since the information is almost certainly not copyright, and since website proprietors plaster copyright notices about with unthinking abandon, it is hard to credit that this alone is sufficient to make the access unauthorized. This is especially the case since the copyright notice did not speak to the issue of access. The Court of Appeal did not rely on this as the touchstone of authorization. Instead, it looked to the unusual fact that EF's ex-employee, now a vice president of Explorica, had signed a confidentiality agreement when an employee of EF. This agreement included a clause to the effect that the employee would not disclose any confidential information to the outside world. As a result, the court concluded Explorica's subsequent access was unauthorized, even though it is quite clear that the website information was in the public domain and therefore, by definition, not confidential.

In all of the cases, the information was publicly accessible, but it was subject to some type of limitation on the uses that this information might be put. Typically this limitation was found in a "terms of use" document, available somewhere on the website, or provided via a clickwrap agreement, though other bases of limitation were present in unusual circumstances. Whatever the basis for the limitation, courts have generally found unauthorized access for the purposes of the CFAA where the cyberspace owner simply objects to access.

The CYBERSPACE AS PLACE metaphor explains why the results in these cases are anything but surprising, even if, as I will argue shortly, they have dire implications. Cyberspace is a place that conforms to our understanding of the real world, with private spaces such as websites, email servers, and fileservers, connected by the public thoroughfares of the network connections. Viewed through the filter of the CYBERSPACE AS PLACE metaphor, computer trespass does not just involve an infringement on my right to use the personal property of my computer system. Instead, the action becomes a trespass against a form of quasi-land that exists online. Trespasses to land have always been considered to be more serious than the equivalent actions against personal property: for example, the action lies for the most trivial trespass to land, whereas trespasses to chattels have always required serious damage.

Anyone who enters my space without authorization is a trespasser. Private spaces may allow people entry under certain conditions, which may be posted on the door, or otherwise communicated to them. This has its cyber-equivalent: conditions of entry on the door are just like the Terms of Use of the website, or information contained in the robot exclusion file.[7] Other mechanisms of removing permission would include direct email contact or requirements in confidentiality agreements as seen in *EF v. Explorica*. As a result of the CYBERSPACE AS PLACE metaphor, anyone using my website appears to be "entering" my place, and should be treated just like an invitee at common law. If the invitee oversteps the bounds of their invitation then they become trespassers, and will be subject to the full force of the criminal law. Courts have had few qualms about applying the metaphor to reach this conclusion.

Laws proscribing computer and network trespass are the obvious starting point for the examination of the legal application of CYBERSPACE AS PLACE. Recently, however, we have seen an example within torts, with the resurrection of the moribund tort of trespass to chattels, and its wholesale application to the Internet. This tort is very similar to the criminal computer trespass material covered here, and indeed courts have applied the two as though they were interchangeable.

Resurrecting Trespass to Chattels

If courts have been speedy in accepting CYBERSPACE AS PLACE within the CFAA, then their reaction to it within tort law has been positively jaw-dropping. Notwithstanding scholars' early suggestions of its inappropriateness, courts have rushed to resurrect the late, largely unlamented, tort of trespass to chattels, and apply it to the new cyberspace arena. The rise of this extraordinary development can be explained by the CYBERSPACE AS PLACE metaphor.

Perhaps the leading case in this area is *eBay v. Bidder's Edge*. eBay, the major web-based auction site, created a huge following for its online auctions. It was, and remains, the major player in online auctions. Bidder's Edge ran a website, "AuctionWatch.com," which aggregated the auction details of many online auction sites, including eBay's. The user was therefore able to see, on one AuctionWatch.com screen, all of the auction sites where a particular type of product was being auctioned, rather than having to search the many auction sites available. In order to provide this service, Bidder's Edge sent out "bots," or automated software query agents, to inquire (or "crawl") on all of the auction sites about the auctions that were being conducted. Similar to the situation in *Verio*, these queries were standard http requests that are individually indistinguishable from a human making the same request. The information from eBay's site was reported back to Bidder's Edge and collected in its database.

eBay was unhappy about these automated enquiries, ostensibly because the requests from the bots placed a heavy load on its servers. It therefore sought to enjoin Bidder's Edge from using its bots to send requests to the eBay site. eBay argued that the trespass to chattels tort should be revived and applied to this new scenario, and the court concurred.

eBay's most striking claim involved an analogy between the real world and the virtual one: "eBay's allegations of harm are based, in part, on the argument that [Bidder's Edge]'s activities should be thought of as equivalent to sending in an army of 100,000 robots a day to check the prices in a competitor's store." This analogy relies directly on the CYBERSPACE AS PLACE metaphor: the idea that real-world physical robots can be mapped onto virtual software "bots" involves an acceptance that the virtual world is a space that can be roamed like the real one. The court, however, did not accept the argument, but not because it rejected the CYBERSPACE AS PLACE metaphor. Rather the court enthusiastically accepted the metaphor, but even after accepting it rejected the argument because the harm was *de minimis*.

The court did, however, grant a preliminary injunction on another of eBay's claims. It concluded that irreparable harm would flow to eBay if the crawling continued. The basis of the court's decision was that Bidder's Edge was engaged in a trespass to chattels, a tort similar to the more familiar tort of conversion, but involving a lesser degree of annexation of the personal property. The court outlined the necessary elements a plaintiff must establish for a chattel trespass claim in relation to computer systems: "(1) defendant intentionally and without authorization interfered with plaintiff's possessory interest in the computer system; and (2) defendant's unauthorized use proximately resulted in damage to plaintiff."[8]

Prior to eBay, there were a number of lesser-known cases on cyberspace chattel trespass, where the issue was whether the sending of unsolicited bulk email via a free email system involved a chattel trespass to the provider of the system. Courts in these cases readily adopted this idea, no doubt as it amounts to one of the few ways to fight the scourge of spam in the absence of legislative assistance. We might have hoped it would be limited to these situations, however, after the *eBay* decision courts have enthusiastically embraced the expansion of the tort, and have been applying it whole-heartedly to situations well beyond the narrow confines of the initial circumstances of spam. In both of the *Verio* and *Explorica* decisions, trespass to chattels was adopted by the courts in addition to the criminal computer trespass provisions discussed above. Most recently, in *Intel v. Hamidi*[9] the California courts have extended the principle to prohibit regular email sent to a corporation that has requested that the email not be sent. Ken Hamidi, a disgruntled ex-employee of Intel Corporation, sent a small number of emails to all employees of Intel, complaining of various injustices. The court focused on the large number of employees this extended to, rather than the character or number of unique emails, and applied the approach of the earlier trespass to chattels cases to this new scenario.

Though some courts have declined the application of this new tort to specific situations—typically where the use of the computer system was *de minimis*—the evolution of the trespass to chattels action, from spam, through web sites, and culminating in regular email, demonstrates the importance of the CYBERSPACE AS PLACE metaphor. There are three pieces of evidence that the metaphor is operating in this new field of cyberspace tort. First, there is the nature of the chattel that is allegedly trespassed upon. Recall that it is the plaintiff's *personal* property that is supposed to be the subject of the tort. Initially, in *eBay* the court did focus on the personal property of the computer or the "computer system." However, later in the decision, the court's analysis magically expanded to suggest the chattel included the plaintiff's bandwidth and server capacity.

The cases struggle with the question of what exactly is the chattel in issue. At times the courts suggest that the chattel is simply the computer, but more often it is a nonspecific combination of computer, bandwidth, capacity, processing power, or network. With the exception of the computer itself, none of these "chattels" are actually chattels at all. There is no private property in bandwidth or processing power or network. It seems here that courts are trying to reconcile the CYBERSPACE AS PLACE metaphor with the personal property basis of the chattel trespass theory. At times the personal property feature emerges, as when the focus is placed upon the computer itself. Other times, the notion of the computer as an aspect of place emerges, as when the focus is placed on the spatial characteristics such as the plaintiff's bandwidth, capacity, processing, or network.

There is a second piece of evidence supporting the argument that CYBERSPACE AS PLACE is used in trespass to chattels. Contrary to the typical conception of torts to personal property, courts in the cyberspace arena have ignored the damage requirement of the tort. When the trespass to chattels action was first mooted as applicable to the cyberspace arena, the consensus was that plaintiff would fail for want of appropriate damage to the chattel in question. As a result, learned commentary suggested that plaintiffs would lose in the cyberspace world. Of course, the opposite is true. In most cases, the courts have ignored the damage requirement or been extremely flexible in determining what damage is sufficient. The conclusion seems to be therefore that the tort is much more like trespass to real property, since real property trespasses have always been considered more serious and as a result the infringement *per se* is actionable without proof of damage. As Dan Burk has concluded, by ignoring the harm requirement the courts that have developed trespass to chattels in the arena "essentially reversed several hundred years of legal evolution, collapsing the separate doctrines of trespass to land and trespass to chattels back into their single common law progenitor, the action for trespass."[10] The reason for this is, I think, that because of the metaphor personal property seems much more like real property. And hence the two different causes of action become conflated.

Then there is the final piece of evidence that CYBERSPACE AS PLACE rules in this old-made-new tort. The language that the courts use often unconsciously reveals how they are thinking about the action. This is, of course, exactly what the Lakoffian metaphor theory teaches us, and the examples in this area are particularly striking. Courts forget that they are supposed to be talking about personal property, that is, a thing, and describe the defendant's actions as though they were trespassing on a place. For example, in *Hamidi*, when the court dismissed defendant's First Amendment claim it noted that the ACLU's amicus brief: "cites cases which confer First Amendment protection in private tort actions, but they differ from the present case in that Hamidi was enjoined from trespassing *onto Intel's private property*."[11] One can only trespass *on* land or a cyberspace equivalent. The court should have said that Hamidi trespassed *against* Intel's personal property, or some other language that indicated that the chattel was misappropriated or abused. Instead the court clearly had the real property action in mind when it dismissed the First Amendment claim. Indeed, this is bolstered when one views the cases it cited in support of the conclusion, most of which involved the posting of material in real property locations such as shopping centers, hardware stores, and so forth. Earlier it had characterized Hamidi's actions as "*invading*" Intel's "internal, proprietary e-mail system," and characterized Hamidi's use of the system as "*entry*" into their system. All of these examples show how the court was conceiving the chattels-based tort in real property terms.

The question then is: So what?

III. The Digital Anticommons

From the fifteenth century onward, land holding in England changed profoundly. Property held by a number of people in common was appropriated in various ways to the exclusive possession of powerful gentry. This was called the "Enclosure Movement" after the fencing and enclosing of the commons by these new private landholders. That most fundamental of property rights, the right to exclude, was used to alter the default position of land tenure from commons property to private property.

Recently we have seen a similar process occurring within intellectual property. Commentators have begun noting the increasing private control of what previously had been intellectual commons property. James Boyle has called this the "Second Enclosure Movement,"[12] and he and other scholars have detailed the enclosure movement within intellectual property and the concomitant erosion of the public domain. The enclosure of the intellectual commons takes many forms: ongoing term extensions for copyright (such that few works have moved from copyright into the public domain for decades), scope extensions for patents, new intellectual property rights for hitherto unprotected collections of facts, the erosion of fair use, and the rise of digital rights management systems. The trend is so obvious that it is no longer confined to earnest scholarly musings, and courts are now being asked to consider the problem.

If intellectual property is the subject of the Second Enclosure Movement, then the Internet is the subject of a related trend, which we might term the "Cyberspace Enclosure Movement." This particular enclosure movement began when online actors, who cheerfully adopted the benefits of the online commons, decided to stake out their own little claims in cyberspace, and used the law to fence off their little cyber-holding, keep out intruders, and "privatized" what once had all the characteristics of a commons.

The previous parts of this chapter have detailed the way in which we think of cyberspace as though it were a place. In law, places become property, and so the next section describes the property-based analysis that is used in regulating cyberspace. This may not initially appear troubling. Private ownership of resources of itself is not problematic; indeed private ownership is generally considered to be the most efficient form of allocation of property resources, and the economic history of the last five hundred years has been characterized by the movement from the public to the private. The quintessential exemplar of the benefits of private ownership is, of course, Garrett Hardin's tragedy of the commons: public resources are overused and destroyed where there is no private property interest in limiting the use. However, as Michael Heller has recently demonstrated, private ownership can lead to the opposite of the tragedy of the commons: the tragedy of the anticommons. Anticommons property occurs when multiple parties have an effective right to preclude others from using a given resource, and as a result no-one has an effective right of use.

The tragedy of the anticommons occurs when these multiple rights of preclusion lead to inefficient underuse of the resource. I argue that this is precisely where the CYBERSPACE AS PLACE metaphor leads. The rise of the Cyberspace Enclosure Movement means that we are headed to a digital anticommons, where no-one will be allowed to access competitors' cyberspace "assets" without licensing access, or other agreeing to some other transactionally expensive permission mechanism.

The Cyberspace Enclosure Movement

If we think of cyberspace as a place, then the legal response is to impose a real property-based regulatory structure on the place. Moreover, because our real-world property system is based on private land tenure, the legal assumption is to use property mechanism to delineate and fence-off these new property entitlements in cyberspace.

However, the Internet initially assumed a number of commons-like features. Indeed, without these commons characteristics, the Internet would not be as we understand it today. Take the most fundamental process of information transfer. There is no centralized server that arranges transfer of packets in the system, and so the transfer of data from one computer to another is entirely dependent on many computers voluntarily transferring packets onto the next machine in the path. This process is called peering, and until recently was performed for free, as a matter of network etiquette and a recognition that the commons benefit of the network was dependent on this process. A similar process initially occurred with email transport. Until the advent of spam, many email servers would maintain an "open relay" for email. The relay provided a means for transfer of email messages from systems that did not have the resources to provide email to its users, or was unable to accommodate the Internet's email protocol. In essence, the email relay server donated their processor and bandwidth to systems less fortunate than themselves. Donating resources was found elsewhere: as Lessig notes, the University of Chicago (circa 1995) allowed anyone to jack into their network and use the Internet. A more recent example is the proliferation of wireless Internet access points, many of which provide their access for free. Free access, free relay, and free peering—for the common benefit of all. Commons property assumptions defined the early architecture of the Net.

Consider then the various services and protocols that work at above the fundamental architecture, starting with the masterstroke that defines the web: the ability to link to every other website. The links are a free and creative endeavor of themselves, and have generated tremendous network externalities. More than this, many individual websites adopt the commons mentality: Project Gutenberg scans and places public domain texts on the web for all to use, Project Perseus translates Ancient Greek and Latin classic texts into English and posts them. Napster, Gnutella, Morpheus, and Kazaa encourage the wholesale sharing of files, much to the anger of the music industry and others.

Any number of examples exists. Free and shared resources, created for the betterment of all, are the online norm, not the exception—so much so that online content companies bemoan the "gift economy" of the Net, and the difficulty of getting anyone to pay for the digital content they want to sell. However, this digital commons is under attack. The Cyberspace Enclosure Movement threatens to privatize out of existence much of the commons character of the network.

Property and Invitees

Let us say that I am a consumer retailer, or a parking garage operator, or an accountant. I own, or via lease have a right of exclusive possession over, the premises where I do business. Of course I want you to come into my premises in order to shop, or park your car, or engage me to shred documents. However I do not want you to have complete freedom of access to the premises, and so I designate you an invitee or licensee, who is entitled to enter the premises under certain conditions. These conditions might include terms that you not steal the stock in the shop, or that you not sue me for scratching your car, or that you not stage a sit-in in my office. I post these conditions on the front door of the premises, so that you can see them and be advised of the conditions of your invitation into the premises.

The basic framework therefore is this: your exclusion from my place is guaranteed by property law, and the terms of your entry into my place are governed by contract law.

As detailed above, this is exactly what we see in cyberspace. The trespass to chattels and computer trespass actions applied to cyberspace operate using precisely this framework. You are forbidden from entering the cyberspace place, except upon conditions that have been set by the proprietor of that space. Sometimes the space involved is a website, sometimes it is an email system. Sometimes the conditions are set using

Terms Of Use of the site, other times they are set by robot exclusion headers, confidentiality agreements, or a letter from a lawyer. However, the approach is the same. You are forbidden from entering my cyberplace unless you agree to my terms. If you access my place in defiance of my terms then you lose your invitee status, and become a trespasser, subject to both civil and criminal action.

Unlike the terms of entry in physical establishments, cyberspace Terms of Use are often extraordinarily broad, and grant rights to the proprietor that are remarkable. The owners of the filesharing system Kazaa, Sharman Systems, outraged many when it was revealed that by downloading the software they had agreed to allow Sharman to turn on their computers and use them in a massive peer-to-peer network. By using Microsoft products and websites you have—at one time or another—(1) agreed to allow them to scan the contents of your PC and download whatever software it deems necessary, (2) agreed not to abuse or flame anyone, (3) given them a license to use your email in any way it sees fit, and (4) agreed not to use their products to create any "objectionable" material.

Some commercial operators of websites go so far as to post Terms of Use that forbid the quintessential web mechanism of establishing a hypertext link to their site, except for certain narrow purposes. For example, the accounting firm KPMG claimed that only those who had negotiated a web-linking agreement with them could include a link to their site. KPMG's lawyers threatened legal action against a commentary site that had been critical of KPMG if it did not remove the link to kpmg.com.

Of course, every site has unique terms, so that users cannot "enter" sites with a reasonable understanding of what to expect, based on experience. The legal expectation is that every user will read every term of use, no matter how complex, no matter how hard-to-find. And users must do this every time they go to an online place.

The Consequences

The enclosure of cyberspace represents a fundamental change in the way the Internet operates. Rules of property to exclude, and rules of contract to provide a limited form of entry, leads to an extraordinary series of splintered interests in cyberspace. Hitherto, cyberspace has flourished because the default rule has been to allow common access and use of the resources. Now, we see the emergence of a default rule of exclusion. This unduly simplifies the situation however, since in fact the proliferation of unique terms of use leads to a situation where there is no default rule at all, but a series of unique rules for access. This gives rise to an explosion of different rights of access and use. The section that follows articulates the theoretical reason why this is a terrible consequence. Before turning to the theory, it is worth considering here a number of very practical implications of the Cyberspace Enclosure Movement.

First, searching the Net will become more difficult and less complete. The search engines that index the Net and the Web will be severely constrained, for two reasons. The "owner" of an indexed site can stop search engines that rely on any kind of competitive business model. These business models might be as simple as placing competitor's advertisements on the page where the indexed site is listed. The site can draft terms of use forbidding this particular type of access. Even if this were not to happen, the costs of assessing terms of use, or other "de-authorizing" device will be prohibitive. We can say goodbye to new types of search engines that affect—in any way—the business models of the sites that they index.

We can also kiss goodbye to aggregation products that were a boon to consumers, and one of the more interesting features of the dot com boom. AuctionWatch was a better product for consumers than eBay, since it covered more auctions. However, eBay succeeded in shutting it down using the trespass to chattels tort. Any type of innovative aggregation product is subject to the same problem. The same is true for comparison shopping agents, which find the most competitive price for a given product. Why should Amazon or CDNow allow comparison agents to "invade" their sites and index them for the benefit of consumers, and not themselves?

Then there is the issue of email. *Hamidi* makes it a tort to email an computer where the proprietor of that system has indicated that it does not wish your email. Does this mean that one is obliged to read the "Terms of Acceptable Email Usage" of every email system that one sends email to in the course of an ordinary day? If the University of Pennsylvania has a policy that sending a joke by email is an unauthorized use of their system, then under the current enclosure movement you have "trespassed" on their system when you email me a Calvin and Hobbes cartoon.

These are just some of the practical problems that the Cyberspace Enclosure Movement leads to. The next section argues that these practical problems are a consequence of a more general theoretical concern. The enclosure movement leads to a tragedy of the digital anticommons.

The Tragedy of the Digital Anticommons

Every first year law student knows Garrett Hardin's "tragedy of the commons." A resource will suffer the tragedy of the commons where it is prone to overuse because too many owners have a right to use the resource, and no one has the right to exclude any other. The exemplars include fisheries that suffer from over-fishing, fields that are over-grazed, forests that are over-logged, and so forth. The tragedy of the commons is among the most compelling arguments given in favor of private ownership of resources, and against forms of commons or state ownership.

Until a short time ago the tragedy of the commons was the only tragedy in town. However, Michael Heller has recently introduced the concept of the "tragedy of the anticommons," and systematically explicated its effect. The tragedy of the anticommons is, in most ways, the mirror image of the tragedy of the commons. Anticommons property exists where multiple owners have a right to exclude others from a scarce resource, and no one has an effective privilege of use. Heller's great insight was not in theorizing of the existence of the tragedy of the anticommons as a theoretical opposite of the commons. Others had already suggested that anticommons property might exist in theory. However, prior anticommons theories relied on the idea that—since this was the exact opposite of the tragedy of the commons—for anticommons property to exist *everyone* in the world must have a right to preclude. Given such difficult preconditions, theorists were hard-pressed to identify a real world correlate, and therefore take the argument further.

Heller's dual contributions were to show how a limited number of exclusory rights would be sufficient to generate anticommons property, and perhaps most important, to provide copious real world examples of anticommons property. His initial example was found on the streets of post-communist Moscow: large numbers of shops stood vacant while vendors hawked their wares from flimsy kiosks lined up in front. Why did these vendors stand around in the cold when they might use the shops immediately behind them? The answer was in the complex series of entitlements to those shops that had been created in the transition to a market economy. There was such a complex hierarchy of divided and coordinated rights that, effectively, no one was able to exploit the resource. There was always someone who could object to the use, or holdout for the entire value of the resource. Once he observed the anticommons in action on the Moscow street, he was able to find other examples that previously had been ignored by the literature, because the concept of the anticommons simply had not existed. A telling example was in the post-earthquake reconstruction of Kobe, Japan. Years after the earthquake, notwithstanding billions in aid, large tracts of Kobe remained in rubble. The reason for this was a "world class" tangle of property interests: "In one block of Kobe, over 300 renters, lessees, landowners, and subletters own often-overlapping claims,"[13] and each one had to agree before rebuilding could begin.

The metaphor of CYBERSPACE AS PLACE, and the enclosure movement that uses the metaphor is leading us to a digital anticommons. The "property" in issue is not individual websites or email systems, but rather the commons property of the network resources: the web or the email system that we all used to share. We used to enjoy a general and untrammeled "right" of access to websites, email systems, fileservers, and so forth. The Cyberspace Enclosure Movement is leading to a default principle of exclusion, with a billion unique terms providing the exceptions governing when we can "enter" these cyber-places. The splintering of rights of access is like the overlapping rights on the Moscow street. We do not have a right to access the commons property any more.

As an example, consider website terms-of-use. With a series of permission rules, it is necessary for me to determine every time I enter a website whether the particular use I make is legal. In a world of zero transaction costs this would not turn the commons into an anticommons. However where transaction costs are real, and, as in the case of reading and understanding long tracts of legalese, where these costs are extremely high, no one has an effective right of use, since the cost outweighs any conceivable benefit. The same is true for email, or any new protocol that the Net can support: the old commons property can easily be transformed into anticommons property.

Until Heller formulated his theory of the anticommons, there had been some literature on the anticommons as the exact symmetrical opposite of the commons. Since the commons was defined as every member having a right to use, the assumption was that the anticommons could only come into existence if every member had the right to exclude. Since "member" in this context meant any person, the requirement was thought to mean that an anticommons would only occur if, and only if, every single human being could preclude other uses. This meant that, practically, the anticommons could never exist. Heller redefined the anticommons to occur where multiple persons (but not everyone) had a right to exclude others such that no one has an effective (as opposed to perfect) right to use. This makes it possible to see that a small number of people may effectively block the best use by others. Heller and Rebecca Eisenberg have argued that this is occurring within the arena of biomedical patents, most notably in the patenting of gene fragments. These fragments may be patented before researchers have identified any corresponding gene, protein, biological function, or potential commercial product. However it is likely that subsequent valuable products, such as therapeutic proteins or gene-based diagnostic tests, will almost certainly require use of multiple fragments. Thus, they argue, a small number of the early owners of gene patents can create an anticommons for all. The same is true for the digital anticommons. It does not take a large number of enclosed cyberspace places to effectively create the digital anticommons. As it stands, the law upholds the right to enclose and create these new forms of private property, on terms dictated by the proprietor's attorneys. The diligent user of the network now must take account of these new entry rules, and consequently the transactions costs for all use rises dramatically. It will not take too many more cases for us to see a significant change in the online behavior of the users. At this point, even though we have only a small number of rights-holders blocking the uses, I believe that we will see the emergence of the anticommons.

This observation leads to the penultimate point for the digital anticommons. Consider again the example of Heller and Eisenberg: patenting individual gene fragments before we understand their use fully (or at all). They note that the anticommons may be real without us being aware of its existence. Empty Moscow shopfronts advertise the existence of the anticommons: indeed were it not so obvious, Heller's theory might never have been born. The anticommons in gene patents is not obvious, because it is impossible for us to know what innovative new commercial product would be developed if the commons did not exist. We are not able to combine the gene fragments in novel ways, because the anticommons owners make it impossible to do so. Hence, the existence of the anticommons precludes the better use of the resource, and at the same time masks the recognition that there might be a better use at all.

The above discussion points to the gradual emergence of the digital anticommons. It is happening slowly, and so we might be tempted to wait, to see what eventuates, before making any broad policy reforms. This is a mistake. Once anticommons property is manufactured, it gets locked in its sub-optimal use pattern. "Once anticommons property is created, markets or governments may have difficulty in assembling rights into usable bundles. After initial entitlements are set, institutions and interests coalesce around them, with the result that the path to private property may be blocked and scarce resources may be wasted."[14] Anticommons property becomes stuck in its low value use, wasting resources, with no prospect of reassembling into higher value use.

Since it is so difficult to put the genie back in the bottle, I suggest that we need to address the anticommons before it emerges completely. In the fuller version of this chapter I discuss how we might try to stop the development of the anticommons, but conclude pessimistically that it will be harder than we might imagine.[15] Elsewhere in this book Mark Lemley suggests that there are appropriate strategies,[16] and his more optimistic conclusions are perhaps the best response. I leave it to him to make out this case.

IV. Conclusion

Joseph Singer explains that property is one of the strongest ordering systems we have in society. The kind of property we have determines much of the society we will have; therefore the social life we want should determine the type of property we admit. In the real world, this means, for example, that we choose to enact and enforce public accommodations statutes because we cannot condone racial segregation. "[P]roperty systems form the overall social context in which individuals live. They describe the limits of allowable social relations and channel interaction into certain patterns."[17]

In the online world, these property interests are even more plastic than those in the real world. We therefore have the opportunity to determine first what sort of online environment we want, and then (and only then) choose what sort of legal regime should apply to it. The Cyberspace Enclosure Movement threatens to reverse this process, by forcing our real world property assumptions on the online environment. However, as described above, these assumptions are unnecessary, harmful, and wrong.

In 1992, the Internet was opened up for commercial exploitation. Relying on the public character of the Net, and the vast public commons that was created before they ever arrived, commercial operators have grown exceedingly fat. They now have successfully exploited the CYBERSPACE AS PLACE metaphor, and convinced judges that it is appropriate to carve out remarkable new property rights online. By tiny, almost imperceptible steps, they are enclosing cyberspace. They have mounted a campaign that has eroded the public commons that the Net has been, and they threaten to create a genuine digital anticommon.

We have been lucky. We have witnessed an unprecedented decade of innovation on the Net. This innovation has flourished in part because of the dot com bubble, but more importantly because of the commons that the Net has provided, and the opportunity that this presents. The Cyberspace Enclosure Movement, dependent on the CYBERSPACE AS PLACE metaphor, has not yet closed this off completely. However, if the current approach is not challenged, then little stands between us and the digital anticommons. The intractable characteristic of anticommons will emerge: low value uses beat out high value ones, and it is almost impossible to change state. We will not be able to rebundle the various commons interests that we once shared. The opportunity will be lost forever.

We may already be past the point where we can do anything about this. I hope we are still a little way off. But unless we do something about it, as we all come to stake out our little claim in cyberspace, the commons that is cyberspace will be destroyed.

Notes

1. William J. Mitchell, *City of Bits—Space, Place and the Infobahn*, 114–115 (1995).

2. A fully argued version of this chapter is to appear as Dan Hunter, *Cyberspace as Place, and the Tragedy of the Digital Anticommons*, 91 CALIF. L. REV. ___ (2003, forthcoming).

3. The capitals used henceforth for the metaphor "CYBERSPACE AS PLACE" are to distinguish it as a conceptual metaphor in the tradition of Lakoff and others. See below.

4. George Lakoff and Mark Johnson, *Metaphors We Live By*, 7 (1980).

5. 126 F.Supp.2d 238 (S.D.N.Y. 2000)

6. 274 F.3d 577 (1st Cir 2001).

7. http://info.webcrawler.com/mak.projects/robots/norobots.html

8. 100 F.Supp.2d 1058, 1069–70. (N.D.Cal 2000).

9. 114 Cal. Rptr. 2d 244, 250 (C.A.1. 2001).

10. Dan L. Burk, *The Trouble with Trespass* (2000) 4 J. SMALL & EMERGING BUS. L. 27, 33.

11. 14 Cal. Rptr. 2d 244, 254 (C.A.1. 2001). (emphasis added)

12. James Boyle, *The Second Enclosure Movement and the Construction of the Public Domain*, at http://www.law.duke.edu/pd/papers/boyle.pdf

13. Michael A. Heller, *The Tragedy of the Anticommons: Property in the Transition from Marx to Markets*, 111 HARV. L. REV. 621, 664 (1998).

14. *Id.* at 659

15. Hunter, *supra*.

16. Mark Lemley, *Place and Cyberspace*.

17. Joseph William Singer, *Entitlement: The Paradoxes of Property*, 146, (2000)

2

Place and Cyberspace[1]

Mark A. Lemley[2]

Introduction

For cyberlibertarians, the other shoe is rapidly dropping. In a curious inversion, those who argued less than a decade ago that cyberspace was a place all its own—and therefore unregulable by territorial governments[3]—are finding their arguments and assumptions used for a very different end. Instead of concluding that cyberspace is outside of the physical world, courts are increasingly using the metaphor of cyberspace as a "place" to justify application of traditional laws governing real property to this new medium. Dan Hunter's excellent article[4] explains how and why this is happening with uncanny accuracy, pointing to the power of metaphor in influencing legal thinking and the particular strength of metaphor in making the new seem familiar.[5] He also quite correctly observes that reliance on the cyberspace as place metaphor is leading courts to results that are nothing short of disastrous as a matter of public policy.[6] Finally, he concludes that there is no way for the Internet to escape the firmly entrenched spatial metaphor, either by substituting another metaphor or by eschewing metaphor altogether. Already, he concludes, the idea of cyberspace as a place is too well established in our minds.[7] The result is a paper that is both extraordinarily important and profoundly depressing.

In this chapter, I do not challenge Hunter's argument that the cyberspace as place metaphor is rampant, nor his conclusion that judicial use of the metaphor has had pernicious consequences. Rather, I focus on the logical steps that courts seep to be missing as they move from metaphor to decision. In part I, I explain why the cyberspace as place metaphor is not a particularly good one. In part II, I suggest some ways courts might take account of these differences between the real world and the Internet. In part III, I observe that even if one accepts the place metaphor *in toto*, it need not follow that everything in this new place must be privately owned. Nor must it follow that private ownership rights include complete rights of exclusion. My conclusion is somewhat more optimistic than Hunter's. While acknowledging the dangers of the cyberspace as place metaphor and the fact that courts have already started down the wrong road, I suggest that courts and commentators who think seriously about the nature of the Internet still have ample room to make reasoned policy decisions. Though we may easily be misled by metaphor, we need not be its slaves.

I. "The Internet Is Just Like a Place"

A. No, It Isn't

We speak of the Internet in spatial terms,[8] and in certain respects users may experience some aspects of the Internet as a location. But even a moment's reflection will reveal that the analogy between the Internet and a physical place is not particularly strong.

As a technical matter, of course, the idea that the Internet is literally a place in which people travel is not only wrong but faintly ludicrous. No one is "in" cyberspace. The Internet is merely a simple computer protocol—a piece of code that permits computer users to transmit data between their computers using existing telephone networks. There were computer networks before the Internet that similarly relied on telephonic exchange of data.[9] The real genius of the Internet was that its simple, end-to-end design allowed many different people to write different programs that everyone could use by adopting the same simple communications protocol.[10] Technologically, anyone can transmit data onto the network. Whether that data arrives at a particular location depends on whether the user at the destination has configured her computer to accept data in that particular form and from that particular source. If so, the data—email, a request to download information, an mp3 file, or a virus—is read by the recipient computer. But regardless of the form the data takes, it is data and not people who are traveling.[11] Data have been traveling on wires and through the airwaves for centuries at the behest of humans, but no one thinks the television, the telegraph or the telephone is "places" within which people travel.[12]

The idea of cyberspace as a physical place is all the more curious because the instantiation that most resembles travel to the casual user, the World Wide Web, is in fact much more like a traditional communications medium. People may speak of "visiting" Web sites, but of course they don't do any such thing. They send a request for information to the provider of the Web site, and the provider sends back data: the Web page itself. Because this process is automated, and because most Web site owners have decided in advance to make their Web pages available to anyone, it may appear that one has simply gotten to a page by clicking a button. But in fact the page is voluntarily sent by its creator to the Web user, a fact that should have significance in a number of the Internet property cases.[13]

The cyberspace as place metaphor seems instead to act at some more conceptual level. Even if we understand somewhere in the back of our minds that we are not really going anywhere, perhaps when we access the Internet it seems so much like we are in a different physical space that for all intents and purposes we accept cyberspace as a "real" place. There are two problems with this description. The first is that most users of the Internet surely do not experience it as anything remotely resembling a real place. While William Gibson wrote beautifully of the visual representations of computer code in cyberspace,[14] it is probably no accident that he did his writing on a manual typewriter.[15] Despite the utopian dreams of early cyberspace "pioneers,"[16] the Internet is hardly replete with black ICE palaces[17] and visual data structures. It is a medium that transmits mostly text, images, and more recently sounds, just as television does. People may speak occasionally of being "lost in" or "transported" by a television show, a movie, or even a book, but we hardly surrender our understanding that "television space" is merely a series of images being transmitted to us. Nor do we think that when catalogs or letters are delivered to our door, we have magically entered into a store or a friend's house.

Second, what's really different about the Internet is interconnection: the fact that links can "take" us from one Web page to another. We aren't limited to visiting the stores in our town or reading only the mail that comes to us. With a click of the mouse, we can see information offered on billions of Web pages by millions of people and companies from all over the world. Further, we can move from a page in Switzerland to one in Swaziland merely by following a link. Perhaps it is this automatic connection to pages that come from distant lands that makes us feel as though we are traveling through cyberspace. But if so, it is surely the supreme irony of the cyberspatial metaphor. For it is precisely this automatic interconnection between data offered by different people in different places that makes the Internet so *different* from the real world. And indeed, it is this very interconnection that courts using the cyberspace as place metaphor threaten to eliminate by treating the Internet "just like" the physical world.[18] In short, we may instinctively feel that cyberspace is a place, perhaps because we long to make new things seem familiar. But there is no particular reason why we should think this way. Certainly it is not the way the Internet really works.

B. It's Like a Place, Except . . .

Perhaps Hunter is right, and we have already conditioned ourselves to think of the Internet in spatial terms to such an extent that there is no going back.[19] Even if this is true, it does not follow that we must blindly accept a one-for-one correspondence between cyberspace and the real world. There are obvious differences between the way things work in the real world and the way they work online. Here are just a few examples:

- While in the physical world I can occupy only one place at a time, on the Internet I—or at least my data—can be everywhere at once (and indeed it is often hard to avoid doing so).
- Physical stores have spatial constraints that limit the number of customers who can enter the store. While there are some constraints on simultaneous usage of a Website or the Internet itself, for most users and for most purposes bandwidth is effectively infinite.
- Physical places exist in proximity to one another, and human senses can perceive what is happening next door. In cyberspace, by contrast, there is no "next door." Nor is there a public street or sidewalk from which one might observe behavior that occurs in a particular Internet space.
- The Internet consists only of information, and information is a public good. A Web site is trivial to copy, and copying it does not deprive its creator of the use of the original site.[20] By contrast, chattels are much harder to copy, and real property is by definition impossible to duplicate. In order to make use of someone else's real property, I would have to deprive them of some control over it.

We may turn to the cyberspatial metaphor out of familiarity or ignorance, or even because we consciously decide it resembles in certain ways what the Internet is or should be. But it is implausible to argue that the Internet is "just like" the physical world. At most, the Internet is like the physical world except in certain respects in which it is different.

II. Getting to "Except"

Courts can and should take the differences between the Internet and the real world into account.[21] They can do so without rejecting the metaphor of cyberspace as place, simply by using the metaphor as a point of departure. But the departure—the recognition that the Internet is not just like the real world, and that the ways in which it is different may matter to the outcome of cases—is critical.

A. Courts Misled by Metaphor

Hunter correctly points out that a number of courts have made the mistake of overlooking the differences between the Internet and real space in a variety of contexts. In particular, courts applying the doctrine of trespass to chattels to email and Web site access[22] have shown a remarkable lack of sensitivity to these differences. As Dan Burk has observed, while these courts nominally apply the doctrine of trespass to chattels to cyberspace, they are in fact using the different and more expansive doctrine of trespass to real property.[23] They ban third parties from "entering" a Web site without permission, sometimes on the grounds that the third party will fill up the site,[24] sometimes because they assume that Internet bandwidth, like real property, should be inherently inviolate.[25] An even more serious problem is the judicial application of the Computer Fraud and Abuse Act,[26] designed to punish malicious hackers, to make it illegal (indeed, criminal) to seek information from a publicly available Web site if doing so would violate the terms of a "browsewrap" license.[27]

These courts have failed to understand how the Internet is different from the physical world. They have not understood that no one "enters" Web sites. Rather, defendants in these cases merely sent requests for information to a Web server the plaintiff had itself made open to the public, and the plaintiff's own server sent information in return.[28] They have not understood that the requests for information that Verio or Bidder's Edge sent did not exclude others from using the site. They have not understood that cases of this sort were really efforts to control the flow of information to or from a site. Because they had land rather than information in mind, these courts forgot that the information at issue in these cases is a public good to which we have never applied the "inviolability" rules of real property.[29] The courts did not understand these things, and so they got the cases wrong, creating a general tort of stunning breadth.

B. Courts That Understand the Limits of Metaphor

But courts could understand these things—could get the cases right—even within the framework of the cyberspatial metaphor. In other contexts, courts have proven receptive to the idea that Internet law can both rely on a framework designed for the real world and yet modify that framework to take account of the peculiarities of cyberspace.

Personal jurisdiction is one area where courts have demonstrated such agility. Rote application of personal jurisdiction rules and the metaphors of the physical world would lead inexorably to the conclusion that anyone who puts up a Website is amenable to suit anywhere on the planet, on the theory that they have sent their "products" into each and every forum. While a few early cases took that position,[30] most courts quickly recognized its failings.[31] A number of courts developed an Internet-specific "interactivity" test for jurisdiction: passive Websites didn't confer jurisdiction wherever viewed, but interactive Web sites did.[32] This test has its problems, and courts have started to move away from it. Courts seem instead to be moving toward a test that uses traditional standards for determining personal jurisdiction, but applies them with sensitivity to the nature of the Internet, recognizing that not every Web site is necessarily a purposeful availment of the benefits of every forum state.[33] Thus, acceptance of the traditional due process framework for personal jurisdiction has not prevented courts from considering the practical differences between the Internet and the world that existed before its advent.

A second example concerns dormant commerce clause jurisprudence. The Supreme Court has repeatedly held that states are not free to regulate interstate commerce in a way that imposes undue and potentially conflicting burdens on those who sell products or services nationwide.[34] Most dormant commerce clause cases in the physical world have focused on intentional state efforts to burden out-of-state providers in order to benefit local ones.[35] On the Internet, by contrast, courts have applied the dormant commerce clause somewhat differently. On the one hand, courts are more likely to invalidate state regulation of the Internet under the dormant commerce clause, because the inherently interstate nature of Internet communications burdens a larger class of people with understanding and complying with a multitude of regulations.[36] In fact, because a single Web page is accessible in all 50 states, the burden of complying with inconsistent regulations is often greater than it is in the physical world.[37] On the other hand, because Internet communication is more malleable than physical goods are, it may be easier to comply with some sorts of state regulations. Courts have taken this ease of compliance into account in determining that state anti-spam statutes do not violate the dormant commerce clause,[38] while striking down other statutes that reach too broadly.[39] They have, in short, adapted a constitutional doctrine based on the physical world to accommodate the somewhat different character of the Internet.

Freedom of speech is the third area where courts have demonstrated their ability to adapt to the virtual world. Government regulation of indecent speech on the Internet has been especially fertile ground for metaphoric debates. When Congress passed the Communications Decency Act in 1996, it sought to preclude Internet sites not only from disseminating constitutionally unprotected obscene speech, but also speech that was merely "indecent," "lewd" or "harmful to minors."[40] Courts were asked to choose among analogies: was the Internet more like a bookstore or library, in which indecent speech must be permitted, or like a radio or television broadcast in which otherwise protected speech could be forbidden?[41] In *Reno v. ACLU*,[42] the Court concluded that the Internet was deserving of full First Amendment protection, not the lesser protection afforded broadcast media. In so doing, the Court considered how well each metaphor actually applied to the Internet. It distinguished the Internet from broadcast media on five grounds: first, in broadcast, warnings could not protect listeners and viewers from undesired content, whereas on the Internet such warnings were effective; second, the risk of encountering indecent material by mistake online was remote; third, unlike the broadcast spectrum, Internet sites were not scarce; fourth, there was no extensive history of Government regulation over the Internet; and fifth, the Internet did not share broadcast's invasive qualities.[43]

The Internet was not entirely like either a bookstore or a television station, but the Court was able to use both metaphors as points of departure. It considered how the particular characteristics of existing media were thought to justify different regulatory regimes, and compared the characteristics of the Internet to determine what level of regulation should be permitted in what was clearly a new medium. Further, Justice O'Connor's concurrence suggested that the Court was sensitive not only to how the Internet differed from any of the existing media offered as analogies,[44] but also to how the nature of the Internet might change over time in ways that affected its regulability.[45]

In all of these cases, courts have avoided becoming prisoners of the metaphors they used. They have demonstrated their ability to adapt laws and metaphors constructed with the physical world in mind to take account of the rather different world of the Internet. In short, the brief history of Internet law to date suggests that courts have the ability to escape the confines of a metaphor when they need to. The problem is therefore a more limited one than Hunter suggests: so far, courts have simply not succeeded in doing so where the cyberspace as place metaphor is concerned. Hunter is right to worry about the consequences of this failure, but I believe he is wrong to suggest that it is global or inevitable.

III. The Equation of Place and Property

The Internet trespass cases make another mistaken conceptual leap. Courts have assumed not only that cyberspace is a place akin to the physical world, but further that any such place must be privately owned by someone who has total control over the property. This is a common assumption these days; it sometimes seems as though our legal system is obsessed with the idea that anything with value must be owned by someone.[46] But as any scholar of real property will tell you, not all land is privately owned, and even land that is privately owned frequently does not fall totally within the dominion of the owner. To reach the results in cases like *eBay v. Bidder's Edge*,[47] *Register.com v. Verio*[48] and *Intel v. Hamidi*,[49] a court must conclude that a particular type of property law is appropriate for the Internet.

A. The Internet and Public Space

While we often think of physical space as being privately owned, much of it is not. Our society could not exist without abundant public space. Not only would we be poorer if there were no parks, no wilderness, and no public libraries and museums, but the market economy would grind to a halt without the constant support provided by roads, bridges, airports, and the other infrastructure of modern government. And life as we know it would be impossible if we did not reserve the air and water as a public commons.[50]

Public spaces sometimes provide a subsidy to the poor: anyone can enter a city park, while a private garden would exist only if it could charge enough to be self-supporting. More importantly for our purposes, public infrastructure serves a vital economic purpose. Roads, lighthouses, and indeed our system of government have some of the characteristics of a public good: it is difficult to exclude nonpaying users.[51] Further, coordination is important to the efficient use of this infrastructure. While it is possible to imagine privatizing all the public infrastructure in our economy, from roads to a postal service to jails to courts,[52] it is not likely that the result will be both convenient and competitive. Imagine paying a different toll every time you turned a corner, navigating roads that only worked with certain types of cars, and living with police forces that obeyed no limits except those set by a private employer. For such a world to be theoretically efficient, we would have to have competitive choices in each of these areas. But the very fact of that competition would create dramatic inefficiencies in production and require a breathtaking number of transactions to "clear" the rights necessary to live one's life. We have public space in the physical world not by accident, or because it is left over space that no one wants, but because it is a necessary part of a functioning system of property.[53]

So too with the Internet. Even the staunchest advocates of propertization on the Internet tend to take for granted all sorts of public "spaces" online. We assume that telephone companies will pass our data along on a nondiscriminatory basis,[54] even as we deregulate telephony[55] and resist any kind of nondiscrimination obligation for cable modems.[56] We assume that no one will own the top level domain names that we all use to communicate, even as we grant property rights in second-level domains.[57] We assume that the protocols that make up the Internet are free for the world to use, even as we permit the patenting of improvements to those protocols.[58] And we assume that search engines and other data collectors will enable us to cull information from the vast archive we have collectively created, even as we begin to impose liability on search engines for finding things we don't like.[59] In short, we rely on public "space" on the Internet, just as we do in the real world. Indeed, even at this early stage in the Internet's development, the public accessibility of its key features is so deeply ingrained that we simply take it for granted. Only when that accessibility is under attack do we even become aware of the baseline assumption of openness.

The cyberspatial metaphor, then, does not determine whether something will be privately owned. Courts that apply the metaphor still have a choice to make: is this the sort of space that should be public or private? On the Internet, there are good reasons to think that the balance should be tilted in favor of public space in many contexts. The economic rationale underlying much privatization of land, the tragedy of the commons,[60] simply does not apply to information goods. It is possible to imagine physical bandwidth or server capacity being overconsumed, though the danger of that currently seems remote.[61] But it is not possible to imagine overconsumption of a nonrivalrous thing like data. My use of your data does not deplete it or prevent your use in the same way that my use of your land might.[62] From an economic perspective, the more people who can use information, the better.[63]

Further, some of the differences between the Internet and the real world—notably the absence of physical proximity online—suggest that we should be more worried about the consequences of privatizing online space. If an antiques dealer kicks me out of his auction house, I can stand on the street outside, observe who enters and who leaves with a Chesterfield sofa, and ask people to share information with me. If eBay kicks me off their site, I have no similar power online. So even if the goal were to mimic the rights that private physical property provides in the online world, granting the equivalent of real property rights online wouldn't necessarily achieve that goal. In *eBay*, the rights that the court granted eBay exceed anything it could have obtained offline.

This is not to suggest that there should be no ownership of Internet "spaces" at all. Rather, the point is that private and public spaces must coexist on the Internet, just as they do in the real world.[64] As a result, one cannot look at any given part of the Internet and assume that it must or should be private property. It might be appropriate to declare that space private, but it might not.[65]

B. How Private Is Private?

Even were we to decide not only that the Internet is like a physical place, but also that it is a physical place that should be privately owned, that still wouldn't justify the results in the trespass cases. It is wrong to think of private property as a unitary phenomenon. Rather, to rely on a time-honored metaphor, property itself is a "bundle" of rights that attach to a particular piece of property. Which sticks are in the bundle and who holds them may vary from case to case. For example, the rights we normally grant to owners of intellectual property and owners of personal property differ in significant respects from the rights we give to owners of real property.[66] This is particularly important in the Internet trespass context both because the courts have nominally applied the law of personal rather than real property[67] and because the underlying issue in many of these cases was really one of intellectual property, not the sanctity of a Web site. Intellectual property rights are notably incomplete, limited in a variety of ways in order to advance the ultimate public good.[68]

Physical resources are also subject to different rules depending on their nature. We have different sets of rights for air,[69] minerals, land and water. Water is a particularly interesting analogy to the electrons that are at issue in the Internet trespass cases, as both flow according to the laws of physics. As Blackstone put it, water "is a movable, wandering thing, and must of necessity continue common by the law of nature so that I can only have a temporary, transient, usufructary property therein: wherefore, if a body of water runs out of my pond into another man's, I have no right to reclaim it."[70] On one view, the Internet trespass cases are all about chasing down electronic "water" in order to reclaim it.[71]

Even within the narrower context of private land, not all rights are uniform. Private property is held in a variety of forms; the fee simple absolute is only one extreme example.[72] Property may be held subject to reversionary interests, or only for a period of years, or without a right to make certain kinds of uses of the land.[73] Different parties may own land and the buildings thereon.[74] Parties may jointly own overlapping rights to use the same piece of property. Carol Rose and Elinor Ostrom have both written of "limited commons property," regimes in which property is held in common by a subset of the general public or in trust for the benefit of a particular group.[75] Property interests of all sorts may be limited by easements or covenants, both those recorded with the property itself and those implied for some public purpose. And property interests are hardly immutable; the fundamental legal rights associated with ownership have changed over time.[76]

Remedies for incursions upon property interests also vary depending on the nature of the interests on both sides. While the normal rule is that property is protected by injunctive relief, that is not always the case. In some cases, courts have permitted infringement of a property right to continue and required the interloper to pay damages to compensate the property owner.[77] Further, courts have not always concluded that an unwanted incursion upon real property is illegal. Under the doctrine of trespass, physical intrusion onto the land is itself actionable. But under the true law of trespass to chattels—as opposed to the mutant version courts have applied to cyberspace—a trespass is not actionable absent evidence of harm to the chattel itself or to the owner's use of it. And under the law of nuisance, certain more intangible intrusions onto private space—the playing of loud music next door, say, or the emission of pollutants—are only actionable if the harm they cause the property owner exceeds the benefits associated with the conduct.[78]

The nuisance cases are particularly instructive because they show that even in the context of real property, there is room to focus on the defendant's conduct. Nuisance law permits us to weigh the costs and benefits of exclusion. It is a standard rather than a rule. The balancing approach of nuisance lawmay be administratively more costly than an absolute rule,[79] but it is almost certainly the right approach when considering the creation of a fundamentally new right that would change the established patterns of behavior on the Internet.[80] In a nuisance analysis, the cyberspatial metaphor would not impel us inexorably toward an absolute right of exclusion. Rather, it would allow us to ask whether, in the context of the Internet, the defendant's conduct intrudes on some fundamental right we want to confer on the owner of a Web server. Even if we accept the metaphor of cyberspace as real property, therefore, we are left with a variety of legal means to implement that idea. As Dan Burk has persuasively argued, conferring an absolute right to exclude has been the wrong choice as a policy matter in the Internet cases that have come up so far.[81]

In the Internet trespass cases, the defendants' conduct has fallen into two basic categories: attempts to acquire information and attempts to convey information.[82] *eBay, Register.com*, and *Oyster Software v. Forms Processing*[83] all involved efforts to download unprotected information from a publicly accessible Web site. This was unexceptional conduct that it made little sense to enjoin. Acquisition of information is normally a social good, so long as the information is available in a public place and is not itself protected by intellectual property law. In Burk's terms, access to eBay's public data by those who would promote competition was "locally objectionable but globally beneficial."[84] By contrast, the downloading of copyrighted songs, text or software from a Web page without authorization can have market-destructive effects. Similarly, acquiring information from a nonpublic source by hacking into a private computer system is conduct that deserves to be prohibited. But we do not need a broad doctrine of trespass or even nuisance to reach that result. Intellectual property and computer crime laws already punish the improper acquisition of information, without also punishing socially beneficial uses.[85]

Dissemination of information can also be either good or bad, depending on the context. Dissemination of unprotected speech, such as obscenity, true threats, defamation, and false statements of fact, serves no social function and has great capacity for mischief. By contrast, dissemination of other kinds of information is generally desirable as a social matter, except when a recipient is overwhelmed with large quantities of undesired information. In cases like *Intel v. Hamidi*, efforts to disseminate protected speech of relevance to a particular targeted audience deserve to be protected, even if a recipient like Intel does not like the message.[86] The *Cyber Promotions* cases, by contrast, involved bulk, unsolicited commercial email that in many cases also falsely represented its source. While there is some speech value to spam, its social harm outweighs its value, and so it should probably be prohibited. But once again, we don't need a broad doctrine of trespass to reach this result. Statutes that prohibit spam, obscenity, defamation and libel already exist, and they do not also punish desirable social conduct like Hamidi's.

In short, to call something property is only to begin the inquiry, not to end it. Our society has many different rules of property to account for many different situations. The rights and remedies we give to private property owners depend in part on the social value of allocating control to the property owner and the social value of the use that defendants make of that property. When we apply these principles to the Internet, we find that existing tort law already does a rather good job of punishing undesirable conduct. Adding a particular form of strong property protection into the mix threatens to deter a good deal of valuable use of the Internet without doing much more to stop bad uses.

Conclusion

Metaphors exist to help us think through new problems by analogizing them to old ones. The cyberspace as place metaphor can be valuable. Thinking about the Internet by reference to the real world is fine, if for no other reason than that courts must apply a host of real-world laws to the Internet. But blind application of the metaphor to reach a particular result obscures more than it illumines. The metaphor will serve its purpose only if we understand its limitations—the ways in which the Internet is not like the physical world. Courts must also understand that metaphor is no substitute for legal analysis. "Property" is a doctrinal tool that we use to create a just society. To reify it—to make it a talisman whose very invocation renders us incapable of thinking through the optimal social result—is to exalt form over substance. Choosing form over substance is rarely a good idea; and certainly not on the Internet, where the form itself is nothing but a metaphor.

Notes

1. ©2003 Mark A. Lemley. This is a shortened version of an article that first appeared in volume 91 of the California Law Review in March 2003, and discussion of many relevant sources has been truncated. I encourage you to read the full article.

2. Professor of Law, Boalt Hall, University of California at Berkeley; of counsel, Keker & Van Nest LLP.

3. See, *e.g*, David Johnson & David Post, *And How Shall the Net Be Governed? A Meditation on the Relative Virtues of Decentralized, Emergent Law, in* Coordinating the Internet 62 (Brian Kahin & James Keller, eds., 1997); David R. Johnson & David Post, *Law and Borders—The Rise of Law in Cyberspace,* 48 Stan. L. Rev. 1367 (1996); David G. Post, *Governing Cyberspace,* 43 Wayne L. Rev. 155, 159 (1996); David G. Post, *Anarchy, State and the Internet: An Essay on Law-Making in Cyberspace,* 1995 J. Online L. art. 3, ¶ 34, http://www.wm.edu/law/publications/jol/articles/post.shtml (1995). Johnson and Post would doubtless be appalled by the use to which their cyberspace as place metaphor is currently being put.

4. Dan Hunter, *Cyberspace as Place and the Tragedy of the Digital Anticommons*, 91 Calif. L. Rev. __ (2003) [hereinafter Hunter, *Cyber-space As Place*].

5. Hunter, *Cyberspace As Place*, *supra* note 4, at 460–72. As Post puts the problem, "[t]hese are new things . . . and thinking about new things is hard, because we only really can understand old things. David G. Post, *"The Free Use of Our Faculties": Thomas Jefferson, Cyberspace, and the Language of Social Life*, 49 Drake L. Rev. 407 (2001) [hereinafter, Post, *Jefferson*]. See also Harmeet Sawhney, *Information Superhighway: Metaphors as Midwives*, 18 Media, Culture & Soc'y 291, 291–92 (1996).

For analogous work on the role of metaphor in law in another context, see Thomas W. Joo, *Contract, Property and the Role of Metaphor in Corporations Law*, 35 U.C. Davis L. Rev. 779 (2002).

6. Hunter, *Cyberspace As Place*, *supra* note 4, at 509–13.

7. Hunter cites ample evidence of the use of the metaphor by courts and commentators. He is certainly correct that it seems pervasive. There are, to be sure, other metaphors that have been offered in the Internet context, though many of them are at root about place. See, e.g., Sawhney, *Information Superhighway*, at 304 (noting the multiplicity of transportation-related metaphors for the Internet: superhighway, pipeline, etc.). Most notably (and perhaps ironically, given his role in developing the cyberspace as place metaphor), David Post analogizes the Internet to a language. See Post, *Jefferson*, at 409. But these alternative metaphors have not caught on as the place metaphor has.

Indeed, even those whose main message is to warn of the dangers of centralizing control over the Internet fall back on spatial metaphors to help make their point. See, e.g., Lawrence Lessig, Code and Other Laws of Cyberspace (1999) (framing his argument in terms of the "architecture" of the Internet). Personally, my inclination is to resist the idea that we cannot think through new problems for ourselves without reliance on imperfect analogies to the ideas of the past. Thus, I would be happiest in a world in which we looked to context and effect, not metaphor, to help us decide what to do. Orin Kerr has argued that this may still be possible, pointing to a minority of courts that take what he calls an "external" perspective on the Internet—looking at the Internet as it actually exists rather than how it is perceived from within. Orin S. Kerr, *The Problem of Perspective in Internet Law*, 91 Geo. L. J. __ (forthcoming 2003). I think there is value in that approach, because it may help courts to reach results that are based on real-world consequences rather than perceived similarities. But for purposes of this article, at least, I will accept Hunter's conclusion that metaphor will continue to influence our treatment of the Internet.

8. See Hunter, *Cyberspace As Place*, *supra* note 4, at 453–54 (citing copious examples).

9. For a discussion of the prevalence of private "bulletin board systems" in the late 1980s and early 1990s, see, e.g., Debra B. Burke, *Cybersmut and the First Amendment: A Call for a New First Amendment Standard*, 9 Harv. J. L. & Tech. 87, 91–92 (1995).

10. On the importance of the end-to-end design of the network, see Mark A. Lemley & Lawrence Lessig, *The End of End-to-End: Preserving the Architecture*

of the Internet in the Broadband Era, 48 UCLA L. Rev. 925, 928 (2001); J. H. Saltzer et al., *End-to-End Arguments in Systems Design*, *available at* http://web.mit.edu/Saltzer/www/publications/endtoend/endtoend.pdf (Apr. 8, 1981), *reprinted in* Innovation in Networking 195 (Craig Partridge ed., 1988).

11. *Cf.* Voyeur Dorm v. City of Tampa, 265 F.3d 1232 (11th Cir. 2001) (holding that a live sex show broadcast over the Internet from a house in Tampa did not violate a local zoning ordinance prohibiting adult entertainment because the entertainment was not physically provided at that location but sent to remote users).

12. Indeed, the analogy to the telephone is more exact than it might at first appear. Most Internet users even today access the Internet through a dial-up modem, which takes data from a computer and converts it to analog sounds that can be sent over a telephone line just like the human voice. For these users, the technical reality of Internet communication is essentially identical to telephonic communication. Only what is being "said" and the way it is perceived differ.

13. See *infra* notes 28–29 and accompanying text.

14. See William Gibson, Count Zero (1985); William Gibson, Mona Lisa Overdrive (1988); William Gibson, Neuromancer (1984). *Cf.* Neal Stephenson, Snow Crash (1992) (offering a similar spatial conception of "the MetaVerse").

15. See, e.g., Interview with William Gibson, *at* http://books.guardian.co.uk/authors/author/0,5917,96528,00.html (last visited Sept. 29, 2002).

16. See, e.g., Michael Benedikt, *Cyberspace: Some Proposals, in* Cyberspace: First Steps 119 (Michael Benedikt, ed., 1991); Marcos Novak, *Liquid Architectures in Cyberspace, id.* at 225; Alan Wexelblat, *Giving Meaning to Place: Semantic Spaces*, *id.* at 255.

17. Intrusion countermeasure electronics, in Gibson's terminology. See Gibson, Neuromancer, *supra* note 14; http://burks.brighton.ac.uk/burks/foldoc/76/59.htm (defining the term).

18. See *infra* notes 22–27 and accompanying text (discussing these cases).

19. See Hunter, *Cyberspace As Place*, *supra* note 4, at [143–45] (suggesting that the cyberspatial metaphor is ineradicable); *but* see Kerr, *supra* note __ (arguing that not all courts accept this metaphor).

20. Copying may have other pernicious effects, of course, particularly on incentives to create. Intellectual property protection exists not because information is like physical property (it isn't), but because unrestrained copying leaves creators with too little incentive to develop new works.

21. See O'Rourke, *Analogy*, *supra* note 4, at 561 ("[R]ather than searching for analogies, courts and legislators could more profitably devote their energies to understanding how the Internet differs from physical space, evaluating whether those differences call for new legal rules . . .").

22. For courts applying the doctrine of trespass to chattels to the Internet, see, e.g., America Online v. National Health Care Discount, Inc., 174 F. Supp. 2d 890 (N.D. Iowa 2001); Oyster Software, Inc. v. Forms Processing, 2001 WL 1736382

(N.D. Cal. 2001); eBay, Inc. v. Bidder's Edge, Inc., 100 F. Supp. 2d 1058 (N.D. Cal. 2000); Register.com, Inc. v. Verio, Inc., 126 F. Supp. 2d 238 (S.D.N.Y. 2000); CompuServe, Inc. v. Cyber Promotions, Inc., 962 F. Supp. 1015 (S.D. Ohio 1997); Intel Corp. v. Hamidi, 114 Cal. *But* Rptr. 2d 244 (Cal. Ct. App. 2001), *review granted*, 43 P.3d 587 (Cal. 2002). see Ticketmaster Corp. v. Tickets.com, 2000 Copyright L. Decs. ¶28,146 (C.D. Cal. Aug. 10, 2000) (refusing to apply the doctrine of trespass to chattels to the Internet in the absence of proven harm); Express One Int'l v. Steinbeck, 53 S.W.3d 895 (Tex. App. 2001) (finding no cause of action for conversion of information posted online, because information was not property). Not surprisingly, more claims are coming. See, e.g., *Homestore Files Lawsuit Against Bargain Network*, http://www.rebuz.com/retech02/0402/homestore_files_lawsuit_against_.htm (discussing suit by Homestore.com against Bargain.com for collecting real estate listings from its site and making them available to bargain.com users).

Many commentators have discussed the application of trespass to chattels to the Internet. Most are critical, especially of cases that apply the doctrine to web sites rather than spam and that disregard the requirement of damage.

23. See Burk, *The Trouble with Trespass*, *supra* note 22, at 34 (arguing that because courts addressing the issue in cyberspace have ignored the damage requirement of trespass to chattels, they are actually applying the trespass to land tort).

24. E.g., *eBay*, 100 F. Supp. 2d at 1064–66.

25. *Cf. eBay*, 100 F. Supp. 2d at 1071–72 (articulating an alternative theory to its holding of likely injury to eBay's servers: that any use of eBay's servers inherently deprives eBay of its property). While the discussion of inherent injury in *eBay* was clearly dictum, it is this dictum and not the court's holding that actual injury was likely that courts have pointed to in subsequent cases. See, e.g., *Register.com*, 126 F. Supp. 2d at 250; *Oyster Software*, 2001 WL 1736382, at *12; *Intel*, 114 Cal. Rptr. 2d at 252.

26. 18 U.S.C. § 1030 (1996).

27. See, e.g., EF Cultural Travel BV v. Explorica, Inc., 274 F.3d 577 (1st Cir. 2001); *Register.com*, 126 F. Supp. 2d 338.

28. This fact distinguishes the trespass and CFAA cases I discuss here from true cases of unauthorized access, in which crackers exploit software bugs to gain access to a computer system or part thereof that the owner never intended to open to the outside world. It is this latter set of cases that the CFAA and other computer crime statutes were designed to deal with. See Orin S. Kerr, *The Troubling Trigger of Cybercrime* (working paper 2002) (noting the amorphous nature of "unauthorized access" under the CFAA).

29. The fact that information is what is really at stake in these cases is most clear in *Intel v. Hamidi*, 114 Cal. Rptr. 2d 244, in which Intel objected to email from a former employee to current employees because of its content, and in *eBay v. Bidder's Edge*, 100 F. Supp. 2d 1058 (N.D. Cal. 2000), in which eBay filed a variety of intellectual property claims in an effort to get control over its uncopyrightable data before finally prevailing on the trespass-to-Web-server theory.

30. See, e.g., Inset Systems, Inc. v. Instruction Set, Inc., 937 F. Supp. 161, 165 (D. Conn. 1996). For criticism, see Dan L. Burk, *Jurisdiction in a World Without Borders*, 1 Va. J. L. & Tech. 3, ¶ 54 (1997), http://www.vjolt.student.virginia.edu/graphics/vol1/home_art3.html.

31. See, e.g., Cybersell, Inc. v. Cybersell, Inc., 130 F.3d 414 (9th Cir. 1997); Bensusan Restaurant Corp. v. King, 937 F. Supp. 295, 301 (S.D.N.Y. 1996), *aff'd*, 126 F.3d 25 (2d Cir.1997).

32. See, e.g., Zippo Mfg. Co. v. Zippo Dot Com, Inc., 952 F. Supp. 1119, 1124–25 (W.D. Pa. 1997). For a discussion of these cases, see Michael A. Geist, *Is There a There There? Toward Greater Certainty for Internet Jurisdiction*, 16 Berkeley Tech. L.J. 1345 (2001).

33. See, e.g., GTE New Media Servs. v. BellSouth Corp., 199 F.3d 1343 (D.C. Cir. 2000) (rejecting the *Zippo* approach); Millenium Enters. v. Millenium Prods., 33 F. Supp. 2d 907 (D. Or. 1999) (rejecting the *Zippo* approach); ALS Scan v. Digital Serv. Consultants, 293 F.3d 707 (4th Cir. 2002) (modifying the *Zippo* approach). Geist chronicles the rejection of the *Zippo* test in favor of a test based on purposeful availment of the benefits of doing business in the forum, coupled with a reliance on the jurisdiction in which the effects are felt in the case of intentional torts. See Geist, *Is There a There There?*, *supra* note 32, at 1371–80.

34. See, e.g., Quill v. North Dakota, 504 U.S. 298, 312 (1992); City of Phila. v. New Jersey, 437 U.S. 617, 626–27 (1978); Pike v. Bruce Church, Inc., 397 U.S. 137, 144–45 (1970). For a good discussion of the Internet cases, see Dan L. Burk, *Federalism in Cyberspace*, 28 Conn. L. Rev. 1095, 1123–34 (1996) [hereinafter Burk, *Federalism*].

35. This was arguably the effect of the regulations at issue in *Quill* and *Healy*. See Healy v. Beer Institute, 491 U.S. 324, 336–37 (1989) (striking down a Connecticut statute that required beer merchants not to charge a different price in a neighboring state than they were in Connecticut).

36. See Am. Library Ass'n v. Pataki, 969 F. Supp. 160, 173–75 (S.D.N.Y. 1997) (striking down Internet antipornography statute on dormant commerce clause grounds); Burk, *Federalism, supra* note 34, at 1131–32 (suggesting that states are limited in their ability to regulate Internet commerce because of the burdens this would place on out-of-state companies). *But* see Hatch v. Superior Court, 94 Cal. Rptr .2d 453 (2000) (holding that a California statute that made it a crime to transmit harmful matter over the Internet to a child does not violate the Commerce Clause); People v. Hsu, 99 Cal. Rptr. 2d 184, 190 (2000) (same); Jack L. Goldsmith & Alan O. Sykes, *The Internet and the Dormant Commerce Clause*, 110 Yale L.J. 785, 823 (2001) (suggesting that most state statutes won't raise dormant commerce clause problems).

37. In particular, it may be impossible for a Web site owner to comply with two inconsistent state regulations, since the same page will be viewed in both states, while an offline company could presumably ship different products to different states.

38. See Ferguson v. Friendfinders, Inc., 115 Cal. Rptr. 2d 258 (Cal. Ct. App. 2002); State v. Heckel, 24 P.3d 404 (Wash. 2001).

39. See ACLU v. Johnson, 194 F.3d 1149 (10th Cir. 1999) (striking down state law regulating indecent speech); American Booksellers' Foundation for Free Expression v. Dean, 202 F. Supp. 2d 300, 304 (D. Vt. 2002) (striking down an antipornography statute that punished conduct communicated on the Internet without regard to where it occurs); PSINet, Inc. v. Chapman, 167 F. Supp. 2d 878, 890–91 (W.D. Va. 2001) (same); Cyberspace Communications v. Engler, 142 F. Supp. 2d 827, 831 (E.D. Mich. 2001) (indedent speech statute struck down); *Pataki,* 969 F. Supp. 160 (indecent speech statute struck down).

40. 47 U.S.C. § 223(a) (Supp. 1997).

41. See, e.g., FCC v. Pacifica Found., 438 U.S. 726 (1978). The special First Amendment rules for broadcast media have been devastatingly criticized on their merits, see, e.g., Thomas G. Krattenmaker & Lucas A. Powe, Jr., Regulating Broadcast Programming 203–36 (1994), but the law clearly treated the two media differently, *id.* at 203.

42. 521 U.S. 844 (1997).

43. *Id.* at 867, 868–69 (distinguishing *Pacifica Found.*, 438 U.S. 726 (1978)).

44. *Id.* at 889–90 (O'Connor, J., concurring): A minor can see an adult dance show only if he enters an establishment that provides such entertainment. And should he attempt to do so, the minor will not be able to conceal completely his identity (or, consequently, his age). Thus, the twin characteristics of geography and identity enable the establishment's proprietor to prevent children from entering the establishment, but to let adults inside.

The electronic world is fundamentally different. Because it is no more than the interconnection of electronic pathways, cyberspace allows speakers and listeners to mask their identities. Cyberspace undeniably reflects some form of geography; chat rooms and Web sites, for example, exist at fixed "locations" on the Internet. Since users can transmit and receive messages on the Internet without revealing anything about their identities or ages . . ., however, it is not currently possible to exclude persons from accessing certain messages on the basis of their identity.

45. *Id.* at 890 (O'Connor, J., concurring) ("Cyberspace differs from the physical world in another basic way: Cyberspace is malleable. Thus, it is possible to construct barriers in cyberspace and use them to screen for identity, making cyberspace more like the physical world and, consequently, more amenable to zoning laws. This transformation of cyberspace is already underway"). O'Connor's concurrence has been criticized as a rote application of the cyberspace as place metaphor, however.

The issue of the Internet's changing nature returned in 2002, when the Court considered the constitutionality of the CDA's successor statute, COPA. A fractured Court held that the statute's command to apply contemporary community standards to the Internet was not itself unconstitutional, though the Court remanded the case for consideration of other constitutional problems. Ashcroft v. ACLU, 122 S. Ct. 1700 (2002).

46. See, e.g., Jessica Litman, *Breakfast with Batman: The Public Interest in the Advertising Age*, 108 Yale L.J. 1717, 1725 (1999) ("There has been inexorable pressure to recognize as an axiom the principle that if something appears to have substantial value to someone, the law must and should protect it as property").

Rochelle Dreyfuss describes this instinct as "if value, then right." Rochelle Cooper Dreyfuss, *Expressive Genericity: Trademarks as Language in the Pepsi Generation,* 65 Notre Dame L. Rev. 397, 405 (1990). The idea that there is an inherent connection between value and ownership is a pernicious one, as scholars have recognized. See, e.g., Wendy J. Gordon, *On Owning Information: Intellectual Property and the Restitutionary Impulse,* 78 Va. L. Rev. 149, 167 (1992) ("A culture could not exist if all free riding were prohibited within it"); Mark A. Lemley, *The Modern Lanham Act and the Death of Common Sense,* 108 Yale L.J. 1687, 1715 (1999) (decrying this trend); Mark A. Lemley, *Romantic Authorship and the Rhetoric of Property,* 75 Tex. L. Rev. 893 (1997) (same).

47. 100 F. Supp. 2d 1058 (N.D. Cal. 2000).

48. 126 F. Supp. 2d 238 (S.D.N.Y. 2000).

49. 114 Cal. Rptr. 2d 244 (Cal. Ct. App. 2001), *review granted*, 43 P.3d 587 (Col. 2002).

50. For a description of a world in which one company has monopoly control of the air we breathe, see Total Recall (Tristar Pictures 1990).

51. As Ronald Coase has pointed out, public goods are sometimes provided privately. See R.H. Coase, *The Lighthouse in Economics*, 17 J.L. & Econ. 357, 374–76 (1974). But the fact that a public good can sometimes be provided privately, usually in some form of limited commons arrangement, does not mean that private ownership is necessarily the most efficient form of provision.

52. For an innovative effort along these lines, see Stephenson, *supra* note 14.

53. For a discussion of a related problem, "anticommons" property, in which ownership in necessary inputs is too divided to permit efficient usage, see Michael Heller, *The Tragedy of the Anticommons: Property in the Transition from Marx to Markets*, 111 Harv. L. Rev. 621 (1998); Michael A. Heller & Rebecca S. Eisenberg, *Can Patents Deter Innovation? The Anticommons in Biomedical Research*, 280 Sci. 698 (1998). The anticommons problem can be solved either by making the property in question a public resource or by concentrating ownership in fewer hands. On the Internet, the latter option doesn't seem realistic. A search engine needs to access billions of Web sites owned by tens of millions of different parties; there is no way to consolidate those rights.

54. For a discussion of the question of nondiscriminatory carriage of Internet traffic, see James B. Speta, *A Common Carrier Approach to Internet Interconnection*, 54 Fed. Comm. L.J. 225 (2002). For a discussion of Internet "peering" arrangements among providers, see Jay P. Kesan & Rajiv C. Shah, *Fool Us Once Shame On You—Fool Us Twice Shame on Us: What We Can Learn From the Privatizations of the Internet Backbone Network and the Domain Name System*, 79 Wash. U.L.Q. 89 (2001) [hereinafter Kesan & Shah, *Fool Us*].

55. See Telecommunications Act of 1996, Pub. L. No. 104–104, 110 Stat. 56 (1996).

56. The Federal Communications Commission has so far refused to compel cable companies to provide broadband Internet access on nondiscriminatory

terms, though the Federal Trade Commission did require a form of such access as a condition of the AOL-Time Warner merger. See Daniel L. Rubinfeld & Hal J. Singer, *Open Access to Broadband Networks: A Case Study of the AOL/Time Warner Merger,* 16 Berkeley Tech. L.J. 631 (2001). For discussion of the open access issue, see Lemley & Lessig, *End of End-to-End, supra* note 10; Christopher S. Yoo, *Vertical Integration and Media Regulation in the New Economy* [cite] Yale J. Reg. 171 (2002); Jim Chen, *The Authority to Regulate Broadband Internet Access over Cable* 16 Berkeley Tech. L.J. 677 (2001); Phil Weiser, *Paradigm Changes in Telecommunications Regulation*, 71 U. Colo. L. Rev. 819 (2000).

57. For a discussion of the domain name allocation process and its flaws, see A. Michael Froomkin, *Wrong Turn in Cyberspace: Using ICANN to Route Around the APA and the Constitution*, 50 Duke L.J. 17 (2000); Jonathan Weinberg, *ICANN and the Problem of Legitimacy*, 50 Duke L.J. 187 (2000); Jonathan Weinberg, *ICANN as Regulator* (working paper 2002). See also Graeme Dinwoodie & Lawrence Helfer, *Designing Non-National Systems: The Case of the Uniform Domain Name Dispute Resolution Policy*, 43 Wm. & Mary L. Rev. 141 (2001); Kesan & Shah, *Fool Us, supra* note 54; Jessica Litman, *The DNS Wars: Trademarks and the Internet Domain Name System*, 4 J. Small & Emerging Bus. L. 149 (2000); Timothy Wu, *Application-Centered Internet Analysis*, 85 Va. L. Rev. 1163 (1999).

58. The Internet runs on a set of open, nonproprietary protocols in large part because the Internet Engineering Task Force (IETF), the standard-setting organization that controls the TCP and IP protocols, had a long-standing policy that it would not adopt proprietary standards. That policy has now changed. The World Wide Web Consortium (W3C) also recently considered changing its policy to permit proprietary Web standards, prompting a firestorm of criticism. See, e.g., Janice Mueller, *Patent Misuse Through the Capture of Industry Standards*, 17 Berkeley Tech. L.J. 623, 629–30 (2002) (describing this debate); Wade Roush, *Web Tolls Ahead?*, Innovation 20 (Jan/Feb. 2002). At this writing, the W3C appeared likely to adhere to its royalty-free patent-licensing policy. See Margaret Kane, *W3C Retreats From Royalty Policy*, News.com, Feb. 26, 2002, *at* http://news.com.com/2100–1023–845023.html. Assertions by IBM and Microsoft of patents that allegedly cover ebXML and SOAP have also caused huge controversy in the Internet standards community. See, e.g., David Berlind, *IBM, Microsoft Plot Net Takeover,* Enterprise, April 11, 2002. IBM later backed down on its ebXML patent. See email from Robert Sutor to the ebXML Joint Coordinating Committee, April 18, 2002 (on file with author). It remains to be seen whether the open nature of the Internet will survive this shift to proprietary standards. For an argument that the Net is moving away from openness, see Lawrence Lessig, The Future of Ideas: The Fate of the Commons in a Connected World (2001) [hereinafter Lessig, Future].

59. Plaintiffs have filed a number of suits challenging the propriety of linking to content or displaying search results. See, e.g., Kelly v. Ariba Soft, 280 F.3d 934 (9th Cir. 2002) (collection of images by search engine to display in search results was fair use, but linking to the images on the searched site was illegal); Anick Jesdanun, *Lawsuits Nip at the Heart of the Web*, S.F. Chron., June 10, 2002, at

E1 (detailing more such claims). *Cf.* Yahoo!, Inc. v. La Ligue Contre le Racisme et L'Antisemitisme, 169 F. Supp. 2d 1181 (N.D. Cal. 2001) (refusing to enforce French criminal conviction of Yahoo! for serving as a clearinghouse in which others auctioned Nazi paraphernalia). For prescient discussion of this issue, see Edward A. Cavazos & Coe F. Miles, *Copyright on the WWW: Linking and Liability,* 4 Rich. J. L. & Tech. 3 (1997); Maureen A. O'Rourke, *Fencing Cyberspace: Drawing Borders in the Virtual World*, 82 Minn. L. Rev. 609, 631 (1998).

60. See *generally* Garrett Hardin, *The Tragedy of the Commons*, 162 Sci. 1243, 1244 (1968).

61. It is significant that in none of the Internet trespass cases was there any real threat of such physical overuse. In *eBay*, the only case that focused significant attention on the issue, the use in question never consumed more than two percent of eBay's server capacity. eBay, Inc. v. Bidder's Edge, 100 F. Supp. 2d 1058 (N.D. Cal. 2000). Subsequent cases posed even less of a threat. Indeed, the courts in Register.com, Inc. v. Verio, Inc., 126 F. Supp. 2d 238 (S.D.N.Y. 2000); Oyster Software, Inc. v. Forms Processing, 2001 WL 1736382 (N.D. Cal. Dec. 6, 2001); and Intel Corp. v. Hamidi, 114 Cal. Rptr. 2d 244 (Cal. Ct. App. 2001) all acknowledged that the defendant's conduct would not harm the servers or impose capacity constraints. They nonetheless found trespass to chattels because they followed *eBay*'s dictum in concluding that such harm need not be proven.

62. See, e.g., Rose, *Romans*, *supra* note 53, at [draft at 2].

63. A separate concern is the "public goods problem," which occurs when creators have insufficient incentive to develop new ideas because it is cheap and easy to copy what they have created. But this problem hardly justifies the application of real property models to the Internet. Intellectual property law is designed to deal with the public goods problem and already provides substantial incentives to develop new works. Indeed, because intellectual property protection is expanding day by day, reliance on real property law as a supplement seems less appropriate today than ever before.

64. Carol Rose suggests that the Internet might be divided into private and public spaces with very different characteristics, sharing different needs. Rose, *Romans*, *supra* note 53, at [154].

65. Many have argued for the creation of public space online, often on a public trust model. See, e.g., Maureen Ryan, *Cyberspace as Public Space: A Public Trust Paradigm for Copyright in a Digital World*, 79 Or. L. Rev. 647 (2001); Molly S. van Houweling, *Cultivating Open Information Platforms: A Land Trust Model*, 1 J. Telecom. & High Tech. L. 309 (2002). See *generally* Lessig, Future, *supra* note 58 (articulating the metaphor of the Internet as an "information commons").

66. For example, patents and copyrights expire. The law sometimes compels licensing of those rights at a mandated rate. And copyright, trademark and trade secret law prevent only some types of uses of a protected work. Indeed, the differences are sufficiently great that there is a substantial debate over whether IP is really property at all.

67. The doctrine of trespass to chattels traditionally required actual harm to the chattel, while trespass to land was actionable whether or not the owner's interest in the land was injured.

68. The Supreme Court has repeatedly invoked the limited and instrumental nature of intellectual property rights, not hesitating to limit those rights when the public interest has so required. See, e.g., Graham v. John Deere Co., 383 U.S. 1, 9 (1966) ("The patent monopoly was not designed to secure to the inventor his natural right in his discoveries. Rather, it was a reward, an inducement, to bring forth new knowledge"); Mazer v. Stein, 347 U.S. 201, 219 (1954) ("The economic philosophy behind the clause empowering Congress to grant patents and copyrights is the conviction that encouragement of individual effort by personal gain is the best way to advance public welfare"); see also Fogerty v. Fantasy, Inc., 510 U.S. 517, 524 (1994) ("The primary objective of the Copyright Act is to encourage the production of original literary, artistic, and musical expression for the good of the public"); Feist Publications, Inc. v. Rural Tel. Serv. Co., 499 U.S. 340, 349–50 (1991) (stating that the "primary objective of copyright" is to promote public welfare); Stewart v. Abend, 495 U.S. 207, 224–25 (1990) (noting the Copyright Act's "balance between the artist's right to control the work . . . and the public's need for access"); Bonito Boats, Inc. v. Thunder Craft Boats, Inc., 489 U.S. 141, 167 (1989) (noting the "careful balance between public right and private monopoly to promote certain creative activity"); Sony Corp. of Am. v. Universal City Studios, Inc., 464 U.S. 417, 429 (1984) (stating that the limited monopoly conferred by the Copyright Act "is intended to motivate creative activity of authors and inventors . . . and to allow the public access to the products of their genius after the limited period of exclusive control has expired").

Various statutory provisions reflect this balancing, see, e.g., 15 U.S.C. § 1125(c)(4) (1994) (protections for news reporting and noncommercial use in federal dilution statute); 17 U.S.C. § 102(b) (idea expression dichotomy); *id.* § 107 (fair use doctrine); *id.* § 108 (right to make library copies); *id.* § 110 (right to make certain miscellaneous copies and performances); *id.* § 117 (right to copy computer software); 35 U.S.C. § 112 (requirement of public disclosure of patents), and numerous commentators have alluded to it. See, e.g., 1 Paul Goldstein, Copyright § 1.14, at 1:40 (2d ed. 1995); L. Ray Patterson & Stanley W. Lindberg, The Nature of Copyright 120–22 (1991); Julie E. Cohen, *Reverse Engineering and the Rise of Electronic Vigilantism: Intellectual Property Implications of "Lock-Out" Programs*, 68 S. Cal. L. Rev. 1091, 1198 (1995); Dennis S. Karjala, *Federal Preemption of Shrinkwrap and On-Line Licenses*, 22 U. Dayton L. Rev. (1997); Pierre N. Leval & Lewis Liman, *Are Copyrights for Authors or Their Children?*, 39 J. Copyright Soc'y 1, 11 (1991); Jessica Litman, *The Public Domain*, __ Emory L.J. __, 967–68 (1990); Margaret Jane Radin, *Property Evolving in Cyberspace*, 15 J.L. & Com. 509, 515 (1996). These are only a few of the innumerable citations on this point.

69. It is, if you will forgive the expression, as "free as the air to common use." See Int'l News Serv. v. Associated Press, 248 U.S. 215, 250 (1918) (Brandeis, J., dissenting).

70. 2 William Blackstone, Commentaries *18.

71. I don't want to push this analogy too far. There are significant differences between water and information that would require greater elaboration than I can give here. My point is only that if we look to the physical world for analogies, the analogy to land is not necessarily the most apt.

72. For a discussion of those forms, see Thomas W. Merrill & Henry E. Smith, *Optimal Standardization in the Law of Property: The Numerus Clausus Principle*, 110 Yale L.J. 1 (2000).

73. See *generally* Jesse Dukeminier & James Krier, Property (4th ed. 1998).

74. For example, employees of Stanford University may buy houses on the Stanford campus, but the university owns the land on which the houses are built, and restrictive covenants prevent the houses from being resold except to other members of the Stanford community. Similarly, much of the land in Hawaii is owned not in fee simple, but in long-term leasehold interests.

75. See, e.g., Elinor Ostrom, Governing the Commons: The Evolution of Institutions for Collective Action 23 (1990); Carol M. Rose, *The Several Futures of Property: Of Cyberspace and Folk Tales, Emissions Trades and Ecosystems*, 83 Minn. L. Rev. 129, 139–43 (1998).

76. The clearest example is the enclosure movement, which Hunter references, see Hunter, *supra* note 4, at 500–08, in which the law went from recognizing the right of an animal owner to graze the animal on another's private but unfenced land to the opposite assumption. A number of scholars have referred to the unchecked growth of intellectual property rights as a "second enclosure movement."

77. In intellectual property law, for example, there are a number of compulsory licenses mandated by the copyright statute. See, e.g., 17 U.S.C. §§ 111, 114, 115, 119 (1994). In addition, the Supreme Court has on several occasions suggested that injunctive relief may not be appropriate against certain types of infringement. See, e.g., Campbell v. Acuff-Rose Music, Inc., 510 U.S. 569, 578 n.10 (1994); New York Times Co. v. Tasini, 533 U.S. 483, 505 (2001). Courts have taken analogous action in cases involving real property. See, e.g., Boomer v. Atlantic Cement, 257 N.E.2d 870 (N.Y. 1970) (refusing injunction in nuisance case in favor of permanent damages where injunction would wreak social harm); Raab v. Casper, 51 Cal. Rptr. 590 (1975) (ruling that damages but not an injunction were available to an infringed-upon landowner from a neighbor who had in good faith began building past the boundaries of his property).

78. See, e.g., Burk, *The Trouble With Trespass*, *supra* note 22, at 51–54.

79. *Cf.* Edward Lee, *Rules and Standards for Cyberspace*, 77 Notre Dame L. Rev. 1275 (2002).

80. More generally, I think it is fair to say that in uncharted areas rules require more justification than standards, precisely because it is hard to know what approach is optimal. The courts in the Internet trespass cases are effectively creating new legal doctrine. My inclination is to be cautious rather than bold in creating such a new right, and to give courts an opportunity to evaluate the consequences of the new claim.

81. *Id.* at 51–54. See also Hunter, *Cyberspace As Place*, *supra* note 4, at 510–13.

82. Tim Wu suggests that we might distinguish between these two on the grounds that an attempt to acquire information from outside requires a different sort of conduct than an attempt to send information to the outside. See Wu, *supra* note 57, at 1176. But in both cases the defendant has engaged in some affirmative conduct; the real question is whether the law should prohibit that conduct on the basis of its effects rather than the form it takes. *Accord* Orin S. Kerr, *Are We Overprotecting Code? Thoughts on First-Generation Internet Law*, 57 Wash. & Lee L. Rev. 1287 (2000) (suggesting that the law is wrongly regulating computer code based on its form rather than what it does).

83. 2001 WL 1736382 (N.D. Cal. 2001).

84. Burk, *The Trouble With Trespass*, *supra* note 22, at 52.

85. At least, they do if they are interpreted properly. Orin Kerr has pointed out that the CFAA has been interpreted too broadly because of its vague definition of unauthorized access to a computer.

86. See, e.g., Hill v. Colorado, 530 U.S. 703, 728 (2000) ("the First Amendment protects the right of every citizen to 'reach the minds of willing listeners and to do so there must be the opportunity to win their attention'"); Kreimer, *supra* note 22 at 147 ("a great deal will turn on the question of whether, in noncommercial cases, the courts will give priority to the metaphor of property rights in the servers or to" the First Amendment).

3

Will the Real Internet Please Stand Up? An Attorney's Quest to Define the Internet

Robert Cannon[1]

I. Introduction

We start with a problem. The problem is that so frequently while discussing legal or regulatory issues related to The Internet, we have no idea what *it* is that we are talking about.

Countless courts, legislative bodies, and packs of pundits have sought to produce definitions of The Internet. Their attempts have at times been poor and misconstrued. Many definitions do not agree. Some talk about applications. Some talk about packets. And others are simply too vague to be of value. And yet, not comprehending the object of the policy, the lawyers go merrily forward constructing on faulty foundations new, imaginative, and perhaps misdirected policy and law.

Is it possible to formulate a good definition of The Internet? Is The Internet the technical specs that currently make up The Internet? Is The Internet based on the experience of The Internet by the user? Can a definition of The Internet reflect its robust and flexible nature, in all of its diversity, as it exists today and what it might become tomorrow? Is there something wrong with this simple question about the definition of The Internet that makes it so hard to answer?

This chapter sets forth on a modest quest: make fun of previous legal definitions of The Internet; attempt to provide a foundational exploration of what a good understanding of The Internet might look like; and finally, look at why none of this (usually) should or does matter.

A response to this modest quest might be that what The Internet is depends upon why you are asking. There may be a degree of truth to this. The cubist will argue that there are multiple perspectives of a horse, all of which are valid. The modernist will argue that how we perceive the horse depends on where we ourselves stand. Nevertheless, there is a horse there. While there may be multiple perspectives of a horse, looking at the bird sitting on the horse is not looking at the horse. Part of the mission here is to say to the legal community, stop looking at the bird.

II. What It Ain't

We have perhaps all heard the popular and tedious definition of The Internet as "a network of networks."[2] Boring. This tells us little about The Internet. The telephone network is a network of networks. The cable television system is a network of networks. The lawyers in Washington, D.C., circulate in networks of networks. This definition provides us little useful information about The Internet (but perhaps this concept will have to be revisited later in the chapter).

Nor is The Internet a network of "interoperable packet switched data networks."[3] This is the first statutory definition, codified as a part of the Communications Decency Act.[4] The first two statutory attempts to define The Internet come from the first two statutory attempts to censor The Internet—both of which have been declared unconstitutional.[5] The definitions remain in the wreckage of those laws. With regards to this first statutory definition, there are lots of "packets switched networks" using other protocols known as Asynchronous Transfer Mode (ATM), Frame Relay, or the ITU's X.25. There can also be IP networks that are not themselves a part of The Internet. The Internet is just one packet switched network in a family of packet switched networks. A good definition of The Internet is something more than "a network of networks" and it is something more than "a network of packet switched networks."

The second attempt to censor The Internet, the Children's Online Protection Act, brought the second statutory definition.[6] And although the CDA and COPA are codified in sections directly next to each other, the two definitions are different.

Sec. 230 is far broader than Sec. 231. The set of networks covered by Sec. 230 is the over inclusive all "interoperable packet switched data networks." Sec. 231 is better—but not by much. The problem with Sec. 231 is that all it has done is narrow the set of qualifying networks to all IP networks. There are IP networks other than The Internet.[7] Merely look-

Table 3.1
Comparison of CDA and COPA definitions

47 U.S.C. § 230(f)(1) (the Communications Decency Act)	47 U.S.C. § 231(e)(3) (the Children's Online Protection Act)
Internet: The term "Internet" means the international computer network of both Federal and non-Federal interoperable packet switched data networks.	The term "Internet" means the combination of computer facilities and electromagnetic transmission media, and related equipment and software, comprising the interconnected worldwide network of computer networks that employ the Transmission Control Protocol/Internet Protocol or any successor protocol to transmit the information.

ing at a networking and saying, "I spy with mine own eye . . . IP" is not enough to conclude that you have the genuine product in front of you.[8]

The Internet is defined in federal statutes six times. These definitions follow in the footsteps of Sec. 230[9] and Sec. 231.[10]

While legislators struggled with a definition of The Internet in terms of the network, some judges have blundered in a different direction. In a domain name trademark dispute, one district court judge concocted the following definition:

> The Internet, or the World Wide Web, is a network of computers that allows people to access information stored on other computers within the network. . . . "Information on The Internet is lodged on files called web pages, which can include printed matter, sound, pictures, and links to other web pages. An Internet user can move from one page to another with just the click of a mouse."[11]

What is wrong with this definition? This mistaken definition occurs with unsettling regularity.[12] This is a definition of The Internet network in terms of an application. There are many applications that can be used over The Internet: email, USENET, IRC, chat, games, streaming media, IP telephony, file transfer protocol, peer-to-peer file transfer (aka Napster), and many others. The Internet network was intentionally designed so that the network transmission capacity would be separate from, be disinterested in, and have no involvement in the higher layer applications. The network just moves the bits creating an open platform for applications. What you do with that platform is your choice. You make this choice at your end of the network where you use the application of your choice.[13]

And as this stupid network[14] is not optimized for any application, it means that the end user is empowered to use any application of desire.

Furthermore, there is a significant limitation in understanding The Internet in terms of a single application. Each application behaves differently. To conceive of how the network behaves by referencing one application is to fundamentally miscomprehend how the network will behave with another application. The interactivity of USENET is different than the interactivity of the World Wide Web, which is different than the interactivity of telephony.

Finally, these applications are not particular to The Internet. Email, chat, telephony, even the World Wide Web can be provisioned over networks that are not The Internet.

In some ways it is an understandable mistake. People comprehend a thing by what they can observe about a thing. People comprehend The Internet by what they observe, which is the applications. An end user interacting with a computer is conscious not so much of The Internet as the end user is conscious of the email application being used or the web page from which the user is gathering information. The Internet is made manifest through the applications that are visible on a computer screen. However, like the man behind the curtain, The Internet is not, in fact, those applications.[15]

III. What It Is

We are not without a good definition with which to start. In 1995, the Federal Networking Council (FNC) came up with the following definition:

On October 24, 1995, the FNC unanimously passed a resolution defining the term Internet. This definition was developed in consultation with members of The Internet and intellectual property rights communities.

RESOLUTION:

The Federal Networking Council (FNC) agrees that the following language reflects our definition of the term "Internet."

"Internet" refers to the global information system that—

(i) is logically linked together by a globally unique address space based on The Internet Protocol (IP) or its subsequent extensions/follow-ons;

(ii) is able to support communications using the Transmission Control Protocol/Internet Protocol (TCP/IP) suite or its subsequent extensions/follow-ons, and/or other IP-compatible protocols; and

(iii) provides, uses or makes accessible, either publicly or privately, high level services layered on the communications and related infrastructure described herein.[16]

This is a good definition. It contains many of the general characteristics one might think of when one ponders The Internet. It includes (1) interconnection, (2) IP, and (3) the end-to-end policy. It can appropriately, therefore, be set up as a straw man for scrutiny.

IP

Perhaps the preeminent defining characteristic of The Internet is The Internet Protocol. This is where The Internet gets its name. Where you have IP, you can have The Internet; where you do not have IP, you cannot have The Internet (right?).

But The Network did not always use IP. Starting in the early 1970s, The Network, at the time known as the ARPANet, used The Network Control Protocol (NCP).[17] In 1974, Robert Kahn and Vint Cerf introduced a new network protocol, the Transmission Control Protocol (TCP).[18] However, TCP had technical limitations and was eventually un-packed into two separate protocols, TCP and IP.[19] In time, this won favor, and the U.S. Government declared that The Network would migrate from NCP to TCP/IP (aka IP version 4) on January 1, 1983 (an early Y2K drill). But the role of The Internet protocol in transforming communications was not yet assured. Further incidents helped to place this protocol in its current preeminent role. TCP/IP was incorporated into the UNIX operating system, resulting in its propagation throughout the academic research community. In 1985 the government decided to build NSFNet in response to a need for a national backbone network to connect academic and research networks. The government made the crucial decision that TCP/IP would be the protocol for NSFNet.[20] Finally, Microsoft elected to include the TCP/IP in the Windows operating system. In a sinfully reductionist history, this is how IP grew to take over the world. Currently The Network uses IP version 4 and is in the process of migrating to IP version 6.[21]

The Internet network is a virtual network at the logical network layer,[22] overlayed on top of physical networks (i.e., Ethernets, telephone networks, cable networks, wireless networks). Designed to be provisioned over a diversity of physical networks without the need for altering those physical networks, the overlayed logical network provides a means for these networks to exchange information. The Internet supports in the higher layers a diversity of applications and services. TCP/IP is indifferent to the physical network and the applications used. It is intentionally a thin protocol that leaves the intelligence for the higher layers, at the edge, where the end user is empowered to communicate at will. The Internet network is simply in the business of passing the bits without acting on the content.

Internet packets are not products of the network. The Internet protocol stack is within the computing device of end users. End users interact with applications. The computer then takes that content and breaks it into Internet packets, with each packet individually addressed with an IP number. These packets are then injected into the network and are transported until they reach the next point, the next router, in the network. The router reads the IP number address on the packet (it does not interact with the content), consults an IP number routing table, and routes the packet based on what that router believes to be the best known route to the IP number destination. The packet will in this way hop through the network from router to router until it (we hope) reaches the device with that IP number. Different packets can travel different routes through the network until they are reassembled at the destination. All that the network is doing is processing the packet based on the IP number and, where appropriate, engaging in error correction.

So how does all of this implicate a legal definition of The Internet? The use of IP is a crucial characteristic of The Internet. Years ago some of the early proprietary commercial computer services (i.e., Prodigy, Compuserve, and AOL) did not use IP in their networks. It was common practice to refer to these services as *online* service providers but not *Internet* service providers.[23]

Next, consider that the FNC indicated that TCP/IP was a key defining characteristic. However, as noted, TCP is used for error correction and control of transmission rates. For some applications, the behavior of TCP is not only useless but also destructive. Traditional Internet applications are not real time. If the packets arrive now or a few moments from now, it does not matter. However, newer applications, such as telephony, are time sensitive. If a packet of data does not arrive now, with the voice that is a part of the conversation, it is no longer useful. Retransmitting it merely gets in the way. Therefore, many real time applications have migrated from TCP to UDP (User Datagram Protocol). Thus, while the

FNC defined the network protocol in terms of TCP/IP, it might be more appropriate to define it solely in terms of IP (unless we want to go in the opposite direction and conclude that any application failing to use TCP is not on The Internet, even though it is fully interconnected, reachable, and visible using the IP number addressing system).

Finally, the FNC nailed the definition of The Internet to a particular protocol. There is some degree of difficulty being so technically specific. The network protocol is in flux, changing at different moments in The *Network's* history (i.e., NCP, TCP, TCP/IP, Ipv4, Ipv6). There are independent networks that use IP that are not The Internet. What happens when the technical specification changes? Does each alteration of the technical specification result in an identity crisis where the question must be asked, "is this thing still The Internet?" The Internet protocol itself is almost 30 years old. Might it be replaced with something novel tomorrow, meeting the demands of the new network? Nailing the definition to a particular technical specification is apt to leave us in a bit of a pickle.

Extensions, Follow-ons, Successors

In response to the above discussion, an imaginary proponent of the FNC definition might argue, "well yes, but the FNC definition was not limited to TCP/IP. It anticipated evolution of the protocol with the phrase 'subsequent extensions/follow-ons.' The FNC was not so daft as to think that the network protocol was static; but recognized that it evolved with time and that the heirs of The Internet protocol ought properly be considered likewise The Internet."

While this is certainly true, it is also saying that The Internet is The Internet regardless of how it mutates or transforms. As long as the mutant has some historical basis in IP, well then it is The Internet.

It creates a bit of a problem if the object of our definition is permitted to evolve into areas unknown. Assume that we have created regulation X for The Internet. And assume that in the best of all possible worlds that regulation X is the best of possible regulations. Now the defenders of regulation X will tell you that this regulation of The Internet was not created in abstraction with no regard for the actual nature of The Internet; rather this regulation, as was required of it, took into account economics, societal values, the public interest, freedoms, and other interests. This regulation was created specifically with a set of assumptions concerning the context of The Internet.

Now change those assumptions. If The Internet morphs into areas unknown, shedding and distorting those things that led to those assumptions, then the best of all possible regulations, regulation X, has just become a problem.

The law has a need. In order to effectively address something, it has to know what that something is. If the definition of that something is vague, then the law's effective intercession with that thing experiences problems. This is a crucial point. Many purists would prefer that The Internet remain undefined so that the law lacks a target and The Internet can remain a libertarian paradise. But this is divorced from reality. Reality is that legislators and policy makers are making decisions on a day-to-day basis concerning The Internet and activity over The Internet. Senator Exon, who reportedly had never been online, introduced the infamous and ill-fated Communications Decency Act.[24] Congress has every year sought to criminalize Internet gambling. And there are countless programs where funding is being provided for Internet projects. Lacking solid definitions of the object of these policies, these policies potentially may go astray and in the end may cause more damage than good. One who professes to desire to protect The Internet by ensuring that policy makers have before them only vague and misconstrued definitions defeat their own desires. Good law is based on a good foundation.

Defining a thing for what it might become is problematic. One might think of a definition of beer. According to Mr. Webster, beer is defined as "an alcoholic beverage usually made from malted cereal grain (as barley), flavored with hops, and brewed by slow fermentation."[25] Now consider the suggestion that beer is beer and anything it might become. Well, it might become Welsh Rarebit.[26] But there is hardly a drunk Welshman who might confuse a pint and a plate of gooey cheese poured impossibly over some toast. A thing is properly defined as what it is and not what it might possibly become.

After seven years of trying, let us assume that Congress finally passes a law regulating Internet gambling. Now this law assumes that The Internet is this vibrant global information system where borders mean nothing and 15 year old boys can gamble away their college savings with a guy named Lou operating a somewhat tilted server on some island. Everyone assumes that this Internet gambling law is the best of all possible laws and prevents 15 year olds from getting cyberscammed. Now assume that in five years, the mega corporation AOLTWCBSMCDONALDS builds a super new network that is "an extension of or follow-on" of the network. This new network is everyone's dream come true, a network of highly connected super artificially intelligent toasters.[27] These toasters have one desire, to toast. They do not gamble. And by the time they are 15 years old they have become obsolete. According to our definition, this mutant evolutionary off shoot is The Internet. But now our best of all possible regulations designed to protect 15-year-old boys (or perhaps to protect Lou) now protects a pack of hyper artificially intelligent toasters.

Jurisprudence disdains vagueness. Vagueness leads law into unanticipated and unintended applications.[28] To define something is the act of "describing, explaining, or making definite and clear" the thing as it exists.[29] This does not include the rather large set of things that it might potentially become.

Interconnected

The FNC lists as a definition criterion that the network is logically linked. The Internet is a "network of networks." Ack! Having previously denigrated use of that phrase, how can I now use it? The phrase does get to an important concept (although in and of itself the phrase is entirely insufficient and vague). The concept is interconnection. IP creates a virtual or logical network over the different physical networks, logically linking together edge devices and creating the ability for these computers to interact with one another.

There certainly was a time when a perception of The Internet as a "networks of networks" was valid. The problem, in the early 1970s, that Robert Kahn and Vint Cerf set out to solve was to develop a network protocol to permit interconnection between otherwise incompatible networks. They set out with the purpose of connecting ARPANet to SATNET, PRNet, and ALOHANet.[30] Thus, in the 1970s, the perception was of multiple networks that themselves needed somehow to be networked. Now, I would suggest, the exact opposite perception is true. No longer are the underlying networks the primary concern. Now the end-to-end Internet itself is the key concern, and what underlying network will be used to access it has become the secondary concern. The Internet has gone from being the glue of other networks to being The Network itself.

The physical networks are interconnected at the logical network layer. An ethernet communication goes to an ethernet address at the other side of the network and is terminated at a black box. There is not necessarily an ethernet at the other side of the black box. It could be any network (telephone, wireless, cable . . .). Any sense of this continuing to be an ethernet communication ceases. What goes from one edge to the other, traveling over multiple interconnected networks, is The Internet. To be "On Net" or to have Internet access is to be linked with this unique network of interconnected devices and users.[31]

This criteria is useful. Confronted with the question, "is a long distance telephony network that uses IP in the backbone a part of The Internet," the answer would be no. The long distance IP telephony network is not interconnected.

There is, however, a degree of vagueness here. What is meant by "logically linked" or interconnected? If someone interconnects three IP networks, would that be The Internet? This network of networks has IP and is interconnected. The answer, as suggested by FNC definition, is that not just any interconnection will do. There is a specific historical networked community known as The Internet that is interconnected.

Now a hypothetical test: Imagine, if you will, a huge intranet. Imagine that a class of users had become dissatisfied with the performance of The Internet and therefore, without changing The Internet protocol and using the same IP address system, constructed a private intranet for their own purpose. This intranet, in order to ensure top performance, was constructed with top of the line electronics, premium design, and sufficient bandwidth. Having constructed this high performance intranet and in order to maintain its high performance nature (i.e., uncongested), access was restricted. Only those on the intranet could communicate with others on the intranet. You could not communicate from the intranet to The Internet and you could not communicate from The Internet to the intranet (though the end user may hardly notice as intranet traffic is simply properly routed over the intranet and outside traffic is simply routed over The Internet). Would it be fair to conclude that this intranet should not properly be called The Internet—it is not logically linked (interconnected) to the rest of The Internet (you cannot get there from here)? Would it be fair to conclude this when the intranet is called Internet2?

Address Space

The FNC definition states that The Internet "is logically linked together by a *globally*[32] *unique address space* based on The Internet Protocol (IP)."

It is compelling to think of a network in terms of its address space. One of the tell-tale signs of what network one is using is how the network is addressing the traffic. The Public Telephone System uses the North American Numbering Plan, in other words, telephone numbers.[33] To use the public telephone system, one needs a telephone number. This number is entered into the network in order to transmit communications from point A to point B. If you are communicating with people using telephone numbers, this is a good indicator that you are on the Pubic Telephone Network. If an end point lacks a telephone number, it is not on the Public Telephone Network.

A point that is reachable by a network's addressing system is on that network. If it is not reachable, if it cannot be "seen," then it is not on that network.

IANA

What is the "unique address space" of The Internet? It is the unique set of IP number addresses. Each device on the *Network* is assigned an IP number. Internet packets traveling from one device to another are addressed with the IP number of the destination and the IP number of the originating device. To be visible a device must be associated with an address; if a device is not associated with an address, it cannot be reached and cannot be said to be on the network.

In the early days of the *Network*, Jon Postel at UCLA and then later at ISI volunteered to administer the total set of IP numbers. This administration of IP numbers came to be called The Internet Assigned Number Authority (IANA). With the formation of The Internet Corporation of Assigned Names and Numbers, the IANA function, by contract with the Department of Commerce, is now under ICANN.[34]

Domain Names

The Internet has an interesting network addressing system. The devices on the network are reachable using IP numbers. Those are the network addresses. But for all practical purposes, the humans are unaware of those network addresses. Instead, recognizing years ago that the IP numbers are difficult for the humans, the domain name system (DNS) was developed.

The DNS is a database that maps letter strings to IP numbers.[35] The domain name cybertelecom.org is mapped to a particular IP number, for example 192.16.0.192. An individual who wishes to view the Cybertelecom website can remember the domain name and enter it into a web browser. The first thing the browser does is a DNS look up, asking a DNS server what IP number that domain name is mapped to. Having the IP number, only now can the browser make a query to the cybertelecom.org web server for the content. The web server responds by sending packets of content to the IP number of the device that originated the query. In other words, while the human may know only the domain name, the devices on the network communicate with each other with only the IP number.

There are two wrinkles with this. First, while the DNS is the compelling way with which humans communicate with each other over the network, it is not necessary. Communication can be achieved with only IP numbers or a different alias system can be mapped to IP numbers, such as instant messaging names. While the DNS is fantastically convenient, it is still an adjunct to and not a necessary component of the network address system.

Second, domain names have the added advantage of remaining constant regardless of what IP number they are mapped to. The website cybertelecom.org has changed web hosts several times. The humans are not aware of this. They simply enter the domain name, do a DNS look up, get the new IP number of the new host, and they are happy. In this way, destinations on The Net can have smooth portability. A user can move from service to service with the domain name remaining constant while the IP address changes.

It is also true that humans on intranets do not always have the same IP number even though they might retain the same domain name. Joe may log on to the network with his email address joe@cybertelecom.org. When Joe logs on, the intranet, using dynamic host control protocol, assigns Joe a new IP number from its pool of IP numbers. Thus, Joe's network address is changing on a regular basis within the confines of the pool of IP numbers allocated to that network, while again, his domain name email address remains constant.

This demonstrates how powerfully the DNS compliments the IP number system. But Joe is still reached by IP numbers. Fred, trying to reach Joe, enters Joe's email address in the application. A DNS look up is conducted and, magically, the *same* IP number is produced every time. But doesn't Joe's number change on a daily basis? Yes, but the IP number produced for Fred is the IP number not of Joe but of the Cybertelecom email server. Fred's email application uses that IP number to send the email, not to Joe's computer, but to Cybertelecom email server where it is stored until the next time the Joe logs on.

The point is not to get too technical with the functionality of the DNS. The point is that ultimately the DNS is an adjunct to the IP number system that makes the IP number system useful to humans. The network itself communicates with IP number addresses. So close is DNS to the addresses system of The Internet that it is difficult to talk about Internet addresses without talking about DNS. But DNS is an adjunct, it is not necessary, and packets are routed through the network using IP numbers, not domain names.

E2E

Another defining characteristic as noted by the FNC is the end-to-end design of the network. The Internet religious among us will suggest that end-to-end is an essential characteristic of The Internet.[36] The Internet protocol is a very thin and stupid protocol. It does very little beyond pass packets back and forth. It does not care what physical network the packets originate on. It does not care what application they are from. The Internet Protocol virtual network simply transmits the packets back and forth to destinations on the network without change in form or content of the data transmitted by the network itself. There is no interactivity or interference with content at this layer.

As described originally by J. H. Saltzer, David Clark, and David Reed, application functionality is reserved for higher protocol layers in The Internet protocol stack.[37] Application functionality is removed from the logical network layer so that all that the logical network does is pass packets without interactivity.

This means that instead of having a network with intelligence in the core where the network is optimized for a single application, like the telephone network, the intelligence resides at the end. End users are confronted with a virtual network not optimized for any application. All packets from applications are equal and are transmitted with best effort. The end user can select any application and indeed must select some application before communication can be initiated. The processing and interactivity involved in that application occurs at an edge and not within the network itself.

End-to-End Religion

But a muddling question with this element of the definition is as follows. The architecture of The Internet creates the *potential* for end-to-end. But must that potential for end-to-end be *actual* in order to have The Internet. If The Network is not fully end-to-end, do I have a definitional problem? What if I insert a firewall? What if I use a spam filter? What if I block certain application ports? What if I elect to use a pocket device simply for Internet email? In all of these examples, the potential end-to-end design is fully present (for example, you could turn off the firewall) but is not fully actual. If I say that end-to-end must be fully actual, does that mean that the only time I have The Internet is with an unobstructed pipe (no firewalls et al) connected to a computer that can perform any function—thus precluding my pocket Internet email device from being a part of The Internet?

The end-to-end design of The Internet is what permits robust and flexible implementations of The Internet. It is what permits people to build Internet picture frames that simply display pictures of the grandchildren or robust computer networks that grace users with full versatility of opportunity on The Internet. While the potential for end-to-end is a characteristic of The Internet, so is the fact that given the unique circumstances of implementation, an individual can elect to actualize only a portion of the potential end-to-end design. Both The Internet picture frame and a fully functional Internet computer are what comprise The Internet and it is not coherent to suggest otherwise.

To suggest otherwise, to advocate the argument that The Internet exists only with full end-to-end leads, reductio ad absurdum, to the conclusion that any limitation of potential end-to-end design results in departure from The Internet (and while with an Internet picture frame we clearly have something, we are no longer sure what that thing is). If, with the full potential of my end-to-end Internet computer, I limit myself by electing to use a given application, I have limited end-to-end and would no longer have The Internet. This is not coherent. Part and parcel of the end–to-end principle is not only that I can elect to use any application over The Internet but also that I can elect *not* to use any application. Thus, the end-to-end defining characteristic of The Internet lies in its potentiality although not necessarily in its actuality.[38]

The Internet to the Ends

This highlights an additional perspective. The Internet is The Internet all the way to the end. At the end generally is a computer. It is in that computer that the higher layer intelligence (applications, services, and content) exist and is created. The computer uses applications to create content and injects this content as packets into the network. The network itself does not interact with the content. The content is created and processed at the end. Thus it is inaccurate to say that, for example, an ISP gives an end user access to The Internet, as if The Internet were some far off and remote thing. Rather, the ISP provisions Internet connectivity. Every device and end user that has Internet connectivity is "On Net" and is a part of The Internet.

What About Lower Layers?

Finally, note that while the FNC definition gives recognition to the higher layer applications that The Internet supports, it gives no recognition to the physical networks over which The Internet is provisioned. This is important as The Internet is regularly confused with the underlying physical networks over which it is provisioned.

Historical

Of course The Internet is not just any network; it is a unique network in history. Unlike, for example, cable, where there is a wide assortment of different cable networks out there, there is one unique network in history with the name The Internet. The Internet was born on an October day in 1969. On that day, a pack of geeks located at UCLA sent the first transmission over The ARPANet to another pack of geeks at Stanford. By the end of that year, there were four nodes on *The ARPANet*. By 1971, there were 15 nodes.[39] This unique networked community continued to grow, adding networks, devices, and users.

In 1983, the U.S. Government decided that The Network would migrate from the existing NCP network protocol to the TCP/IP protocol. This unique network came to have the name The Internet. And The Internet continued to grow.

The National Science Foundation recognized a need for a new backbone to service the community and thus established the NSFNet. The

NSFNet however had an Acceptable Use Policy that prohibited commercial traffic (driving the creation of alternative commercial backbones). As demand for commercial use of The Internet grew, commercial networks, which could not use NSFNet to exchange traffic, set up the Commercial Internet Exchange. Eventually NSF altered the NSFNet AUP to permit commercial traffic, established The Network Access Points (NAPs) for traffic exchange, and privatized the NSFNet.

This Internet continued to grow but was nevertheless largely hidden from popular sight. Then, in 1990, Tim Berners-Lee invented the World Wide Web and in 1993, Marc Andreesen invented the Mosaic Web Browser, and The Internet broke into popular use.

The Internet is a unique networked community that traces its history back to a particular day in history. It would be hard to identify this network with a specific technology as the technically has evolved over the years and is used in other networks. What has been unique and identifiable has been the networked community. It grew from a small community of users experimenting with novel packet switched communications, to a global community of millions of users engaged in an information revolution.

In Conclusion, It's the Wrong Question

So enough of making fun of other people's definitions. What is *The Internet*?

After hammering on this question long enough, one realization is that it is not so much a *definition* as it is a *proper name* (this is why Internet is capitalized). The Internet is the proper name of a unique network.

Definitions of The Internet generally collapse into circular statements. The Internet is defined as the network with the name The Internet. Consider that one defines[40] terms[41] but not necessarily proper names.[42] A term is defined by a set of differentiating criteria; things that meet those criteria are identified with that term. If a *flying disc* (aka Frisbee) is a "a plastic disk several inches in diameter sailed between players by a flip of the wrist,"[43] and if that thing there on the ground is a "a plastic disk several inches in diameter sailed between players by a flip of the wrist," then we can joyfully conclude that the thing on the ground (and all other things that meet the definition) is a *flying disc*. Definitions provide us with differentiating criteria that allow us to determine whether a thing in question is properly labeled with a term.

Now consider proper names and the question, "who is *Bob*." The answer is that "*Bob* is the person with the name *Bob*." Things with proper names are not so much defined as they are described. "*Bob*" is that particular tall person over there with a beard and blonde hair who generally wastes too much time on Internet law issues. But it is not true that all tall blonde people with beards are *Bob*s. And if Bob's description changes, if Bob shaved his beard and dyed his hair green, he would still, in fact, be *Bob*.

There is only one computer network with the name The Internet. Other networks that meet the technical description of The Internet nevertheless are not The Internet (these may properly fall under the definition of *an internet*, a computer network that is logically linked using IP but which is not The Internet). And the technical description of The Internet can change over time (for example, from NCP to TCP to TCP/IP to IPv4 to IPv6) and still be The Internet. This is true because The Internet is a name given to a particular network rather than networks that meet certain technical specifications.

Even if "what is the definition of The Internet" is the wrong question, when confronted with a bunch of wires and boxes and blinking lights, one would like to have the ability to understand whether these things are The Internet and how this Internet thing is distinguished from anything else. The Internet is the name of a unique networked community that saw its birth in October 1969.

- On October 25, 1969, The Network was comprised of two end points. At the end of the year it had four end points. Slowly this networked community grew and in the year 2002 it is estimated that there are 580 million individuals on the network.
- In 1983, this Network migrated to The Internet Protocol, from which it got its name.
- The Network is logically linked using the unique address space based on IP, currently managed by IANA.
- The Network is a computer network overlayed over physical networks.
- The Network supports higher layer applications following the end-to-end principle.

This is not intended to be a final word. Instead it is intended to be a first step, or perhaps a shove, in the right direction. It is not far from the

FNC definition—only now it is a description. The primary revision is to change the perspective from a focus on technology to a focus on a unique networked community. Also, this describes the network relationally both down and up. It describes The Internet in relationship to the physical networks over which The Internet is provisioned (something that the FNC did not do). It then describes The Internet in relationship to the applications and services in the higher layers that it supports (the end-to-end principle). In this way it clarifies that The Internet is a particular network that is neither the underlying infrastructure nor the overlying applications. It is a particular logical network, created over existing physical networks, creating a global network of interconnected devices and users.

IV. Postscript: Why It Doesn't Matter

It is prudent to struggle for a proper understanding of The Internet so that when Internet legal issues are addressed, things do not get mucked up. But in struggling so hard for a definition, and then realizing that perhaps seeking a definition was the wrong endeavor, perhaps there is another realization; perhaps the definition does not or should not matter.

The Internet is a specific network. The network can change. The technology can evolve. Use of the network can change. Users can become fickle and prefer the next pet-rock fad in communications. There are other networks. Is it prudent to create law particularly based on one specific network? The struggle to define what The Internet is, begs the question, "why are you asking."

Consider this. Two guys break the law playing three card Monty. One is running a scam on a webserver based on some off-shore island. The other has a beat on the corner of Bleeker and Sullivan. Would it make sense for the law that applies to be different for guy A and guy B? Should one spend two years in jail and another spend four?

The argument here is for neutrality. The application of law should probably not vary depending upon technology.[44] Suggestions of The Internet as a unique and mythical medium where business plans defy the need for making a profit, where criminal activity defies jurisdictional boundary, and where The Network for some exceptional reason defies definition, are (thankfully) gone. The mythology of The Internet having been dispelled, perhaps we can return to the conclusion that a policy goal is a policy goal online or off. Online and offline activity should generally be treated equally.

Returning to the example of Internet gambling, here is a phenomena that has created the opportunity for anyone with a computer and a credit card to easily waste away vast sums of money, engaging the offerings of off-shore gambling companies. The cyber casinos are not regulated. There is no assurance that the game is not rigged. There are no background checks on the operators. Scams can fold up shop at night fall. There is no money going into a state's coffers. There are no state citizens being employed. Removed from the highly social glitz and glamour of Las Vegas or Atlantic City, Internet gambling is a solitary activity between individual and computer screen, where compulsive gamblers can become entranced and minors can act like adults. It is a disruptive innovation to the business plans of established incumbents' gambling businesses.

Every Congress since the 104th Congress has introduced legislation to combat Internet gambling.[45] And they have come close to success. Congress has generally sought to meddle with the Wire Act, which prohibits the business of gambling on sports over wire communications.[46] While the Department of Justice has supported clarification that The Wire Act applies to all situations including Internet gambling, it has opposed Congressional attempts to solve the problem by creating separate and inconsistent legislation. Get nailed under the Wire Act and get treated one way; get nailed under the new Internet gambling law and get treated a different way. DOJ repeatedly advocated that Congress lose the technology bias:

> [The Internet Gambling Prohibition Act of 1999] is not technology-neutral, but applies only to Internet gambling while leaving the existing prohibition on gambling over "wire communication facilities" in general unchanged. While the Department is generally concerned about legislation designed for particular technologies such as The Internet, it is specifically troubled here by the creation of two inconsistent gambling prohibitions—one expressly for The Internet and a different one for the use of wire communication facilities (which includes The Internet).

Indeed, any effort to distinguish Internet transmission from other methods of communication is likely to create artificial and unworkable distinctions. For example, we expect digital Internet telephony to grow in popularity over the next few years. How would we deal with gambling that occurred over this technology, which would use The Internet or other packet-switched networks for pure voice communications? "Would it be under the proposed Internet Gambling Prohibitions Act, specific to the Internet, or under The Wire Act, which deals with wire communications in general (but also includes the Internet)? This is especially problematic as these statutes would have different standards and punishments."

The Department urges Congress to identify the conduct that it is trying to prohibit and then to prohibit that conduct in technology-neutral terms. The fact that gambling, an age-old crime, has gone high-tech and can now be done through The Internet is no reason to pass new laws that specifically target The Internet for regulation. Passing laws that are technology-specific can create overlapping and conflicting laws prohibiting the same activity, but with different legal standards and punishments. "This would be the result if the Internet Gambling Prohibition Act were enacted. We would have both the Wire Act, which has been used to prosecute Internet gambling, and a new Internet Gambling Prohibition Act which would prohibit some, but not all, types of Internet gambling." This overlap in the statutes can only complicate law enforcement's efforts on The Internet gambling front.[47]

In response to the endeavor to define The Internet comes the begged question, "why are you asking." Perhaps the answer is, "you should not be asking." If the legal concern is for activity in the application or content layers (gambling), then focus the legal solution on the layer where the problem exists, the application or content layer. The underlying network layers can be swapped arbitrarily, and yet the policy concern with the activity (the application) likely remains the same. Laws inconsistently applied in different spaces ought to be greeted with a raised eyebrow.

When Does It Matter?

But wait! Having said that The Internet should not matter, I backpedal to suggest that there are times when it does. Circumstances where a proper understanding of The Internet is necessary include (1) First Amendment law, and (2) communications law. There are certainly other occasions where the unique environment of online communications leads to unique legal concerns (but even here, the law may wish to consider general online concerns as opposed to specific Internet concerns).

In the first instance, in different circumstances the federal government has greater ability to restrict speech than in others. On television and radio, the government has been able to restrict the broadcast of "The Seven Dirty Words," a monologue by George Carlin.[48] Conversely with printed matter such as newspapers and books, the government has significantly reduced ability to restrict the dissemination of adult content. According to the Supreme Court,[49] the First Amendment analysis depends upon a proper understanding of the medium. Broadcast spectrum is scarce and held in the public trust, therefore the FCC has authority to restrict the broadcast of mature content to later hours. Printed matter is a highly competitive medium, with great diversity, which is not invasive in your home. With a TV, you set the channel and passively absorb whatever pours from the set; however, reading material must be affirmatively acquired by the reader. Likewise, with The Internet, the Supreme Court has stated that the nature of the medium is crucial.[50] The Internet is not scarce. It is not held in the public trust. It is not invasive. Rather The Internet is a broad and diverse medium where the user goes out and selects the content of that user's choosing. Given the nature of the medium, censorship by the government would be inappropriate.

In the second instance, communications law is not uniform but is rather broken into silos. Title II of the Communications Act regulates the telephone network. Title III regulates wireless services. Title VI regulates cable. While convergence of the mediums is making these silos arcane, nevertheless, for the time, regulation applicable to a given communications medium is dependent on what that communications medium is. In communications terms, it is important that The Internet is The Internet, an information service, and not a telephone service or a cable service. This is at the center of several FCC proceedings.[51]

Notes

1. Views expressed are likely those of the author but could not conceivably be the views of anyone else, including but not limited to his employers, the FCC, TPRC, or his dogs. Robert Cannon can be reached at cannon@cybertelecom.org.

2. See *Reno v. ACLU*, 521 U.S. 844, 850 (1997) ("The Internet is an international network of interconnected computers that enables tens of millions of people, if not more, to communicate with one another and to access vast amounts of information from around the world"). *Cited by Bihari v. Gross*, No. OO Civ. 1664 (SAS), 1 (SDNY Sept. 25, 2000); *OBH, Inc., v. Spotlight Magazine, Inc.*, 86 F.Supp.2d 176, 178–79 (WDNY 2000); *Brookfield Communications, Inc. v. West Coast Entertainment Corp.*, 174 F.3d 1036, 1044 (9th Cir. 1999). See also In Re

Implementation of the Local Competition Provisions in the Telecommunications Act of 1996, Inter-Carrier Compensation for ISP-Bound Traffic, CC Docket No. 96–98, CC Docket No. 99–68, Declaratory Ruling ¶ 3 (Feb. 26, 1999); *In re* GTE Telephone Operators GTOC Tariff No. 1 GTE Transmittal No. 1148, Memorandum Opinion And Order, CC Docket No. 98–79 ¶ 2 (1998), recon. denied (1999). Several courts have put forth their own versions of this definition. See *Name.Space, Inc. v. Network Solutions, Inc.*, 202 F.3d 573, n. 1 (2d Cir. 2000) ("The Internet is a vast system of interconnected computers and computer networks"); *Jews for Jesus v. Brodsky*, 993 F.Supp. 282, 287 n. 2 (D.N.J.1998) ("The Internet is not a physical or tangible entity, but rather a giant network which interconnects innumerable smaller groups of linked computer networks. It is thus a network of networks"); *Zeran v. AOL*, Civil Action 96–952–1, n. 1 (ED VA Mar. 21 1997) ("The 'Internet,' as the term is used here, refers to the immeasurable network of computers interconnected for the purpose of communication and information exchange").

3. 47 U.S.C. § 230(f)(1). See also *Blumenthal v. Drudge*, CA No. 97–1968, n. 6 (DDC Apr. 22, 1998) (citing Sec. 230); *In re* Application of WorldCom, Inc. and MCI Communications Corporation for Transfer of Control of MCI Communications Corporation to WorldCom, Inc., Report and Order, CC Docket No. 97–211, ¶ 143 (Sept. 14, 1998) ("The Internet is an interconnected network of packet-switched networks").

4. The Telecommunications Act of 1996, Pub. L. No. 104–104, 110 Stat. 56, Sec. 501 (1996).

5. See *Reno v. ACLU*, 521 U.S. 844 (1997); *Ashcroft v. ACLU*, No. 00–1293, __ U.S. __ (2002).

6. 47 U.S.C. § 231(e)(3).

7. Examples of IP networks that are not The Internet include international telephone networks that use IP.

8. There is also a problem with the phrase "Extensions, Follow-ons, Successors," which will be discussed on page 61.

9. See 47 U.S.C. § 1127 ("The term 'Internet' has the meaning given the term is section 230(f)(1) of the Communications Act of 1934 (47 U.S.C. 230(f)(1))"); AntiCybersquatting Consumer Protection Act, Sec. 3005. DEFINITIONS ("The term 'Internet' has the meaning given that term in section 230(f)(1) of the Communications Act of 1934 (47 U.S.C. 230(f)(1))").

10. *See* Internet Tax Freedom Act, Pub. L. No. 105–277, Div. C, tit 11, § 1101(e)(3)(C) ("The term 'Internet' means collectively the myriad of computer and telecommunications facilities, including equipment and operating software, which comprise the interconnected world-wide network of networks that employ the Transmission Control Protocol/Internet Protocol, or any predecessor or successor protocols to such protocol, to communicate information of all kinds by wire or radio"); The Children's Online Privacy Protection Act, Pub. L. No. 105–277, Div. C, tit 13, § 1302(6) ("The term 'Internet' means collectively the myriad of computer and telecommunications facilities, including equipment and operating software, which comprise the interconnected world-wide network of

networks that employ the Transmission Control Protocol/ Internet Protocol, or any predecessor or successor protocols to such protocol, to communicate information of all kinds by wire or radio").

11. *Morrison & Foerster LLP, v. Brian Wick and American Distribution Systems, Inc.*, 94 F.Supp.2d 1125, 1126 (D.Co. 2000).

12. See, e.g., *Sporty's Farm L.L.C., V. Sportsman's Market, Inc.*, 202 F.3d 489, 492 (2d Cir. 2000) ("The Internet is a network of computers that allows a user to gain access to information stored on any other computer on the network. Information on The Internet is lodged on files called web pages, which can include printed matter, sound, pictures, and links to other web pages. An Internet user can move from one page to another with just the click of a mouse"); *Bihari v. Gross*, No. OO Civ. 1664 (SAS), 1 (SDNY Sept 25, 2000) ("The Internet is an international network of interconnected computers that enables tens of millions of people, if not more, to communicate with one another and to access vast amounts of information from around the world. Information on The Internet is housed on webpages").

13. See discussion of End-to-End, *infra*, page 67.

14. See David Eisenberg, Rise of the Stupid Network, Computer Telephony, August 1997, pg 16–26, *available at* www.rageboy.com/stupidnet.html.

15. Consider that The Internet predates most applications. In the case of the World Wide Web, The Internet predates the Web by decades.

16. FNC Resolution: Definition of "Internet" 10/24/95 *at* www.itrd.gov/fnc/Internet_res.html.

17. TCMHC: Internet History 1962–1992 *at* computerhistory.org/exhibits/internet_history/index_page (hereinafter "TCMHC History"); Barry Leiner, Vinton Cerf, David Clark, Robert Kahn, Leonard Kleinrock, Daniel Lynch, Jon Postel, Lawrence Roberts, Stephen Wolff, *The Past 4and Future History of The Internet*, Communications of the ACM, 102, 103 (Feb. 1997) (hereinafter "ACM History"); PBS Nerds 2.0.1 Timeline www.pbs.org/opb/nerds2.0.1/timeline/ (hereinafter "PBS Timeline").

18. Vint Cerf and Robert Kahn, *A Protocol for Packet Network Interconnection*, IEEE Transactions on Communications Technology (1974).

19. The Internet Protocol provides for the creation and transmission (routing) of packets of data. The Transmission Control Protocol provides for error correction in order to ensure that packets get to their destination (retransmitting packets that do not make it or slowing down the transmission rate where packets are being transmitted at a rate faster than can be received).

20. Richard T. Griffiths, The Origin and Growth of The Internet and the World Wide Web (Oct. 2001); PBS Timeline, *supra* note 17; ACM History, *supra* note 17, p. 105; TCMHC History, *supra* note 17.

21. See Ikedam and Yamada, Glocom, *Is IPv6 Necessary?* (Mar. 2002) *at* www.glocom.org/tech_reviews/tech_bulle/20020227_s2/index.html; ISOC Briefing Paper 6: The Transition to IPv6 (Jan. 2002) *at* www.isoc.org/briefings/006/isocbriefing06.pdf.

22. The Layered Model of Regulation sets forth a policy framework that divides communications problems into (1) the physical network layer, (2) the logical network layer (aka CODE or transport layer), (2) the application and service layer, and (4) the content layer. See Kevin Werbach, *A Layered Model for Internet Policy*, presented at Telecommunications Policy Research Conference (Fall 2000) *at* www.tprc.org/abstracts00/layered.pdf; Robert M. Entman, Rapporteur, *Transition to an IP Environment*, The Aspen Institute (2001); Michael L. Katz, *Thoughts on the Implication of Technological Change for Telecommunications Policy*, The Aspen Institute (2001); Douglas C. Sicker and Joshua L. Mindel, *Refinements of a Layered Model for Telecommunications Policy* (unpublished 2002).

23. See, e.g., *In re* Applications for Consent to the Transfer of Control of Licenses and Section 214 Authorization by Time Warner Inc and America Online, Inc., Memorandum Opinion and Order, CS Docket No. 00–30, 16 FCCR 6547, 6551 (2001).

24. Robert Cannon, *The Legislative History of Senator Exon's Communications Decency Act: Regulating Barbarians on the Information Superhighway*, 49 FCLJ 51 n. 6 (Nov. 1996) *available at* law.indiana.edu/fclj/pubs/v49/no1/cannon.html.

25. Merriam-Webster Online, Collegiate Dictionary, Beer (2002) *at* www.m-w.com/cgi-bin/dictionary.

26. See Merriam-Webster Online, Collegiate Dictionary, Welsh Rarebit (2002) *at* www.m-w.com/cgi-bin/dictionary (Rarebit is "melted often seasoned cheese poured over toast or crackers").

27. See Red Dwarf, Talkie Toaster *at* www.reddwarf.co.uk/.

28. It can also lead to difficulties under the 1st and 5th Amendments.

29. Merriam-Webster Online, Collegiate Dictionary, Definition (2002) *at* www.m-w.com/cgi-bin/dictionary.

30. See Nerds 2.0.1, Surfing the Net (1998) *at* www.pbs.org/opb/nerds2.0.1/networking_nerds/tcpip.html; TCMHC History, *supra* note 17.

31. *Compare*, David Crocker, IETF Informational RFC 1775, *To Be "On" The Internet* (Mar. 1995) *at* ietf.org/rfc/rfc1775.txt.

32. As Vint Cerf is fond of pointing out, The Internet has passed beyond the reach of the globe. Vint Cerf, Cerf's Up: Interplanetary Internet (2002) *at* www1.worldcom.com/global/resources/cerfs_up/interplanetary_internet/.

33. See 47 C.F.R. Part 52.

34. See Milton Mueller, Ruling the Root (MIT Press 2002).

35. The DNS can go far beyond this but for purpose of this chapter the inquiry is restricted to Internet addressing.

36. See, e.g., Larry Lessig, The Future of Ideas (2002). *Compare* Christian Sandvig, Communications Infrastructure and Innovation: The Internet as End 2 End Network that Isn't (unpublished Nov. 2002), *at* www.dcn.davis.ca.us/~csandvig/research/Communication_Infrastructure_and_Innovation.pdf.

37. See Saltzer, Reed, and Clark, *End-to-End Arguments in System Design,* ACM Transactions on Computer Systems (Nov. 1984) *available at* www.ana.lcs.mit.edu/anaweb/PDF/saltzer_reed_clark_e2e.pdf. See *also* Carpenter, IETF Information RFC 2275, *Internet Transparency* (Feb. 2000) *at* www.ietf.org/rfc/rfc2775.txt.

38. End-to-End design is the subject of much policy debate. Among religious, a break in the end-to-end design gives rise to policy conundrums that should be solved by intervention. But seen in the light noted in the paper, the policy question is altered. The cause for alarm is no longer that end-to-end design might be broken. See Sandvig, *supra* note 36. Rather, this perspective acknowledges that the potential end-to-end design will inevitably be limited in actuality; thereby asking not if but my whom. Will end-to-end be limited by the end user? Will end-to-end be limited by the LAN building a firewall? Will end-to-end be limited by a cable network that declares no streaming media will compete with its cable content? The crucial end-to-end design question becomes *who decides.*

39. See Robert H'obbes' Zakon, Hobbes' Internet Timeline v5.6 (2002) *at* www.zakon.org/robert/internet/timeline/.

40. See Merriam-Webster Online, Collegiate Dictionary, Define (2002) <www.m-w.com/cgi-bin/dictionary> (" to discover and set forth the meaning of [as a word]").

41. See Merriam-Webster Online, Collegiate Dictionary, Term (2002) <www.m-w.com/cgi bin/dictionary> ("a word or expression that has a precise meaning in some uses or is peculiar to a science, art, profession, or subject").

42. See Merriam-Webster Online, Collegiate Dictionary, Proper Noun (2002) <www.m-w.com/cgi-bin/dictionary> ("a noun that designates a particular being or thing, does not take a limiting modifier, and is usually capitalized in English—called also *proper name* ").

43. See Merriam-Webster Online, Collegiate Dictionary, Frisbee (2002) <www.m-w.com/cgi-bin/dictionary>. "Frisbee" is the trademark of Whamo for a flying disc.

44. There are other factors that have led to differentiating The Internet. In the FCC's Computer Inquires, it was not the technology of computer networks that set them apart and led to separate policy; it was the markets. The computer network market was highly competitive with low barriers to entry, in contrast to the telephone network market, which was a consolidated bottleneck market. See Robert Cannon, *Legacy of the Computer Inquiries*, 52 FCLJ __ (2003).

45. See Washington Internet Project, Legislation *at* www.cybertelecom.org/legis/legis.htm.

46. 18 USC § 1084.

47. Testimony of Kevin V. Di Gregory, Deputy Assistant Attorney General, Addressing Internet Gambling Before the Subcommittee on Crime of the House Committee on the Judiciary (Mar. 9, 2000) *at* www.cybercrime.gov/kvd0309.htm.

48. See *FCC v. Pacifica Foundation*, 438 U.S. 726 (1978).

49. See *Sable Communications of California, Inc. v. FCC,* 492 U.S. 115 (1989).

50. See *Reno v. ACLU,* 521 U.S. 844 (1997).

51. See *In re Inquiry Concerning High Speed Access to The Internet over Cable and other Facilities,* CS Docket 02–52, Declaratory Ruling and Notice of Proposed Rulemaking (Mar. 15, 2002) *available at* hraunfoss.fcc.gov/edocs_public/attachmatch/FCC–02–77A1.pdf; *In re* Appropriate Framework for Broadband Access to The Internet over Wireline Facilities, CC Docket 02–33, Notice of Proposed Rulemaking (Feb. 14, 2002) *available at* hraunfoss.fcc.gov/edocs_public/attachmatch/FCC–02–42A1.doc.

4

Governance in Namespaces

Stefan Bechtold

1. Introduction

In the fall of 2000, a web site offered a new service allowing politicians, individuals, and corporations to bid on and buy political votes from citizens. The first Internet auction site for real votes had opened. The election in question was the U.S. Presidential Election of 2000, a memorable event for many reasons. The web site in question, which described itself as "satirical," was located in Austria. It bore the name "voteauction.com."

After the Chicago Board of Election Commissioners had filed a lawsuit against voteauction.com, on October 18, 2000, the Circuit Court of Cook County, Illinois, issued an injunction against the web site. The U.S. registrar, who had registered the domain name, had been named as a co-defendant in the lawsuit. After the injunction was issued, the registrar cancelled the domain name, effectively shutting down the web site all over the world.

About a week later, the web site appeared again under the new domain name "vote-auction.com." This time, the domain name had been registered with a Swiss registrar. A few days later, it was cancelled as well. However, no court had issued any injunction ordering the cancellation. No official authority had addressed the question of whether a domain name registered in Switzerland and located in Austria is subject to U.S. jurisdiction. Rather, the domain name was cancelled after some telephone and e-mail discussions between the Chicago Board of Election Commissioners and the Swiss domain name registrar. The Swiss registrar, a private entity, exercised its power over an asset, the domain name space, to exclude this domain name from the Internet.[1]

In September 1998, a freshman at Northeastern University in Boston began working on a software program that would revolutionize online music business. Only two and a half years later, the Napster network had over 70 million users who downloaded up to 2.8 billion music files per month. In July 2000, the District Court for the Northern District of California issued a preliminary injunction effectively ordering Napster to shut down its service. The Court of Appeals for the Ninth Circuit later affirmed the injunction with some modifications.[2]

Voteauction.com and Napster each raise different problems. Voteauction.com is a case about election fraud, freedom of speech, and personal jurisdiction. Napster is a case about copyright infringement and innovation policy. At the same time, both cases are very similar. They illustrate how technical control over a particular component of a network can be used as leverage for legal and policy control. Voteauction.com lost its domain names because private entities—the domain name registrars and, ultimately, the domain name registry—could exclude its domain names from an authoritative list recognized by all computers connected to the Internet. Music files could no longer be shared over the Napster network because Napster could exclude them from an authoritative list of files recognized by all computers connected to the Napster network. In both cases, the network component that enabled this control was a "namespace."

While namespaces may seem an obscure concept of computer science, we are in fact surrounded by them. In the world of computers, the domain name system, public key infrastructures, Yahoo! Categories, Usenet newsgroups, and computer file systems are all examples of namespaces. Yet, namespaces are not confined to computers. Telephone numbers, Social Security numbers, the "International Standard Book Number" (ISBN), zip codes, bar codes, and bibliographic classification schemes form namespaces, too.

Both Voteauction.com and Napster show that, in cyberspace, the capability for legal regulation often depends on the technical control over a namespace. Technical namespaces are not unalterable, given facts. Rather, technology is a social construct. The cultural and societal structure of those producing technology shapes the technology itself.[3] Conversely, technology enables, shapes, and limits social, legal, and political relationships among citizens, businesses, and the state. As Lawrence Lessig has shown, this interrelation between technology, law, and society implies that technology is not a neutral artifact, but can be shaped according to conscious design decisions that originate from external value systems.[4]

This article analyzes the interrelation between technology and law for namespaces in general. It attempts to highlight a common feature of namespaces: designing namespaces and exercising control over them is not a mere technical matter. The technical control over a namespace creates opportunities for the intrusion of politics, policy, and regulation. The very technological architecture of a namespace may encompass a regulation of speech, access, privacy, content, copyright, trademark, liability, conflict resolution, competition, innovation, and market structures. Therefore, legal and policy considerations should be taken into account even during the design stages of a namespace.

The analysis of such questions is not novel. The best-known namespace on the Internet is the domain name system (DNS), which maps domain names to numerical IP addresses. Since 1998, the DNS has been managed by the "Internet Corporation for Assigned Names and Numbers" (ICANN), a private nonprofit corporation under California law. The status of ICANN is highly disputed. While some proponents assert that ICANN is a mere technical standardization and coordination body, critics argue that it unjustly uses its control over the technical DNS infrastructure as leverage to control policy aspects of Internet communications such as trademark and copyright issues, surveillance of Internet users, regulation of content, imposition of tax-like fees, and the regulation of the domain name supply industry.[5]

Nevertheless, this is not an article about the governance of the domain name system. Although many issues addressed in this article are discussed in the context of ICANN and the DNS, these discussions often fail to recognize that these issues are not unique to the DNS. Rather, they are general governance problems of namespaces that can be found in other namespaces—from peer-to-peer systems to instant messaging systems—as well. They are not even confined to the computer world. In real space, many namespaces—from bibliographic classification schemes to Social Security numbers—exhibit the same problems. No literature exists that identifies and discusses governance dimensions of namespaces on such an abstract, general level.[6] This article attempts to fill that gap. Its findings can be applied to a wide range of namespaces both in cyberspace and real space.

The article proceeds as follows. In section 2, a more precise definition of namespaces is provided. Section 3 develops several dimensions of namespace governance that can be applied to namespaces in general. It shows the legal and policy implications of design decisions made along these dimensions. In section 4, a more abstract account of the relationship between namespace design and the law is provided. Section 5 concludes the article.

2. What's In a Name?

Names are important tools for identification and communication both in real space and cyberspace. From a legal and social science perspective, personal names are a crucial aspect of personal identity and dignity. A complex mix of social norms, memories, connotations, and shared experiences influences the esteem of personal names, in particular first names.[7] From an economic perspective, commercial names and trademarks facilitate identification and thereby reduce consumer search costs.[8] From a computer science perspective, the definition of "name" is even more sober: a name is a string of bits or characters that refers to a resource.[9] In communication networks, some method to identify and locate the networked resources must exist. Names provide such a method to facilitate sharing and communication.

Computer science, in particular the theory of distributed systems,[10] has developed a rather rigorous theory of naming that proves helpful for the following analysis of namespaces. In general, different kinds of names exist. An "address" is a special type of name that identifies the location of an object rather than the object itself. The IP address of a computer and the number of a telephone are addresses in this sense. Addresses are not well-suited to persistently identify objects. Once an object is moved to another location, its address changes. If a computer connected to the Internet, for instance, is moved to another location, its IP address often has to be changed as well.

In many communication networks, these shortcomings of addresses are resolved by adding a layer of location-independent names on top of the addressing scheme. While addresses *locate* resources, location-independent names *identify* them.[11] The domain *name* of a computer, for example, identifies a computer, while its IP *address* reveals its logical location.

Location-independent names and addresses do not exist separately. Rather, names are resolved to addresses by so-called "name services." Name services allow users and software programs to look up, add, change, and remove names.[12] The aforementioned domain name system (DNS) is a name service that resolves domain names to IP addresses.

The collection of all valid names in a particular system forms a "namespace."[13] Some namespaces are designed for human use, while other namespaces are accessed by computers only. Names used by human beings should usually be mnemonically useful, while the critical feature of names used by computers is that they are unambiguously resolvable. In such a namespace, names must be unique.[14]

Namespaces are pervasive, both in cyberspace and in real space. In cyberspace, namespaces are mainly used to identify four different kinds of resources: computers (or more generally: devices), users, files, and applications (or more generally: services).[15] Device namespaces include the domain name system, the telephone number system, ENUM, as well as IP and Ethernet addresses. User namespaces are Microsoft Passport, the Liberty Alliance Project, public key infrastructures as well as user identification systems on eBay and in instant messaging systems. Uniform Resource Locators (URLs), peer-to-peer (P2P) systems, Yahoo! Categories and the different computer file systems available are examples of file namespaces. Service namespaces are created, for instance, by TCP/UDP port numbers and the "Universal Description, Discovery and Integration" (UDDI) service in the context of Web services. Some technologies even use multiple namespaces. Digital rights management (DRM) systems, for example, employ device, user, and file namespaces at the same time.[16] The list of namespaces used by computers and computer networks is endless.[17]

In real space, telephone, credit card, bank account, passport and Social Security numbers as well as tax identifiers are namespaces that identify devices, natural persons or corporate entities. People, streets, cities, countries, species, diseases, and celestial objects are all identified by namespaces as well. Other examples include P.O. boxes, natural languages, and the system of longitude and latitude. The travel industry uses several namespaces to identify travel agencies, hotels, airlines, car rental companies, travel insurance companies, and consumers worldwide. The Dun & Badstreet Data Universal Numbering System (D-U-N-S) is used to identify

62 million business entities around the world. The system of bar codes that is used for product identification is another example how widely namespaces are used today. Traditional media can be identified by different namespaces such as the "International Standard Book Number" (ISBN) and the "International Standard Recording Code" (ISRC).[18] Bibliographic classification schemes, the frequency spectrum, the ISO 3166 list of country codes, as well as the names of all chemical compounds, may complete this listing of namespaces. To put it succinctly: namespaces are important and ubiquitous.

The article uses various namespaces, mostly from cyberspace, to illustrate the presented theoretical framework. Nevertheless, the framework should also be applicable to namespaces that are not explicitly studied in this article.

3. Dimensions of Namespace Governance

By analyzing the means, intensity and scope of namespace governance as well as the possible namespace topologies, this section identifies several dimensions of namespace governance that illustrate the close intertwining of technology, law and policy.

3.1. Means of Namespace Governance

In general, namespace providers have varying interests to regulate the use of and access to their namespace. They may, for example, want to grant access to the namespace only under certain conditions. They may also grant third-party service providers, who use the namespace in their own services, access to the namespace only after payment of a fee.

3.1.1. Governance by Contract

Namespace providers can condition access to and use of their namespace upon the prior conclusion of a contract. Often, they attempt to bind all end users and service providers by contract. A web of contracts laid over the namespace is intended to protect various nontechnical interests of the namespace provider.

The domain name system (DNS), for example, uses such a web of contracts to govern the domain name space. All registrants, registrars and registries of domain names in generic top level domains (gTLDs), such as

.com, .biz, .net, and .org, have to enter into contractual agreements that either directly or indirectly originate from ICANN, the entity that currently controls the DNS.[19] In order to resolve conflicts between domain name registrations and trademark law, a dispute resolution mechanism was created. This "Uniform Dispute Resolution Policy" (UDRP) enables a trademark holder to challenge the registration of a domain name and potentially gain control over it. As part of the contracts between ICANN and the gTLD registrars, ICANN requires the registrars to impose the UDRP on everyone who wants to register a domain name.[20] As a result, on the one hand, ICANN binds all registrars to the UDRP as a condition of their accreditation. On the other hand, a consumer who wants to register a domain name under the .com TLD, for example, will only be able to register it if he agrees to the terms of the UDRP as well. Through a hierarchical web of contracts that originates from ICANN, ICANN has accomplished the result that every registrar and every registrant is bound to the UDRP.[21] ICANN has effectively enveloped the domain name space with a web of contracts that is used to protect, among other things, interests of trademark holders.

Other namespaces in which contractual webs are used include Microsoft Passport[22] and digital rights management systems.[23] In general, the webs of contracts surrounding namespaces bind both service providers that depend on the namespace and individual namespace users. They may be used by namespace providers to regulate various legal and policy aspects of namespaces, ranging from intellectual property and privacy protection to competition issues.

3.1.2. Governance by Technology

Contractual webs would not be a very promising means of namespace governance if the contracts were, as a practical matter, difficult to enforce. In namespaces, however, it is the technology that enables the automatic enforcement of such contracts and policies. By threatening to exclude namespace users and service providers that do not adhere to namespace contracts or policies, namespace providers can enforce their interests in an über-efficient manner.

This phenomenon can be observed in most namespaces. ICANN can enforce its web of contracts, as described above, through its technical control over the domain name space. By withdrawing or reassigning a domain name, any decision under the UDRP can be enforced in a very effective and inexpensive manner: through technology.[24] Public key infrastructures (PKIs) are another namespace that uses technology as a governance tool. PKIs enable the secure, convenient, and efficient discovery of public keys in asymmetric encryption systems. By resolving public keys to individual persons or corporate entities and vice versa, PKIs create user namespaces. In PKI namespaces, various key revocation mechanisms exist by which compromised public keys can be excluded from further use of the namespace.[25] In general, technology enables the namespace provider to control which names are assigned, modified and revoked in a namespace. It is the most important governance tool in namespaces.

3.2. Governance by Whom?

Namespaces can be created and governed by governments, by private entities, or by hybrid coalitions. Particularly in namespaces governed by private or hybrid entities, interests of third parties and the general public might become under-represented. Private regulation of namespaces may clash with public values. Namespaces must be supported by sufficient accountability structures.

The ICANN debate is a prime example of this governance dimension. To what extent ICANN should exercise control over the domain name space and what accountability structures are appropriate is fiercely contested in Internet policy circles.[26] The UDRP has come under criticism for being biased toward the interests of trademark holders.[27] ICANN has been accused of creating a new body of international, but private trademark law that lacks any of the accountability structures under which traditional statutes operate.

The IP and Ethernet address spaces, the TCP/UDP port number space as well as the Microsoft Passport and P2P namespaces are examples of namespaces that are subject to purely private governance.[28] Bibliographic classification schemes, which are namespaces as well, are usually either sponsored by governments or by private consortiums of interested parties and users.[29] Who is governing a namespace determines in part what values and whose interests are protected by the namespace.

3.3. Namespace Topology

In a namespace, system functions can be positioned in a central location or distributed along a vertical or horizontal axis. Choosing a topology[30]

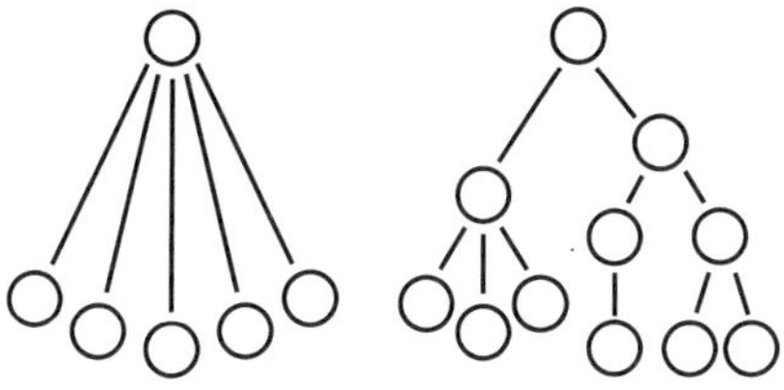

Figure 4.1
Flat versus Hierarchical Namespaces[31]

along these axes has numerous policy and legal implications, as this subsection will illustrate.[32]

3.3.1. Vertical Distribution of Namespaces

Namespace functions can be distributed along a vertical axis in various ways. Whereas a namespace without any such distribution is a "flat" namespace, a namespace with full vertical distribution is a "hierarchical" one (see figure 4.1).

In a flat namespace, a single entity provides the full name service and thereby operates the full namespace. Therefore, a single point of control exists. Flat namespaces can be easily regulated, be it by the namespace provider, by the government, or by hackers. Flat namespaces also have a single point of knowledge: one database stores the names of all objects as well as their locations and other attributes. If the database misuses this knowledge for data mining and marketing purposes, flat namespaces pose a privacy risk.

Hierarchical namespaces have different characteristics. In a hierarchical namespace, the name service is distributed over a hierarchy of different entities. Each entity is responsible for a different subset of names. No single entity exercises direct and perfect control over the whole namespace. Rather, different parts of the namespace can be managed by different entities, and, occasionally, governed by different policies. Hierarchical namespaces therefore enable some competition to occur within the namespace.

The DNS may again exemplify this governance dimension. It consists of a hierarchically organized network of databases (operated by a network of so-called "registries"). Therefore, domain names under the TLD .de are assigned and administered by a different registry than domain names under the TLD .com. The registries have at least some discretion in the way they assign domain names. Many country-code top level domain

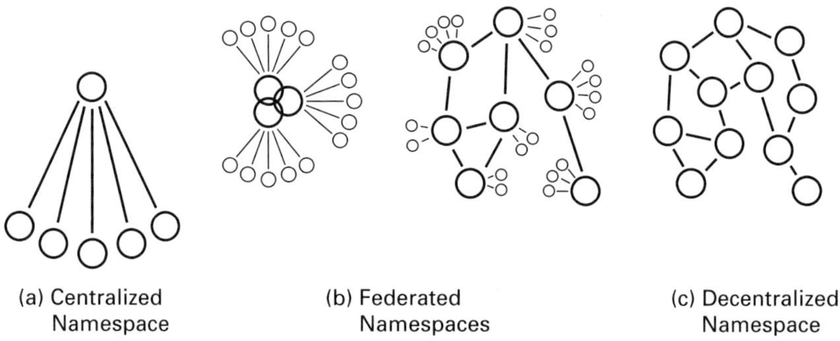

Figure 4.2
From Centralized to Decentralized Namespaces

(ccTLD) registries, for example, do not impose ICANN's UDRP upon domain name registrars and registrants. To some extent, responsibility for assigning domain names and for maintaining the name service is distributed throughout the hierarchical DNS network.[33] Thereby, the decision as to what policies are appropriate for governing the domain name space is decentralized as well. This decentralization in deciding policy issues could only be achieved by making a technical decision at the design stage of the DNS: the choice of a hierarchical structure as the DNS' topology.

ENUM, IP addresses, and the Library of Congress bibliographic classification are further examples of hierarchical namespaces. Conversely, Microsoft Passport and TCP/UDP port numbers are flat namespaces.

3.3.2. Horizontal Distribution of Namespaces

Besides different vertical distributions, namespace functions can be distributed along a horizontal axis in various ways. Whereas a namespace without any such distribution may be called a "centralized" namespace, a namespace with full horizontal distribution is a "decentralized" one. Between those two extremes lie various forms of "federated" or interconnected namespaces (see figure 4.2). Choosing a namespace topology along the horizontal axis determines its regulability as well as its privacy, liability, and competition implications.

3.3.2.1. Centralized Namespaces

In a centralized namespace, a single entity provides the name service and thereby operates the full namespace.[34]

Regulability Centralized namespaces have a single point of control that can be regulated. This is most obvious in centralized peer-to-peer (P2P) systems. In a P2P file sharing network, files can be shared among the participating peer computers without any intervention by a centralized server. In order to share files, however, the individual peer must know where files are located in the network. Therefore, P2P networks need a namespace in which each file that is available in the network is assigned to the address of the peer computer where the file is located.

Early P2P systems used a centralized namespace for locating files in the network. Until Napster was shut down by a court order in 2001, for example, it used a centralized namespace that was located at a server operated by Napster. P2P systems such as Napster have been criticized for facilitating mass-scale piracy. To suppress such piracy, record companies and other copyright holders have demanded that Napster be shut down. In a P2P network with a centralized namespace, shutting down the overall system is a relatively easy task. Shutting down the central namespace destroys the whole P2P network, because without the namespace, a peer computer can no longer locate any file in the P2P network. Therefore, a centralized topology opens the namespace to regulation of various sorts: the government or courts may order that the namespace be shut down. Also, the namespace may be shut down by the namespace provider or by hackers. Centralized namespaces are prone to regulation.

Privacy A centralized namespace is not only easy to regulate, it may also pose privacy risks. In a centralized namespace, all information about the namespace is located within one entity. This entity assigns names, it knows who is accessing the namespace and which names are looked up. During Napster's operation, for example, Napster was in the unique position to know about every download occurring from every computer connected to the Napster system. Such information can be valuable data for surveillance, data mining, marketing and personalization purposes.

However, centralized namespaces may have ambivalent implications for privacy protection, as the Microsoft Passport user namespace exemplifies. By mapping unique identifiers to individual users, this system allows users to establish lasting digital identities on the Internet. Once a user is registered in this user namespace, he can access all web sites that use Microsoft Passport as their authentication service without having to authenticate himself at each individual web site, as Microsoft Passport will provide the participating web site with the necessary authentication information.[35]

Microsoft Passport is a centralized user namespace as Microsoft is currently the only provider of the namespace. User namespaces can theoretically be used to collect large amounts of personal data. Microsoft Passport can store the name of the user, his credit card information and address, as well as demographic or preference data such as gender, occupation, ZIP code, birthday, and language preference. Passport does not transmit such data to participating web sites without the user's consent. Rather, as a default, Passport only transmits a 64-bit-long unique user identifier. Thereby, users can access third-party web sites—such as eBay or McAffee—without having to provide the web site with any personal information such as the user's name, e-mail address, or phone number. Such namespace design can enhance the privacy of its users in light of the fact that the amount of information a user has to share with a particular web site to gain access can be decreased.

This is not to say that the users' privacy is perfectly or even adequately protected in Microsoft Passport.[36] Indeed, the centralization of information storage may lead to increased privacy risks if the central information storage provider is not trustworthy. However, the Passport example illustrates how different namespace topologies lead to different allocations and sizes of privacy risks. Centralized namespaces may protect privacy interests because services that depend on the namespace do not have to store personal information by themselves. However, they may also threaten privacy interests as the central storage may be insecure or the namespace provider itself may misuse the stored information.

Liability In a centralized namespace, knowledge about all issues relating to the namespace is centralized as well. This centralization of knowledge means that, under certain circumstances, the single namespace provider might be held responsible for the activities that its users engage in with the names. Doctrines of contributory and vicarious infringement can be used against centralized namespaces. The courts, for example, held Napster responsible for alleged copyright violations of its users because, as a provider of a centralized namespace, Napster had knowledge about every event occurring within the namespace.

Competition Choosing a centralized topology for a namespace also influences the competitive framework in which the namespace operates. Namespaces are subject to network effects:[37] the more users and service providers use a particular namespace, the larger and therefore more valuable the namespace becomes to them.[38] Therefore, in communication markets shaped by network effects, the optimal number of namespaces is often one. Network effects can lead to *de facto* standards, or even monopolies in a market. In such markets, switching from one namespace to another may involve such high costs for both consumers and producers ("switching costs") that the market is locked into a particular namespace.

Many centralized namespaces are subject to these effects. Network effects are one of the main reasons why no competitor to the ICANN-administered DNS has succeeded in providing universally accessible alternate top level domains.[39] The refusal of AOL to interconnect its instant messaging systems with competing systems can be explained by network effects as well.[40]

3.3.2.2. Federated Namespaces

Competition Although network effects can lead to a namespace monopoly, this is not inherently bad from an economic perspective. If, in a particular market, having a single namespace is more efficient than having several competing namespaces, then this is desirable.[41] Having a single namespace does not mean, however, that the namespace should be owned by a single company, or that only one company should provide the whole namespace. Rather, such namespaces can be opened to competitors. Several competitors may offer competing namespace services that adhere to one common standard. Open standards reduce the lock-in effects produced by network effects. They shift the locus of competition from competing *for* the market to competing *within* the market, using common standards.[42] Such a market structure may combine the best of both worlds: the efficiency gains of one common namespace pushed by network effects, and the efficiency gains of competition between different providers in this namespace.

Centralized namespaces can be opened to competition by introducing interoperability and interconnection between different namespace providers, that is, by "federating" the namespace (see figure 4.2).[43] Federating namespaces frees them from proprietary control. Microsoft Passport may exemplify the difference between centralized and federated namespaces. Microsoft formerly structured its Passport namespace as a proprietary service. Passport did not interoperate with other competing identification and authentication services. In such a centralized namespace, technical, economic, and policy control are exercised by a single entity. However, in September 2001, Microsoft announced that it would open Passport to other authentication systems. A user with an account at a competing authentication system could still access web sites that use Passport as their authentication service. For Passport would accept the authentication from the competing service and issue a Passport ticket for this user. In other words, Passport would translate the "foreign" identity into a Passport identity.[44]

By creating interconnections between different namespaces, competition between the federated, interoperable namespaces becomes possible. A competing user authentication service, for example, could offer its service under a privacy policy different from Passport's privacy policy. If Microsoft chose to offer Passport only on a high usage fee basis or if it tied the Passport service to another product, a competitor could always offer his authentication service under very different terms, but still interoperate with Passport.[45] By federating user namespaces, they are no longer a proprietary tool for data mining, but rather an open authentication platform upon which other applications can build.

Further examples of federated namespaces are PKI namespaces that employ bridge certification authorities, oligarchy models, mesh architectures and various means of cross-certification,[46] interconnected telephone networks,[47] the Internet,[48] as well as the discussions about interoperable instant messaging systems and about root zone level competition in the DNS.[49]

Privacy The partial decentralization in federated namespaces may be used to enhance the protection of privacy interests. In a centralized user namespace such as the current Microsoft Passport architecture, each user is assigned a globally unique ID. Globally unique IDs always pose privacy risks as they can easily be used to connect personal information gathered from various sources. In the federated user namespace of the Liberty Alliance, a competing project for standardizing a federated user namespace,[50]no globally unique ID that is tied to a particular identity provider exists. Rather, users have different accounts with one or more identity providers as well as with numerous service providers. With the consent of the user, all or some of his identities can be linked together. Even if two identities are linked together, however, no common identity exists. Both services remember the other's handle for the user and communicate with each other only with these handles. This architecture enables the user to decide in a very fine-grained way which identities become linked together and which should stay separate. Thereby, the user can control which providers can exchange information about him.[51] Federating namespaces can enhance privacy protection as the overall namespace becomes modularized.

3.3.2.3. Decentralized Namespaces

Whereas in a federated namespace, a small number of interconnected namespaces exists, in a fully decentralized namespace, the namespace itself is dispersed across the network beyond recognition. Decentralized P2P networks are prime examples of such namespaces. In a fully decentralized P2P system, no single namespace exists. Rather, each peer has a namespace in which all locally stored files are registered. Resolving a name means searching the whole network or at least significant parts of it. The P2P system Gnutella uses such architecture. Other decentralized namespaces include encryption systems—such as the original Pretty Good Privacy (PGP) system—that do not employ a structured PKI architecture, but rather a more anarchical model in which public keys are certified on a peer-to-peer basis.[52] Decentralized namespaces possess interesting features regarding their regulability, privacy protection and the liability of the namespace "providers."

Regulability If copyright holders want to shut down a fully decentralized P2P network, they cannot simply shut down a central namespace, because the namespace is scattered across the individual peers of the P2P network. Shutting down any one of the peers in the network would not impact the overall network. Fully decentralized namespaces are much harder to regulate than centralized namespaces.

Liability and Privacy As no single entity exists that operates the namespace, liability for actions occurring within the namespace is scattered as well; there are only the individual users who could be held liable, but no central entities, as no such entities exist. In a fully decentralized namespace, knowledge for actions occurring on top of the namespace is dispersed throughout the network. In a decentralized P2P network, for instance, no central entity exists that monitors all the transactions occurring in the network. Some of these networks—such as Freenet—are even designed with the explicit purposes of preserving privacy for information producers and consumers and resisting censorship.[53] Surveillance of fully decentralized namespaces is an intricate task.

As this subsection has shown, choosing a topology for namespaces has far-reaching implications from a legal and policy perspective. The more decentralized a namespace becomes, the harder it becomes to regulate, the more it protects the privacy and anonymity of its users, the more difficult, more expensive and more inefficient it becomes to hold someone liable for the actions occurring on top of the namespace, and the more competition it enables within the namespace.

3.4 Intensity of Namespace Governance

Namespaces can be governed with various intensities. Whether a namespace is tightly controlled or merely left to its own impacts various policy aspects of namespace governance, ranging from regulability to innovation issues.

3.4.1. Control versus Coordination

Some namespaces are tightly controlled and coordinated. Some namespaces are coordinated, but not controlled. Other namespaces are neither controlled nor coordinated. In various namespaces, some control or coordination is necessary for technical reasons. If a namespace, for example, provides fewer names than needed, that is, if it is a scarce namespace, coordination mechanisms must exist to assign names in an efficient and resource-saving manner. Coordination, however, is not the same as tight control.

A namespace that illustrates the difference in degree between control and coordination is the IP address space. IP addresses form a distinct namespace that is administered by the "Internet Assigned Numbers Authority" (IANA). Traditionally, such addresses had been assigned entirely on a first come, first served basis. In the early 1990s, however, it became evident that the IP address space would be used up in a few years.

To cope with this scarcity, IP address registries began to impose policies that assigned IP addresses based on demonstrated need and made them subject to annual fees. Thereby, the registries attempted to prevent the stockpiling of IP addresses and conserve the current address space as long as possible. The registries increasingly used their technical control over the IP address space to facilitate rationing and policy enforcement. However, apart from this scarcity problem, the IP address assignment process is still restricted to mere coordination tasks. The IP address registries do not exercise any control over any other policy issues that would be worth mentioning.[54]

If one compares the regulatory philosophy governing the IP address space with the current regulatory philosophy governing the domain name space, the difference in degree between control and coordination becomes obvious. Name scarcity may necessitate a coordination of the name assignment process. It does not, however, necessitate any tight control over other, policy-related issues of the namespace.

3.4.2. Control versus Uncoordination and Decentralized Innovation

In some infinite namespaces, the aforementioned coordination problems are solved by the sheer size of the namespace. Such namespaces are fully "democratized." No coordination is necessary, because no entity in the namespace has more knowledge, control, or responsibility over the namespace than any other entity in the namespace. Such namespaces create open platforms that enable decentralized, uncoordinated innovation.

This governance implication of creating infinite namespaces can be best observed in the TCP/UDP port number space. Port numbers provide a standardized means for different computer programs to communicate with each other over the Internet.[55] In combination with the IP address of a computer, port numbers uniquely identify every program running on any computer connected to the Internet.[56] Therefore, port numbers provide a service namespace that identifies applications running on networked computers.

In total, 65,535 distinct port numbers exist. The first 1,024 of the 65,535 ports are all so-called "well-known ports" that are assigned to programs that are widely used across the Internet. Port numbers in the range from 1,024 to 49,151 are called "registered ports." They are assigned to less common programs. Both well-known and registered ports are listed in a list maintained by the "Internet Assigned Numbers

Authority" (IANA).[57] While IANA exercises some control over the assignment of ports 0 through 49,151, the ports 49,152 through 65,535 are totally unassigned ("private ports"). Every application that wants to communicate with another application running on a remote computer can do so by simply using one of the private ports.

Therefore, 25 percent of the port number space are not only uncontrolled, but also uncoordinated. Leaving the port number space open arguably played a major role in fostering innovation on the Internet. To realize how this value is embedded in the structure of the port number space, one need only imagine a different design. *First*, imagine that IANA assigned every port number to specific programs so that no private ports exist. *Secondly*, imagine that IANA assigned port numbers only according to a set of predetermined rules. It could auction ports or charge an administrative fee for assignment. It could choose to assign no ports to P2P applications because of piracy concerns. It could choose to assign no ports to video streaming software because it did not want the Internet to become a competitor of cable TV. Fortunately, it is unrealistic that IANA would ever assign port numbers based on such criteria. The scenario becomes more plausible, however, if one imagines, *thirdly*, that it was not IANA that assigned the port numbers, but a company such as AT&T or Microsoft. In such a scenario, the control over the port number space could be used to allow the operation of certain kinds of applications on the Internet while shutting down other applications.

By keeping 25 percent of the port number space open and uncoordinated, IANA has chosen a different path. It cannot prevent anyone from writing an application that operates over the Internet using a private port. Since nobody exercises control over the port number space, everybody is free to invent new technologies running atop of the Internet without having to ask anyone for permission. When Tim Berners-Lee developed the Hypertext Transfer Protocol (HTTP), one of the technologies underlying the World Wide Web, he did not have to ask the AT&Ts or Microsofts of this world for permission to use a port number. The port number space was a free resource.[58]

The observation that certain design choices in the Internet architecture foster innovation occurring on the Internet is not novel. Indeed, it lies at the heart of the so-called "end-to-end argument" (e2e). First described by Saltzer, Reed and Clark in a seminal paper dating from 1984,[59] the e2e argument claims that as much intelligence as possible should reside at the "edges" of a network, that is, at applications running on networked computers, not in the network itself.[60] It vests power in end users and disables control by a central actor within the network.[61] E2e thereby ensures that the network is a neutral platform that does not discriminate between different applications or services.[62] Concerning innovation, e2e implies that "innovators with new applications need only connect their computers to the network to let their applications run."[63] By decentralizing control, e2e enables decentralized innovation.[64] E2e is also an architectural principle of how to design a computer network system under uncertainty—uncertainty concerning how the network will be used in the future, and uncertainty as to what kind of applications will be run over the network.[65] Although in a network, no single entity may exist that can anticipate all possible uses of the network, this knowledge may indeed exist, but may be distributed among a myriad of individual actors in the network. E2e provides a mechanism to cope with such extremely dispersed knowledge in a network.[66] If what kind of innovation will occur on a network is not predictable, e2e argues, the network should not be biased by its very architecture toward any specific kind of innovation.[67]

The connection between e2e and innovation is not a novel observation.[68] However, previous analyses of this connection did not notice that, in this regard, e2e was implemented on the Internet by a particular design of a namespace: the TCP/UDP port number space. As was described above, the port number space leaves 25 percent of all port numbers uncoordinated, thereby enabling decentralized innovation. In regards to innovation policy, the openness of the TCP/UDP port number space is the Internet's implementation of the end-to-end argument.

3.5 Scope of Namespace Governance

The governance of namespaces may differ not only in intensity, but also in scope. Determining the scope of namespace governance impacts various policy aspects, ranging from privacy and regulability to innovation issues.

3.5.1. Information-rich versus Information-poor Namespaces

Namespaces can be designed to collect large amounts of personal information about the persons who are accessing and registering with the namespace. They can also be designed to store as little personal information as possible. Whereas information-rich namespaces may lead to privacy concerns, information-poor namespaces may become a tool for privacy protection. The DNS is an example of an information-rich namespace: personal information about the registrants of Internet domain names has traditionally been publicly available through the WHOIS database. In contrast, no global public databases exist that contain personal information about every telephone subscriber. From an outside perspective, the telephone network is therefore an information-poor namespace.

3.5.2. Single-purpose versus Multi-purpose Namespaces

While some namespaces serve specific narrow purposes, other namespaces can be used for many different purposes and accessed by different applications. This has implications for regulating such namespaces and for innovation occurring on top of them.

Regulability The P2P file namespace Napster served a narrowly confined purpose: to identify and locate music files in the network. Conversely, the DNS device namespace serves many different purposes. From the perspective of the DNS, it does not matter whether domain names are resolved in order to locate music, video, text documents, persons or any other resources. The DNS is a multi-purpose namespace. Single-purpose namespaces are more prone to regulation than multi-purpose namespaces. As soon as a court had determined that the Napster namespace was used mainly for illegitimate purposes, the namespace could be regulated. A namespace such as the DNS, which is used for some illegitimate, but also for many more legitimate purposes, would be much harder to shut down under this rationale.

Innovation around Namespaces Whether a namespace serves a single or multiple purposes, also determines to a large extent whether the namespace fosters or hinders innovation. Some multi-purpose namespaces are "vertically innovation-friendly" in the sense that they do not prevent the creation of other namespaces on top of them (see figure 4.3).

Such multi-purpose namespaces facilitate innovation in software applications that need their own namespaces, because such applications can use the existing namespace infrastructure and build their own namespaces on top of it. A single-purpose, not vertically innovation-friendly name-

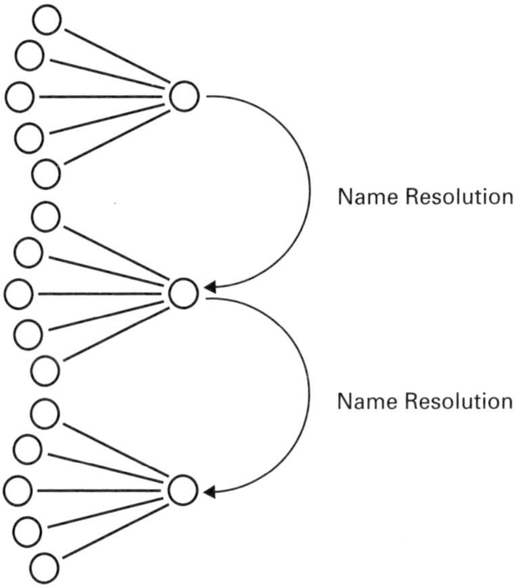

Figure 4.3
Vertically Innovation-friendly Namespaces

space prevents such namespace creation by contractual or technological means. A prime example is the interrelation among the Ethernet address, IP address, and domain name spaces. All three namespaces are vertically innovation-friendly as they are built on top of each other: while the DNS resolves domain names into IP addresses, the "Address Resolution Protocol" (ARP) resolves IP addresses into Ethernet addresses.[69] The Ethernet address space did not prevent the creation of the IP address and domain name spaces on top of it. Other examples of layered namespaces include instant messaging services that build user namespaces on top of the IP address or the domain name space. The Madster network[70] creates a virtual private network on top of the America Online Instant Messenger (AIM) user namespace. It enables users identified by the underlying AIM user namespace to share music and other files identified by the Madster file namespace. This example shows that file namespaces can be built on top of user namespaces that in turn are built on top of several layers of device namespaces. Creating multi-purpose namespaces can therefore foster openness and decentralization—one of the central goals of the e2e argument.[71]

3.5.3. Fixed versus Adaptive Internal Structure

Designing the internal structure of a namespace is complicated by the fact that, to put it simply, history matters. Decisions made at the time of the initial technical design of a namespace may impede its use at a later time when the environment in which the namespace operates has changed. Building a comprehensive, rigid namespace structure at one time does not mean that this structure will be the best possible structure in the future.

Changing Number of Names The most widespread problem in this regard is that the size of a namespace may gradually prove too small. For this reason, the size of the IP and the Ethernet address spaces was enlarged over time. Similar problems arose in the domain name, the Social Security number and the bar code space.[72] Making a namespace too small in the beginning may put it at a disadvantage in the long run.

Changing Kinds of Names A namespace can encode information about the kinds of names that are included in the namespace in its very structure. Because the kinds of names the namespace has to deal with change over time, its structure may become outdated. This is especially important in a particular class of namespaces: bibliographic classifications schemes.[73] For a long time, classification schemes organized knowledge in a strictly hierarchical manner. The Library of Congress Classification (LCC), one of the largest in the world, continues to do so up to the present day. In such a classification scheme, each book or document is assigned one or several numerical classifiers that locate the contained knowledge in a hierarchical representation of all existing knowledge.

However, all bibliographic classification schemes have to grapple with the problem that knowledge is constantly emerging and changing. As new subjects and areas of research emerge, classification schemes become outdated.[74] Although many classification schemes are updated by their editors on a regular basis, it can take years until new fields of science and knowledge are properly reflected in the schemes. Sometimes, classification schemes are even incapable of integrating new subjects into their existing structure. Such classification difficulties impede the organization and processing of new knowledge, which can have detrimental impact on scientific progress.[75] These problems of coding information into the structure of the namespace and of the resulting path dependencies are not confined to bibliographic classification schemes, but can be observed in other namespaces as well.[76]

Regarding bibliographic classification systems, library and information science has invested large amounts of time and effort to solve this problem. Over the last few decades, various forms of "self-perpetuating" classification schemes have been proposed. They are based on an idea developed by the Indian librarian Shiyali R. Ranganathan in the 1930s. As it is beyond the scope of this article to describe his so-called "faceted analytico-synthetic" approach in detail, suffice it to say that it attempts to provide librarians with modularized tools by which they can build classification numbers on their own in a decentralized, yet uniform way. Thereby, even different classifiers working in different libraries should be able to create new subjects without waiting for the next edition of the classification and yet achieve identical results.[77]

The faceted analytico-synthetic approach, which has been adopted to varying degrees by many current classification schemes,[78] exemplifies how namespaces can be designed so that their structure can be adapted in a decentralized, yet uniform way if the kind of names that have to be identified changes over time. By providing tools for modularized and decentralized name creation, namespaces can be dynamically changed in substance and scope without changing their underlying basic modular components. Such ideas can be found and applied in various namespaces.[79]

4. Implications of Governance Dimensions

Hitherto, this article has identified several dimensions along which namespace governance can be studied (means, intensity and scope of governance, namespace topology and governance by whom?). Although these dimensions differ in many respects, they are concerned with two basic aspects, which will be summarized in the following discussion.

4.1. Namespace Architectures Protect and Express Values

As this article has illustrated, technical control over a namespace can be used as leverage for policy and legal control. Namespaces protect policy and legal values. The imposition of the UDRP onto the DNS protects trademark-related interests; the federation of namespaces enables competition between different namespaces; decentralized namespaces can protect privacy interests; uncoordinated namespaces enable an open platform for decentralized, uncoordinated innovation.

At the same time, by protecting certain values, many namespaces communicate a particular *Weltanschauung*. This is particularly noticeable in bibliographic classification schemes. Classification schemes often demonstrate structural biases on the basis of gender, race, ethnicity, culture, or religion.[80] The Dewey Decimal Classification class for religion is biased toward—or, more gently spoken: heavily focused on—Christianity.[81] The Library of Congress Classification exhibits distinct biases "towards the social structure, history, law and cultural concerns of the United States."[82] The plasticity of bibliographic classification schemes can also be used strategically: Chinese classification systems have been deliberately shaped to reflect particular political and ideological beliefs.[83]

This is not the place to criticize particular classification schemes. Indeed, some biases in classification schemes may be unavoidable.[84] Biased bibliographic classification schemes merely illustrate that namespaces are social constructs that reflect the same biases as the culture that creates them.

4.2. Allocation of Knowledge, Control, and Responsibility

While this article has identified several distinct governance dimensions, most of them can be reduced to a single, more abstract dimension. Most governance dimensions described thus far differ in the allocation of knowledge, control, and responsibility within a namespace.

A flat namespace, for example, has a single point of *knowledge*. One database knows all names and their related attributes. Such centralized knowledge can pose a privacy risk. At the same time, centralized knowledge can lead to centralized *control*. If one single entity in a namespace knows about all actions occurring within the namespace, it is an optimal starting point for namespace regulation. The existence of such a control point can lead to a situation in which the flat namespace is held centrally *responsible* for all actions occurring within the namespace. The Napster case is a prime example of such a centralization of knowledge, control, and responsibility.

Table 4.1
Allocation of Knowledge, Control, and Responsibility

Namespace Architecture		Allocation of		
		Knowledge	Control	Responsibility
Vertical Distribution	Flat	c[85]	c	c
	Hierarchical	d	m	m
Horizontal Distribution	Centralized	c	c	c
	Federated	m	m	m
	Decentralized	d	d	d
Intensity	Controlled	c	c	c
	Coordinated	m	d	m
	Uncoordinated	d	d	d
Scope	Information-rich	c	c	c
	Information-poor	d	d	d
	Single-purpose	c	c	c
	Multi-purpose	d	d	d
	Rigid Internal Structure	c	c	c
	Adaptive Internal Structure	d	d	d

In vertically distributed, that is, hierarchical namespaces, different parts of the namespace can be managed by different entities and, occasionally, different policies. Hierarchical namespaces distribute knowledge, control, and responsibility over different hierarchies of the namespace. Other dimensions of namespace governance have similar features. Table 4.1 gives an overview of the allocation of knowledge, control, and responsibility in most of the dimensions of namespace governance identified in this article.

5. Designing Namespace Governance

Designing the architecture of namespaces is not a mere technical matter. It entails decisions about legal and policy questions. Structure has consequences. While the article has provided a descriptive analysis of the close intertwining between technology, law and policy in regards to namespaces, it has not addressed the question what the normative consequences of this analysis are. Should namespaces be designed according to certain principles? What are those principles?

Although answering these questions seems necessary to develop a full-fledged normative theory of namespace governance, this article does not attempt to provide such answers. This would not only be beyond scope of the article, but perhaps even impossible for several reasons. *First*, as namespaces are used in many different areas, it seems hard to draw any general normative conclusions that are applicable to namespaces in general. After all, authenticating users in a PKI system is not the same as developing a method to place books in library shelves in some reasonable order. *Secondly*, developing a normative theory of namespace regulation should be based on a sound general normative theory of technology regulation. Such theory would have to consider the advantages and disadvantages of an *ex ante* regulation (which may be impossible owing to a lack of predictability) and an *ex post* regulation (which may be impossible owing to path dependency). Discussing and developing such a general normative theory of technology regulation is an endeavor that is far beyond the scope of this article. *Thirdly*, a normative theory of namespace governance would have to consider the dynamic interaction between the different governance dimensions described above. And, *finally*, it would also have to take into account the dynamic interactions between different namespaces. If one namespace, for example, is designed to be innovation-friendly, but depends on another namespace that is innovation-hostile, openness and innovation are not preserved in the overall system. When the recording industry wanted to shut down Napster, it could have tried to shut down the TCP port 6699 over which Napster clients communicated. However, the e2e-compliant port number space made such regulation impossible. To achieve its goal, the recording industry turned to another namespace that was more controllable: Napster's own file namespace. As long as an open and decentralized namespace depends on another namespace with a different architecture, keeping the namespace open and decentralized does not necessarily mean that openness and decentralization will ultimately reign.

For all these reasons, the article is confined to presenting a taxonomic structure under which the governance of various namespaces can be analyzed. This taxonomy proves helpful for discussing the legal and policy implications of a namespace during its technical design. If one determines, for example, that a namespace should be open, enable competition, protect privacy and foster decentralized innovation, the taxonomy presented provides answers as to how these legal and policy goals may be implemented in a namespace.

Namespaces are an overlooked facet of governance both in real space and cyberspace. Although we are surrounded by namespaces, policy discussions have not regularly drawn any attention to general policy problems of namespaces. This article has shown that the technical design of namespaces in general has numerous legal and policy implications. While the article has focused mainly on namespaces in cyberspace, many of its findings can be applied to namespaces in real space as well. As we are literally surrounded by namespaces both in cyberspace and real space, governance in namespaces could be a ubiquitous theme as well.

Acknowledgments

I would like to thank Peter Bechtold, Jonathan Greenberg, Jeff Gould, Brian Hemphill, Kurt Jaeger, Lawrence Lessig, Nelson Minar, Wernhard Möschel, Milton Mueller, Markus Müller, Tomas Sander, and the participants at workshops and seminars at Stanford Law School and the University of Tübingen Law School, Germany, as well as the TPRC 2002 participants for their valuable comments and suggestions. A longer version of this article will be published in 36 Loyola of Los Angeles Law Review (forthcoming 2003).

Notes

1. See Henry H. Perritt, *Towards a Hybrid Regulatory Scheme for the Internet*, 2001 U. Chi. Legal F. 215, 241–244.

2. A&M Records, Inc. v. Napster, Inc., 239 F.3d 1004 (9th Cir. 2001), *remanded* 2001 WL 227083 (N.D. Cal. 2001), *aff'd*, 284 F.3d 1091 (9th Cir. 2002).

3. See Manuel Castells, The Internet Galaxy 36–63 (2001).

4. Lawrence Lessig, Code and Other Laws of Cyberspace (1999).

5. See Milton Mueller, Ruling the Root—Internet Governance and the Taming of Cyberspace 267 (2002).

6. For an analysis of the related problems of classification, see Geoffrey C. Bowker & Susan Leigh Star, Sorting Things Out—Classification and Its Consequences (1999).

7. See Douglas A. Galbi, *A New Account of Personalization and Effective Communication* 4, *at* http://papers.ssrn.com/abstract=286288 (2001).

8. See William M. Landes & Richard A. Posner, *Trademark Law: An Economic Perspective*, 30 J.L. & Econ. 265, 269 (1987).

9. See Andrew S. Tanenbaum & Maarten van Steen, Distributed Systems—Principles and Paradigms 184 (2002); John F. Shoch, Inter-Network Naming, Addressing, and Routing, *in* Proceedings of the 17th IEEE Computer Society International Conference 72 (1978).

10. *Cf.* George Coulouris et al., Distributed Systems—Concepts and Design (3d ed., 2001); Tanenbaum & van Steen, *supra* note 9.

11. See Shoch, *supra* note 9, at 72.

12. Coulouris et al., *supra* note 10, at 357; Tanenbaum & van Steen, *supra* note 10, at 183, 194.

13. Coulouris et al., *supra* note 10, at 358.

14. Jerome H. Saltzer, *Naming and Binding of Objects*, *in* Operating Systems—An Advanced Course 99, 121 (Rudolf Bayer et al., eds., 1978); see *also* Mueller, *supra* note 5, at 39.

15. See Coulouris et al., *supra* note 10, at 356; Tanenbaum & van Steen, *supra* note 10, at 184.

16. For an overview, see Stefan Bechtold, *From Copyright to Information Law—Implications of Digital Rights Management*, *in* Security and Privacy in Digital Rights Management 213, 214–216 (Tomas Sander, ed., 2002), *available at* http://www.jura.uni-tuebingen.de/~s-bes1/pub/2002/DRM_Information_Law.pdf (hereinafter Bechtold, *From Copyright to Information Law*). For a more detailed discussion, see Stefan Bechtold, Vom Urheber- zum Informationsrecht—Implikationen des Digital Rights Management 34–75 (2002) (hereinafter Bechtold, Vom Urheber- zum Informationsrecht).

17. Other computer namespaces include variable names in computer languages, character sets, the X.500 directory service, XML namespaces, colorspaces such as RGB or CMYK, databases, and Microsoft Smart Tags.

18. For an overview, see Bechtold, Vom Urheber- zum Informationsrecht, *supra* note 16, at 39–41.

19. Although ICANN is known for managing the DNS, the U.S. government still retains residual authority over the DNS root and has not expressed its intent to give up this authority in the future. See Mueller, *supra* note 5, at 197; A. Michael Froomkin, *Wrong Turn in Cyberspace: Using ICANN to Route Around the APA and the Constitution*, 50 Duke L.J. 17, 91, 105–125 (2000).

20. Internet Corporation for Assigned Names and Numbers, *Registrar Accreditation Agreement* § II.K, *at* http://www.icann.org/nsi/icann-raa–04nov99.htm (Nov. 4, 1999).

21. See Mueller, *supra* note 5, at 192.

22. Microsoft uses paper-based and click-wrap contracts to bind both third-party web sites and consumers who want to access and use the Passport service.

23. In many DRM systems, technology license agreements are used to bind manufacturers of computer electronics and computers. Usage contracts are employed to establish a contractual relationship between the DRM provider and individual consumers. See Bechtold, *From Copyright to Information Law*, *supra* note 16, at 217–222, 227.

24. See Mueller, *supra* note 5, at 191, 232–234.

25. See Russ Housley & Tim Polk, Planning for PKI 107–124 (2001).

26. See *only* Mueller, *supra* note 5, at 192; Froomkin, *supra* note 19; A. Michael Froomkin & Mark A. Lemley, *ICANN and Antitrust*, 2003 U. Ill. L. Rev. (forthcoming 2003).

27. See *only* Michael Geist, *Fair.com? An Examination of the Allegations of Systemic Unfairness in the ICANN UDRP*, 27 Brook. J. Int'l L. 903 (2002); *but* see Annette Kur, *UDRP*, *at* http://www.intellecprop.mpg.de/Online-Publikationen/2002/UDRP-study-final–02.pdf (2002).

28. With Microsoft "Passport," the tension between public and private ordering becomes particularly obvious. As Lawrence Lessig wrote on Slashdot: "When we needed a passport system, we didn't tell Chase Manhattan bank that they could develop the passport system in exchange for a piece of every transaction . . . there was a recognition of the importance of neutral, commons-like, infrastructures upon which others could build neutrally," http://slashdot.org/article.pl?sid=01/12/21/155221 (Dec. 21, 2001).

29. See Allan Wilson, *The Hierarchy of Belief: Ideological Tendentiousness in Universal Classification*, in Classification Research for Knowledge Representation and Organization 389, 393 (Nancy J. Williamson & Michèle Hudon, eds., 1992).

30. In general, the study of a network's topology is concerned with the manner in which the network nodes are interconnected, Roshan L. Sharma, Network Topology Optimization 8 (1990).

31. The following three figures were inspired by Minar, *supra* note 31.

32. Parts of the following analysis build upon the overview of different distributed systems topologies by Nelson Minar, *Distributed Systems Topologies*, *at* http://www.oreillynet.com/pub/a/p2p/2001/12/14/topologies_one.html (part 1); http://www.oreillynet.com/pub/a/p2p/2002/01/08/p2p_topologies_pt2.html (part 2) (2001–2002).

33. See Mueller, *supra* note 5, at 6.

34. Therefore, "flat" and "centralized" namespaces are essentially the same. While the dichotomy between flat and hierarchical namespaces deals with the vertical distribution of a namespace, the dichotomy between centralized and decentralized namespaces deals with its horizontal distribution.

35. Microsoft Passport thereby enables a so-called "single sign-on" (SSO).

36. After a complaint by privacy advocacy groups led by the Electronic Privacy Information Center (EPIC), the FTC conducted an investigation of Microsoft Passport and, in August 2002, proposed a consent order that would prohibit Microsoft from misrepresenting information practices and force the company to implement a comprehensive information security program in Microsoft Passport; see *In the Matter of Microsoft Corporation,* 2002 WL 1836831 (FTC 2002), *available at* http://www.ftc.gov/opa/2002/08/microsoft.htm. In Europe, after an investigation by the European Union's data protection authorities, Microsoft agreed in January 2003 to substantially modify the information flow in the Passport system. See "Microsoft to Alter Online System to Satisfy Europe," *N.Y. Times*, January 31, 2003, at W1; Article 29 Data Protection Working Party,

Working Document on On-line Authentication Services, at http://europa.eu.int/comm/internal_market/en/dataprot/wpdocs/ wp68_en.pdf (Jan. 31, 2003).

37. For some general information about network effects, see Michael L. Katz & Carl Shapiro, *Network Externalities, Competition, and Compatibility*, 75 Am. Econ. Rev. 424 (1985); Mark A. Lemley & David McGowan, *Legal Implications of Network Economic Effects*, 86 Cal. L. Rev. 479 (1998).

38. It is interesting to note that in many communication networks, it is not the network itself that is subject to network effects, but rather the namespace that is underlying the network; see Gerald Faulhaber, *Network Effects and Merger Analysis: Instant Messaging and the AOL-Time Warner Case*, 26 Telecommunications Policy 311, 317 (2002).

39. For an overview of the debate on alternate DNS roots, see *infra* note 49.

40. See Faulhaber, *supra* note 38, at 315–316, 324.

41. Lemley & McGowan, *supra* note 37, at 497.

42. See *id.* 516, 600.

43. See *Consent to the Transfer of Licenses and Section 214 Authorizations by Time Warner, Inc. and America Online, Inc., Transferors, to AOL Time Warner, Inc., Transferee*, 16 F.C.C.R. 6547 ¶ 131 (FCC 2001).

44. For more information about the underlying Kerberos 5.0 security architecture, see B. Clifford Neuman & Theodore Ts'o, *Kerberos: An Authentication Service for Computer Networks*, 32 (9) IEEE Communications Magazine 33, 36 (1994).

45. Of course, this is only true as long as Microsoft would not decide to dissolve the federation with the competitor.

46. See Housley & Polk, *supra* note 25, at 55–66.

47. See Eli M. Noam, Interconnecting the Network of Networks (2001).

48. For more information on backbone peering arrangements, see See Jean-Jacques Laffont et al., *Internet Peering*, 91 Am. Econ. Rev. Papers & Proc. 287 (2001).

49. See Milton Mueller, *Competing DNS Roots: Creative Destruction or Just Plain Destruction?* (2001), *at* http://www.arxiv.org/ftp/cs/papers/0109/0109021.pdf.

50. See http://www.projectliberty.org.

51. For more information, see Liberty Alliance Project, *Liberty Architecture Overview* (Version 1.1) 23-28, at http://www.projectliberty.org/specs/liberty-architecture-overview-v1.1.pdf (Jan. 15, 2003).

52. See Radia Perlman, *An Overview of PKI Trust Models*, 13 (6) IEEE Network 38, 40 (Nov./Dec. 2000).

53. See Ian Clarke et al., *Protecting Free Expression Online with Freenet*, 6 (1) IEEE Internet Computing 40, 41 (2002).

54. See Mueller, *supra* note 5, at 32–39.

55. See Pete Loshin, TCP/IP Clearly Explained 181–210 (3d ed., 1999); Eric A. Hall, Internet Core Protocols 274 (2000).

56. See Loshin, *supra* note 55, at 184–185 (who also provides an explanation of server daemons that complicates this description slightly).

57. http://www.iana.org/assignments/port-numbers.

58. See David P. Reed, Jerome H. Saltzer, and David D. Clark, *Commentaries on "Active Networking and End-to-End Arguments,"* 12 (3) IEEE Network 69, 70 (1998).

59. Jerome H. Saltzer, David P. Reed, and David D. Clark, *End-to-End Arguments in System Design*, 2 (4) ACM Transactions on Computer Systems 277 (1984).

60. See *id.* 286; Lawrence Lessig, The Future of Ideas—The Fate of the Commons in a Connected World 34 (2001); Mark A. Lemley & Lawrence Lessig, *The End of End-to-End: Preserving the Architecture of the Internet in the Broadband Era*, 48 UCLA L. Rev. 925, 930–931 (2001).

61. Hans Kruse, William Yurcik & Lawrence Lessig, *The InterNAT: Policy Implications of the Internet Architecture Debate*, in: Communications Policy in Transition—The Internet and Beyond 141, 150 (Benjamin M. Compaine & Shane Greenstein, eds. 2001).

62. Lessig, *supra* note 60, at 37.

63. Lessig, *supra* note 60, at 36.

64. Kruse, Yurcik & Lessig, *supra* note 61, at 150.

65. Reed, Saltzer & Clark, *supra* note 58, at 70.

66. To some extent, this is reminiscent of Friedrich Hayek's conception of competition as a discovery procedure; see Friedrich A. Hayek, *Competition as a Discovery Procedure*, *in* New Studies in Philosophy, Politics, economics and the History of Ideas 179 (Friedrich A. Hayek 1978); Friedrich A. Hayek, The Political Order of a Free People 67–70 (1979); Friedrich A. Hayek, The Mirage of Social Justice 70–71, 114–115 (1976).

67. See Lessig, *supra* note 60, at 39. A counter-argument against e2e may be that some emerging technologies will need some particular support by the network architecture to reach their full potential.

68. It was clearly formulated by Reed, Saltzer & Clark, *supra* note 58, at 70. Lawrence Lessig builds much of his analysis in his book "The Future of Ideas" on the impact of e2e on innovation.

69. For an overview of ARP, see Hall, *supra* note 55, at 97–134.

70. http://www.madster.com. On October 30, 2002, a district court issued a preliminary injunction ordering Madster to shut down its service. See *In re Aimster Copyright Litig.*, 2002 WL 31443236 (N.D. Ill. 2002).

71. In fact, the call for vertically innovation-friendly namespaces is merely an application of the e2e argument.

72. See Mueller, *supra* note 5, at 73–208; Simson Garfinkel, Database Nation 18–25 (2000); Kate Murphy, *Bigger Bar Code Inches Up on Retailers*, N.Y. Times, Aug. 12, 2002, at C3.

73. For a general overview of the theory and problems of classification, see Rita Marcella & Robert Newton, A New Manual of Classification (1994).

74. *Id.* 30.

75. "A dynamic information society depends on subject access to pioneering literature from the dominant paradigms and literature from the marginal paradigms, as this literature is central for the innovation processes. Classification systems are made from yesterday's concepts of the dominant paradigms. Therefore classification systems are normally not suited to providing subject access to literature from marginal paradigms and pioneering literature in the dominant paradigms," Claus Poulsen, *Subject Access to New Subjects, Specific Paradigms and Surveys: PARADOKS-registration*, 43 (3) Libri 179, 183 (1990).

76. For similar problems in the IP address space (classful versus classless routing), see Mueller, *supra* note 5, at 33–38.

77. Shiyali R. Ranganathan, *Self-Perpetuating Scheme of Classification*, 4 (4) The Journal of Documentation 223, 231 (1949).

78. See Clare Beghtol, *"Facets" as Interdisciplinary Undiscovered Public Knowledge*, 51 Journal of Documentation 194, 201 (1995).

79. The namespace of all chemical compounds with the underlying periodic system shows similar features, see Ranganathan, *supra* note 77, at 232. For attempts to build a facet-oriented search layer on top of the DNS, see John C. Klensin, *A Search-Based Access Model for the DNS*, *at* http://www.rfc-editor.org/internet-drafts/draft-klensin-dns-search–05.txt (Nov. 3, 2002).

80. See Hope A. Olson and Rose Schlegl, *Standardization, Objectivity, and User Focus: A Meta-Analysis of Subject Access Critiques*, 32 (2) Cataloging & Classification Quarterly 61 (2001); see *also* Wilson, *supra* note 29.

81. In the 21st edition of DDC, the class on religion (200) is divided into the following divisions: "philosophy & theory of religion" (210), "the Bible" (220), "Christianity and Christian theology" (230), "Christian practice & observance" (240), "Christian pastoral practice & religious orders" (250), "church organization, social work & worship" (260), "history of Christianity" (270), "Christian denominations" (280), and, finally, "other religions" (290). Over the last years, DDC has undertaken great efforts to reduce systematic biases in its classification scheme.

82. Marcella and Newton, *supra* note 73, at 88.

83. See William E. Studwell, Hong Wu, and Rui Wang, *Ideological Influences on Book Classification Schemes in the People's Republic of China*, 19(1) Cataloging & Classification Quarterly 61, 62, 63–64 (1994).

84. See Robert P. Holley and Robert E. Killheffer, *Is There an Answer to the Subject Access Crisis?*, 1 (2/3) Cataloging & Classification Quarterly 125, 126 (1982); *but* see Wilson, *supra* note 29, at 393.

85. c = fully centralized; m = intermediate between centralized and decentralized; d = fully decentralized.

5

The Geographic Dispersion of Commercial Internet Use

Chris Forman, Avi Goldfarb, and Shane Greenstein

1. Introduction

Advances in frontier technology are only the first step in the creation of economic progress. The next step involves use by economic agents. Adoption by users typically needs time, invention, and resources before economic welfare gains are realized. This principle applies with particular saliency to the Internet, a malleable technology whose form is not fixed across locations. To create value, the Internet must be embedded in investments at firms and households that employ a suite of communication technologies, TCP/IP protocols and standards for networking between computers. Often organizational processes also must change.

The dispersion of Internet use to commercial users is a central concern for economic policy. As a *general purpose technology* (GPT) (Bresnahan and Trajtenberg 1995), the Internet will have a greater impact if and when it diffuses widely to commercial firms. This is particularly so because commercial firms do the vast majority of the investment in Internet infrastructure, and at a scale of investment reaching tens of billions of dollars. Concerns about dispersion are difficult to address, however. Measuring dispersion requires a census of commercial Internet use, which, in turn, requires extensive data and an appropriate framework. This has not been done by any prior research. This study fills this gap.

We construct a census of adoption, the most common yardstick for measuring a new technology's use (Rogers 1995). We use this census to answer questions on the regional distribution of commercial Internet use.[1] How widely dispersed is Internet technology across locations and industries? Which regions adopt often and which do not? How does this measurement of dispersion compare with other ways of measuring the spread of the Internet?

Three themes shape our approach to answering these questions. First, our approach is consistent with standard ruminations about the strategic advantages affiliated with adoption of Internet technology. For example, some investments in Internet technology are regarded as "table stakes"—they are required for companies to be a player in a market—whereas other investments are regarded as the basis of competitive advantage (Porter 2001). Second, our framework extends principles of "universal service" to Internet technology (Compaine 2001, Noll et al. 2001). Third, since there is no preset pattern for the adoption of GPTs, we seek to document differences in adoption between locations.

We propose to analyze the dispersion of use of the Internet in two distinct layers. In one layer—hereafter termed *participation*—investment in and adoption of Internet technology enables participation in the Internet network. Participation is affiliated with basic communications, such as email use, browsing and passive document sharing. It also represents our measure of "tables stakes," namely, the basic Internet investment required to do business. In the second layer—hereafter termed *enhancement*—investment in and adoption of Internet technology enhances business processes. Enhancement uses Internet technologies to change existing internal operations or to implement new services. It represents our measure of investment aimed at competitive advantage.

Our analysis covers all medium and large commercial users, approximately two-thirds of the workforce. We use a private survey of 86,879 establishments with over 100 employees. The survey is updated to the end of 2000. Harte Hanks Market Intelligence, a commercial market research firm that tracks use of Internet technology in business, undertook the survey. We use the County Business Patterns data from the Census and routine statistical methods to generalize our results to the entire population of medium to large establishments in the United States.

We develop three major conclusions: First, we conclude participation and enhancement display contrasting patterns of adoption and dispersion. Overall, we find an average rate of adoption in excess of 88 percent; participation is near saturation in a majority of geographic locations. By any historical measure, such extensive adoption is remarkable for such a young technology. In contrast, though enhancement is widespread across industries and locations, the rate is much lower than that found for participation. Such investment occurs at approximately 12.6 percent of establishments.

Second, we show that Internet technologies displayed geographic usage patterns common to other communication technologies; however we argue different reasons from other authors. Specifically, there is evidence consistent with a mild geographic digital divide in both participation and enhancement. Although participation is high, the average establishment in a small metropolitan statistical area (MSA) or rural area is about 10 percent to 15 percent less likely to participate than one in the largest MSAs. Also, establishments in MSAs with over one million people are one and a half times as likely to use the Internet for enhancement than are establishments in MSAs with less than 250,000 people.

Why do some regions lead and others lag? We offer an explanation that differs sharply with the literature on digital divide. We conclude that the (preexisting) distribution of industries across geographic locations explains much of the difference in rates in enhancement. This is not the entire explanation, but it is certainly important. Hence, we question the prevailing opinion that the dispersion of the Internet sharply benefited a small number of regions. We argue that regional growth policies, in addition to focusing on correcting lack of participation in a few locations, should also focus on understanding how regional growth policies can broaden the foothold that enhancement has across the majority of regions.

Third, existing studies fail to document the dispersion of use by commercial establishments. We establish this by comparing our data with other measures. We find that the geographic dispersion of commercial Internet use is positively related to the dispersion in household and farm use, as documented in previous research, but the relationship is not strong. Hence, we conclude that previous studies provide a misleading picture of dispersion.[2]

2. Background

Our framework builds on microstudies of Internet investment in commercial establishments and organizations.[3] It is motivated by the user-oriented emphasis in the literature on GPTs.[4]

2.1 General Purpose Technologies and the Commercialization of the Internet

The diffusion of the Internet can be viewed in the context of observations about technological convergence (Ames and Rosenberg 1984), which is the increasing use of a small number of technological functions for many different purposes. Bresnahan and Trajtenberg (1995) develop this further in their discussion of GPTs, which they define as capabilities whose adaptation raises the marginal returns to inventive activity in a wide variety of circumstances. GPTs involve high fixed costs in invention and low marginal costs in reproduction. A GPT is adapted for any new use, and this adaptation takes time, additional expense and further invention. Following the literature, we label these as *co-invention expenses*. Studies have found that co-invention influences computing and Internet technology investments by business users (Bresnahan and Greenstein 1997, Forman 2002).

Almost by definition, GPTs have a big impact if and when they diffuse widely, that is, if they raise the marginal productivity of a disparate set of activities in the economy. As a practical matter, "disparate" means a great number of applications and industries, performed in a great number of locations. What stands in the way of achieving wide and disparate diffusion? Barriers arise as a result of users facing different economic circumstances, such as differences in local output market conditions, quality of local infrastructure, labor market talent levels, quality of firm assets or competitive conditions in output markets. Simply put, these barriers are different co-invention expenses.

There is no preset pattern for the dispersion of GPTs. They can diffuse in layers or waves (e.g., Lipsey, Becker, and Carlaw 1998). Below we argue that analysis of the dispersion of the Internet to commercial business requires analysis of distinct layers. We hypothesize that the co-invention costs of certain types of Internet investment were low, whereas other bottlenecks persistently produced high coinvention costs. For low co-invention activities, adoption became a requirement to be in business. When the costs were higher and the benefits variable for other aspects, firms were more hesitant, investing only when it provided competitive advantage.

Consequently, we ignore differences across applications and intensities of use within an establishment. We focus on two layers that vary across location and industry. We label these layers as *participation* and *enhancement.*

The first layer, participation, is a key policy variable. As noted, it represents the basic requirements for being at the table for medium and large businesses. By 2000, participation was regarded as a routine matter. Its emphasis also arises in many studies of ubiquitous communications networks. A ubiquitous network is one in which every potential participant is, in fact, an actual participant. Concerns about ubiquity emerge in policy debates about applying principles of "universal service" to new technologies (Cherry, Wildman, and Hammond 1999, Compaine 2001, Noll et al. 2001). For our purposes, we recognize that many different policies for ubiquity target geographic variance in adoption (e.g., reducing urban/rural differences).

The second layer, enhancement, is also important for policy because its use is linked to the productive advance of firms and the economic growth of the regions in which these firms reside. It usually arrives as part of other intermediate goods, such as software, computing or networking equipment. Implementation of enhancement was anything but routine. Enhancement included technical challenges beyond the Internet's core technologies, such as security, privacy, and dynamic communication between browsers and servers. Organizational procedures usually also changed. Benefits accrue to the business organization employing enhancement through the addition of competitive advantage, but the co-invention costs and delays vary widely.

Participation represents a measure of "table stakes," while enhancement represents a measure of investment for competitive advantage.[5] Both layers of activity are important for economic advance, but each has distinct effects on regional and industrial growth. We do not necessarily presume that the two are closely related, but intend to measure the correlation between them.

3. Data and Method

The data we use for this study come from the Harte Hanks Market Intelligence CI Technology database (hereafter CI database).[6] The CI database contains establishment-level data on (1) establishment characteristics, such as number of employees, industry and location; (2) use of technology hardware and software, such as computers, networking equipment, printers and other office equipment; and (3) use of Internet applications and other networking services. Harte Hanks Market Intelligence (hereafter HH) collects this information to resell as a tool for the marketing divisions at technology companies. Interview teams survey establishments throughout the calendar year; our sample contains the most current information as of December 2000.

HH tracks over 300,000 establishments in the United States. Since we focus on commercial Internet use, we exclude government establishments, military establishments and nonprofit establishments, mostly in higher education. Our sample contains all commercial establishments from the CI database that contain over 100 employees, 115,671 establishments in all;[7] and HH provides one observation per establishment. We will use 86,879 of the observations with complete data generated between June 1998 and December 2000. We adopt a strategy of utilizing as many observations as possible, because we need many observations for thinly populated areas.[8] This necessitates routine adjustments of the data for the timing and type of the survey given by HH.

3.1. Data Description and Sample Construction

To obtain a representative sample, we compared the number of firms in our database to the number of firms in the Census. We calculated the total number of firms with more than 50 employees in the Census Bureau's 1999 County Business Patterns data and the number of firms in our database for each two-digit NAICS code in each location. We then calculated the total number in each location. This provides the basis for our weighting. The weight for a given NAICS in a given location is

$$\frac{\text{Total \# of census establishments in location — NAICS}}{\text{Total \# of census establishments in location}} \cdot \frac{\text{Total \# of establishments in our data in location}}{\text{Total \# of establishments in our data in location — NAICS}}$$

Therefore, each location-NAICS is given its weighting from its actual frequency in the census. In other words, if our data under-samples a given two-digit NAICS at a location relative to the census then each observation in that NAICS-location is given more importance.

Using two survey forms, HH surveyed establishments at different times. To adjust for differences in survey time and type, we econometri-

Table 5.1
National Internet Adoption Rates (in percentages)

	Weighted average	Unweighted average	Northeast	Midwest	South	West
Participation	88.6%	80.7%	88.0%	88.7%	89.0%	85.7%
Enhancement	12.6%	11.2%	12.7%	12.7%	12.4%	12.5%
Enhancement & experimenting with enhancement	23.2%	18.1%	24.0%	23.1%	22.7%	23.3%

cally estimate the relationship between an establishment's decision to participate or enhance as a function of its industry, location, timing of survey and form of survey. We then calculate predicted probabilities of adoption for each establishment *as if it were surveyed in the second half of 2000 and were given the long survey*. Once we weight by the true frequency of establishments in the population, we have information about establishments related to two-thirds of the U.S. workforce. The more observations we have for a given region or industry the more statistical confidence we have in the estimate.

3.2. Definitions of behavior

Identifying participation was simple compared to identifying enhancement. We identify participation as behavior in which an establishment has basic Internet access or has made any type of frontier investment. In contrast, for enhancement, an establishment must have made the type of investment commonly described in books on electronic commerce. We identify enhancement from substantial investments in electronic commerce or "e-business" applications. We look for commitment to two or more of the following projects: Internet-based enterprise resource planning or TCP/IP-based applications in customer service, education, extranet, publications, purchasing or technical support.

In table 5.1 we show the results of these definitions. Participation by establishments within the sample is at 80.7 percent (see Unweighted Average in table 5.1). The sample under-represents adopters. Our estimate of the economy-wide distribution, using the true distribution of establishments from the Census, is 88.6 percent (see Weighted Average in table

5.1). Enhancement has been undertaken by 11.2 percent of our sample and 12.6 percent of the true distribution. We also can estimate the rate of adoption by "experimenters," that is, those establishments with some indication of use, but not much. As one would expect for a technology still in the midst of diffusion, the proportion for experimenters (combined with enhancement) is considerably higher than for enhancement alone, reaching 18.1 percent for the unweighted average and 23.2 percent for the weighted average. We have explored this latter definition and found that it tracks the enhancement definition we use below, so it provides no additional insight about the dispersion of use. We do not analyze it further.

4. Distribution Across Metropolitan Areas

In this section we estimate the dispersion of Internet technology to urban businesses. We identify features of urban leaders and laggards. We also show how the (preexisting) geographic distribution of industries is responsible for Internet technology's geographic distribution.

Tables 5.2a and 5.2b cover the largest economic areas in the United States. In them we list the estimates for both participation and enhancement, organized by MSAs with over one million people and listed by areas in the order of highest to lowest adoption rates.[9]

In tables 5.3a, 5.3b, and 5.3c we summarize results for all MSAs by population and average establishment size. Finally, in tables 5.4a, b, and c we show the estimates for the top ten areas for enhancement by population group as defined above. We also list the standard errors[10] and number of observations to show where we have statistical confidence in the estimates.

4.1. Participation

Table 5.2a shows that participation is high in major urban locations. Virtually all establishments in the major urban areas are participating; they have paid the "table stakes." We estimate that thirty-five of the forty-nine major metropolitan areas (MSAs) are above 90 percent.[11] All but five are within a 95 percent confidence interval of 90 percent. Big differences among metropolitan areas are apparent only at the extreme. The bottom ten areas range from 89.1 percent in Pittsburgh to 84.6 percent in Nashville. Although these are the lower adopting areas, they are not very low in absolute value.

From table 5.3a we see that large MSAs are highest with their average participation of 90.4 percent. Participation in medium MSAs averages 84.9 percent. In small MSAs the participation rates are even lower, 75.5 percent on average.

We examined participation across 320 MSAs in the country (unweighted by population size).[12] The median MSA in the United States has participation at 84.3 percent. The lower quartile is 76.4 percent. Of the 80 MSAs in the lowest quartile, 69 have a population of under one-quarter million. In other words, very low participation in urban settings, when it arises, exists primarily in small MSAs.

4.2. Enhancement

In table 5.2b we examine the use of enhancement at establishments in MSAs with over one million people. We estimate that thirty-eight of the forty-nine areas are above 12.5 percent. All but one are within a 95 percent confidence interval of 12.5 percent. The top ten include a set of areas that partially overlaps with the list in table 5.2a. It begins with the greater Denver area (with 18.3 percent) at number one and the greater Portland area at number ten (with 15.1 percent). In between are the greater San Francisco Bay Area, the greater Salt Lake City area, Minneapolis/St Paul, the greater Houston area, Atlanta, Oklahoma City, Dallas/Fort Worth, and San Antonio. Again, big differences with these leaders are only apparent at the extremes. The bottom ten areas range from 12.4 percent in Phoenix to 9.0 percent in Las Vegas. Even so, these low adopting areas are, once again, not very low relative to the average.

Overall, establishments in urban settings are more likely to adopt enhancement than those located outside major metropolitan areas. Table 5.3a shows the adoption of enhancement in MSAs of different population size, highlighting again that large MSAs are somewhat exceptional. Establishments in large MSAs have adoption rates of 14.7 percent. In medium MSAs, it averages 11.2 percent. In small MSAs the rates are even lower, 9.9 percent on average. Table 5.3b strongly hints at the explanation for these differences. The upper quartile of two-digit NAICS industries with the highest enhancement adoption rates includes management of companies and enterprises (55), media, telecommunications and data processing (51), utilities (22), finance and insurance (52), professional, scientific and technical services (54) and wholesale trade (42).[13] The fraction of the

Table 5.2a
Participation Among Metropolitan Areas with Over One Million People

Rank	City	Rate	Std error	Obs	Population
1	San Francisco–Oakland–San Jose, CA	96.4%	0.4%	2135	7,039,362
2	Denver–Boulder–Greeley, CO	95.9%	0.7%	940	2,581,506
3	Cleveland–Akron, OH	94.8%	0.6%	1099	2,945,831
4	Seattle–Tacoma–Bremerton, WA	93.9%	0.5%	1012	3,554,760
5	Salt Lake City–Ogden, UT	93.5%	0.8%	535	1,333,914
6	San Antonio, TX	93.3%	0.8%	395	1,592,383
7	Providence–Fall River–Warwick, RI–MA	93.0%	1.2%	290	1,188,613
8	Grand Rapids–Muskegon–Holland, MI	93.0%	0.7%	503	1,088,514
9	Minneapolis–St. Paul, MN–WI	92.7%	0.5%	1411	2,968,806
10	Los Angeles–Riverside–Orange County, CA	92.5%	0.4%	4099	16,373,645
11	Kansas City, MO–KS	92.2%	0.6%	753	1,776,062
12	Austin–San Marcos, TX	92.1%	0.7%	344	1,249,763
13	Dallas–Fort Worth, TX	92.1%	0.5%	1720	5,221,801
14	Portland–Salem, OR–WA	92.1%	0.6%	776	2,265,223
15	Houston–Galveston–Brazoria, TX	91.7%	0.6%	1413	4,669,571
16	Phoenix–Mesa, AZ	91.6%	0.7%	988	3,251,876
17	Raleigh–Durham–Chapel Hill, NC	91.6%	0.9%	398	1,187,941
18	Columbus, OH	91.5%	0.9%	574	1,540,157
19	Milwaukee–Racine, WI	91.5%	0.7%	855	1,689,572
20	San Diego, CA	91.5%	0.7%	738	2,813,833
21	Detroit–Ann Arbor–Flint, MI	91.4%	0.6%	1621	5,456,428
22	Indianapolis, IN	91.3%	0.8%	646	1,607,486
23	Greensboro–Winston-Salem–High Point, NC	91.1%	0.9%	570	1,251,509
24	Atlanta, GA	90.9%	0.6%	1426	4,112,198
25	Miami–Fort Lauderdale, FL	90.9%	0.7%	1010	3,876,380
26	Charlotte–Gastonia–Rock Hill, NC–SC	90.7%	0.9%	618	1,499,293
27	Boston–Worcester–Lawrence, MA–NH–ME–CT	90.6%	0.5%	2231	5,819,100

Table 5.2a
(Continued)

Rank	City	Rate	Std error	Obs	Population
28	Chicago–Gary–Kenosha, IL–IN–WI	90.5%	0.4%	3431	9,157,540
29	New York–Northern New Jersey–Long Island, NY–NJ–CT–PA	90.5%	0.4%	4775	21,199,865
30	Washington–Baltimore, DC–MD–VA–WV	90.4%	0.5%	2222	7,608,070
31	Philadelphia–Wilmington–Atlantic City, PA–NJ–DE–MD	90.3%	0.5%	1745	6,188,463
32	Rochester, NY	90.3%	1.0%	373	1,098,201
33	Hartford, CT	90.2%	0.9%	500	1,183,110
34	Oklahoma City, OK	90.2%	1.1%	339	1,083,346
35	Memphis, TN–AR–MS	90.0%	1.0%	437	1,135,614
36	Louisville, KY–IN	89.9%	1.0%	448	1,025,598
37	Cincinnati–Hamilton, OH–KY–IN	89.7%	0.8%	772	1,979,202
38	St. Louis, MO–IL	89.7%	0.7%	936	2,603,607
39	Pittsburgh, PA	89.1%	0.8%	727	2,358,695
40	Buffalo–Niagara Falls, NY	88.5%	1.1%	393	1,170,111
41	Tampa–St. Petersburg–Clearwater, FL	88.4%	0.9%	812	2,395,997
42	Jacksonville, FL	87.6%	1.3%	373	1,100,491
43	Las Vegas, NV–AZ	87.2%	1.2%	417	1,563,282
44	Sacramento–Yolo, CA	87.0%	1.2%	427	1,796,857
45	Norfolk–Virginia Beach–Newport News, VA–NC	86.9%	1.2%	374	1,569,541
46	New Orleans, LA	86.0%	1.1%	386	1,337,726
47	West Palm Beach–Boca Raton, FL	85.9%	1.2%	299	1,131,184
48	Orlando, FL	85.5%	1.0%	622	1,644,561
49	Nashville, TN	84.6%	1.1%	466	1,231,311

Table 5.2b
Enhancement among Metropolitan Areas with Over One Million People

Rank	City	Rate	Std error	Obs	Population
1	Denver–Boulder–Greeley, CO	18.3%	1.3%	940	2,581,506
2	San Francisco–Oakland–San Jose, CA	17.0%	0.9%	2135	7,039,362
3	Salt Lake City–Ogden, UT	16.7%	1.7%	535	1,333,914
4	Minneapolis–St. Paul, MN–WI	15.9%	1.0%	1411	2,968,806
5	Houston–Galveston–Brazoria, TX	15.7%	1.0%	1413	4,669,571
6	Atlanta, GA	15.4%	1.0%	1426	4,112,198
7	Oklahoma City, OK	15.4%	2.0%	339	1,083,346
8	Dallas–Fort Worth, TX	15.3%	0.9%	1720	5,221,801
9	San Antonio, TX	15.3%	1.9%	395	1,592,383
10	Portland–Salem, OR–WA	15.1%	1.3%	776	2,265,223
11	Providence–Fall River–Warwick, RI–MA	14.9%	2.2%	290	1,188,613
12	Austin–San Marcos, TX	14.7%	1.9%	344	1,249,763
13	Cleveland–Akron, OH	14.7%	1.2%	1099	2,945,831
14	Tampa–St. Petersburg–Clearwater, FL	14.6%	1.3%	812	2,395,997
15	Memphis, TN–AR–MS	14.5%	1.8%	437	1,135,614
16	Seattle–Tacoma–Bremerton, WA	14.5%	1.2%	1012	3,554,760
17	Hartford, CT	14.4%	1.6%	500	1,183,110
18	San Diego, CA	14.3%	1.3%	738	2,813,833
19	Cincinnati–Hamilton, OH–KY–IN	14.2%	1.3%	772	1,979,202
20	Washington–Baltimore, DC–MD–VA–WV	14.2%	0.8%	2222	7,608,070
21	Chicago–Gary–Kenosha, IL–IN–WI	14.1%	0.7%	3431	9,157,540
22	Rochester, NY	14.1%	1.9%	373	1,098,201
23	Boston–Worcester–Lawrence, MA–NH–ME–CT	13.9%	0.8%	2231	5,819,100
24	Detroit–Ann Arbor–Flint, MI	13.8%	0.9%	1621	5,456,428
25	Kansas City, MO–KS	13.7%	1.3%	753	1,776,062
26	Raleigh–Durham–Chapel Hill, NC	13.7%	1.7%	398	1,187,941
27	Pittsburgh, PA	13.6%	1.3%	727	2,358,695
28	Indianapolis, IN	13.6%	1.4%	646	1,607,486

Table 5.2b
(Continued)

Rank	City	Rate	Std error	Obs	Population
29	Charlotte–Gastonia–Rock Hill, NC–SC	13.6%	1.5%	618	1,499,293
30	West Palm Beach–Boca Raton, FL	13.6%	2.0%	299	1,131,184
31	Los Angeles–Riverside–Orange County, CA	13.5%	0.6%	4099	16,373,645
32	Miami–Fort Lauderdale, FL	13.5%	1.1%	1010	3,876,380
33	New York–Northern New Jersey–Long Island, NY-NJ-CT-PA	13.5%	0.6%	4775	21,199,865
34	Philadelphia–Wilmington–Atlantic City, PA–NJ–DE–MD	13.3%	0.9%	1745	6,188,463
35	St. Louis, MO–IL	13.2%	1.2%	936	2,603,607
36	Louisville, KY–IN	13.2%	1.6%	448	1,025,598
37	Columbus, OH	13.0%	1.5%	574	1,540,157
38	Buffalo–Niagara Falls, NY	12.9%	1.7%	393	1,170,111
39	Phoenix–Mesa, AZ	12.4%	1.1%	988	3,251,876
40	Greensboro–Winston-Salem–High Point, NC	12.2%	1.4%	570	1,251,509
41	Grand Rapids–Muskegon–Holland, MI	12.0%	1.5%	503	1,088,514
42	New Orleans, LA	11.9%	1.7%	386	1,337,726
43	Milwaukee–Racine, WI	11.7%	1.2%	855	1,689,572
44	Nashville, TN	11.7%	1.5%	466	1,231,311
45	Jacksonville, FL	11.3%	1.7%	373	1,100,491
46	Sacramento–Yolo, CA	11.8%	1.6%	427	1,796,857
47	Norfolk–Virginia Beach–Newport News, VA–NC	10.8%	1.7%	374	1,569,541
48	Orlando, FL	10.5%	1.3%	622	1,644,561
49	Las Vegas, NV–AZ	9.0%	1.4%	417	1,563,282

Table 5.3a
Average Adoption by Size of MSA

Population	Average participation by MSA population	Standard error	Average enhancement	Standard error	Number of areas
> 1 million	90.4%	0.1%	14.7%	0.2%	57
250,000–1 million	84.9%	0.2%	11.2%	0.3%	116
<250,000	75.5%	0.2%	9.9%	0.3%	143

Table 5.3b
Percentage of Establishments in Top Quartile Industry for Enhancement, by Size of MSA

Population in top quartile	Percentage of establishments	# of areas
> 1 million	27.5%	57
250,000–1 million	19.5%	116
< 250,000	19.0%	143

number of these establishments over the total number of establishments in an MSA is highest in large MSAs (27.5 percent). That accounts for much of the difference between larger and smaller MSAs.

Table 5.3c provides a test of this proposition. It shows the results from a simple regression that predicts enhancement for an MSA, illustrating the effect of industry presence controlling for area size and establishment size. The coefficient shows that a ten percent drop in the percent of firms from leading industries (from e.g., 0.3 to 0.2) would lead to the 2 percent drop in the enhancement rate within an MSA. The importance of industry continues to come through even with the addition of MSA size effects, interaction terms, average establishment size, and other measures of laggard industries.[14] In sum, an area is advanced because its establishments happen to come from a leading industry. To be fair, the presence of leading industries is not the only factor, but it is an important one. It alone explains 20 percent of the variance in enhancement. In the last column we also show a similar result for participation. This demonstrates that the presence of leading industries strongly shapes participation as well.

Table 5.3c
Importance of Industry and Population on Enhancement and Participation (standard errors in parentheses)

	(1) enhancement	(2) enhancement	(3) enhancement	(4) enhancement	(5) enhancement	(6) enhancement	(7) participation
(a) Percent firms in top quartile	0.259*** (0.0291)	0.214*** (0.0388)	0.212*** (0.0612)	0.223*** (0.0606)	0.192*** (0.0613)	0.177*** (0.0616)	0.565*** (0.108)
(b) MSA population greater than 1 million	0.0127* (0.00719)	0.0189 (0.0221)	0.0241 (0.0219)	0.0227 (0.0218)	0.0187 (0.0219)	0.177*** (0.0384)	
(c) MSA population between 250,000 & 1 million	0.00331 (0.00479)	0.000163 (0.0132)	0.00335 (0.0131)	0.00462 (0.0130)	0.00248 (0.0131)	0.102*** (0.0230)	
(a)*(b)			-0.0250 (0.105)	-0.0563 (0.104)	-0.0554 (0.103)	-0.0317 (0.104)	-0.458** (0.182)
(a)*(c)			0.0199 (0.0876)	-0.00690 (0.0870)	0.0118 (0.0863)	0.00792 (0.0867)	0.226 (0.152)
% retailing					-0.158*** (0.0630)	-0.186*** (0.0624)	-0.267** (0.111)
% of firms with over 50 employees that have over 500 (census)				0.385*** (0.131)	0.321*** (0.132)		0.419* (0.233)
% of firms with over 50 employees that have over 1000 (census)						0.0823 (0.229)	
Constant	0.0705*** (0.00496)	0.0739*** (0.00538)	0.0742*** (0.00768)	0.0559*** (0.00981)	0.0906 (0.0169)	0.110 (0.0154)	0.719*** (0.0298)
Rsq	0.1995	0.2074	0.2079	0.2292	0.2445	0.2306	0.5338

***significant at 99% confidence level **significant at 95% confidence level *significant at 90% confidence level

Table 5.4a
Leading Adopters of Enhancement among MSAs with Over One Million in Population

MSA	Adoption rate	Standard error	Number of observations	% Establishments in top quartile
San Jose, CA	20.0%	1.6%	638	33.2%
Denver, CO	17.1%	1.4%	778	31.1%
Salt Lake City–Ogden, UT	16.7%	1.7%	535	26.1%
San Francisco, CA	16.5%	1.5%	608	39.4%
Houston, TX	16.2%	1.1%	1320	26.5%
Seattle–Bellevue–Everett, WA	16.0%	1.3%	799	29.1%
Minneapolis–St. Paul, MN–WI	15.9%	1.0%	1411	28.2%
Portland–Vancouver, OR–WA	15.6%	1.4%	683	27.5%
Oklahoma City, OK	15.4%	2.0%	339	19.7%
Atlanta, GA	15.4%	1.0%	1426	32.0%
Average of Top Ten Large MSAs	16.5%			26.6%
Average of Bottom Ten Large MSAs	10.7%			21.7%

In tables 5.4a, 5.4b, and 5.4c we further examine differences in enhancement rates across small, medium and large MSAs, listing the ten leading MSAs for enhancement according to MSA size. In addition we look at the percentage of leading enhancement industries within each MSA. This breakdown of information highlights the differences between large, medium, and small MSAs. These figures reinforce the results in table 5.3, showing that MSAs with leading enhancement rates are not only the larger ones, but also the ones with the highest percentage of leading establishments. Moreover, they show that the difference in enhancement rates between MSA sizes are driven by differences in industry composition at the lower tail of the distribution. Table 5.4a shows the enhancement rates for the top ten and bottom ten large MSAs. The average of the fraction of leading establishments in the top ten large MSAs (26.6 percent) ex-

Table 5.4b
Leading Adopters of Enhancement among MSAs with 250,000 to One Million in Population

MSA	Adoption rate	Standard error	Number of observations	% Establishments in top quartile
Huntsville, AL	19.5%	3.3%	136	27.7%
Appleton–Oshkosh –Neenah, WI	19.4%	3.2%	150	14.4%
El Paso, TX	18.8%	2.8%	185	15.0%
Boulder–Longmont, CO	18.4%	3.4%	121	33.8%
Des Moines, IA	18.0%	2.6%	234	33.7%
Biloxi–Gulfport–Pascagoula, MS	17.8%	4.4%	74	19.6%
Omaha, NE–IA	17.3%	2.1%	343	31.6%
Pensacola, FL	17.1%	4.0%	86	17.1%
Peoria–Pekin, IL	16.2%	3.2%	136	20.3%
Kalamazoo–Battle Creek, MI	16.2%	2.9%	172	15.6%
Average of Top Ten Medium MSAs	17.9%			24.4%
Average of Bottom Ten Medium MSAs	4.4%			16.3%

ceeds the fraction (21.7 percent) in the bottom ten large MSAs (note that the laggards are not shown in this table).

In table 5.4b and 5.4c we can see that the levels of adoption among the leaders of medium MSAs are very similar to those shown in table 5.4a, but the standard errors are much larger owing to smaller sample size. These standard errors make us cautious to emphasize any details about particular locations in these rankings, but we are able to make broad statements. As before, among medium and small MSAs the average fraction of leading industries in the ten leading MSA's (24.4 percent and 16.4 percent respectively for medium and small) exceeds the average fraction of leading industries in the ten laggard MSAs (16.3 percent and 11.1 percent respectively). While leading medium and small MSAs are just as likely to be as advanced as leading large MSAs, however, there are many

Table 5.4c
Leading Adopters of Enhancement among MSAs with Less Than 250,000 in Population

MSA	Adoption rate	Standard error	Number of observations	% Establishments in top quartile
Rapid City, SD	25.6%	6.2%	41	13.5%
Missoula, MT	19.1%	6.1%	32	17.6%
Charlottesville, VA	18.2%	5.5%	47	25.2%
Decatur, IL	17.3%	5.9%	37	16.5%
Cheyenne, WY	17.1%	7.1%	19	14.3%
Dover, DE	17.0%	5.3%	29	20.3%
Jackson, TN	16.9%	4.9%	55	3.7%
Sioux Falls, SD	16.8%	3.9%	86	24.6%
Jackson, MI	16.1%	4.9%	50	8.9%
Casper, WY	16.0%	6.9%	23	14.3%
Average of Top Ten Small MSAs	18%			16.4%
Average of Bottom Ten Small MSAs	2.1%			11.1%

medium and small MSAs with fewer establishments in the top quartile of enhancement adoption. In other words, the difference in distributions arises entirely at the lower tail.

4.3. Comparison with other findings

We compared our findings against the National Telecommunications Information Administration (NTIA) studies of Internet technology use in households for the same year. This study is one among many from NTIA about the digital divide. We aggregated data that appeared in summary form in the NTIA report to the MSA level. We were able to compute household adoption rates for PCs and the Internet for 231 MSAs, a sample weighted toward large to medium MSAs.[15] The correlations between these MSA averages for households and our estimates for commercial establishments in the same location are positive but weak. They range between 0.13 and 0.17. The rank Spearman correlations are mildly higher, between 0.17 and 0.22.

We conclude that the household use of the Internet or computers is mildly informative about the use of the Internet at commercial establishments, as one would expect if the education of the local labor force influences both. However, we also conclude that the correlation is weak within most medium to large MSAs. This is consistent with the view that commercial establishments in urban areas train their workers or simply find mobile technically adept employees. Our findings also support the view that the factors necessary to foster participation and enhancement of Internet business processes did not depend much on local household behavior.

Unlike much previous literature,[16] we find no evidence that this technology is being dominated by a small set of users concentrated in a small region, whether it is in Silicon Valley, along Route 128 outside of Boston, or in any other major urban center. Participation was widespread by the end of 2000, though it tends to mildly favor establishments in heavily populated areas. The use of enhancement to gain competitive advantage spread widely but favored medium and large urban areas with a heavy mix of industries that were high adopter industries. Large MSAs have fewer laggards than medium and small MSAs. We conjecture that the laggard small and medium MSAs may suffer from an inability to achieve scale economies in secondary markets for programmer, technical and other complementary services.

We will say more about the urban/rural divides below, but we speculate at this point that the difference in findings between our study and previous studies arises for four reasons: (1) We are observing medium to large commercial adopters, who have the highest propensity to invest in Internet technology; (2) We are observing their behavior late enough in the diffusion cycle to overwhelm geographic biases associated with very early experimentation (i.e., early experimentation tends to favor areas with a high proportion of technical and scientific users); (3) We are observing business use, which has quite distinct determinants compared with household and farm adoption of computing and internet technology; (4) We are observing use of technology, not production or design of new products, and the latter tends to receive much more attention in public discussion, but leaves a false impression about use.

4.4. Urban dispersion in broader perspective

We close the discussion of MSA adoption by noting that the geographic distribution of establishments largely existed prior to the commercialization and diffusion of the Internet. This leads to three striking observations. First, the preexisting distribution of industries shaped the diffusion of Internet technology.

Second, this technology was adopted across many industries—not all of which share similar geographic distributions. Hence, there are straightforward economic reasons why the use of this technology had a large dispersion over geographic space. It would have taken an implausibly fast and massive relocation of existing establishments and labor markets to concentrate this technology in a small number of places.

Third, concerns about the concentration of use (as emphasized in studies of the digital divide and early development of Internet infrastructure) are out of proportion with the technology's actual pattern of diffusion in business. To be sure, there are leader and laggard regions, but we hardly find it alarming, nor surprising, for an expensive business technology just past its early stages of development.

In this sense, we agree strongly with analysts who argue that geography plays a role in shaping the diffusion and impact of virtual communities.[17] At the same time these findings make us skeptical that this technology's diffusion is headed toward geographically concentrated use. Too many regions have numerous establishments using the Internet for enhancement.

5. Urban/rural divides across states

Tables 5.5a and 5.5b present adoption rates for participation and enhancement for rural and nonrural establishments across all the states in the United States except New Jersey and the District of Columbia.[18] This is also a useful perspective for policy, since many policies for encouraging universal service within rural communities are determined by state regulators and legislatures.

5.1. Participation and enhancement

The estimates for participation in table 5.5a are high in most rural establishments, as expected. One striking feature of the table is its spread.

There are only five states where the rate of participation in rural areas is lower than 80 percent, and eighteen below 87 percent; however, this is still worse than in urban areas. There are two states with urban areas below 80 percent adoption (Vermont and Montana) and only six below 87 percent.

The estimate for rural enhancement adoption in table 5.5b has a distinct distribution. The enhancement rates in the leading states are comparable with the leading metropolitan areas. The lead state is Minnesota with a rate of 15.5 percent. This is followed by Rhode Island, South Carolina, Louisiana, New York, Ohio, West Virginia, Wyoming, Utah and Alaska. In the leading rural states the rates in the urban and rural areas are comparable. However, the differences in the lower tail are large. Twenty-four states have rural enhancement rates below 10 percent, while only three states have urban rates under 10 percent.

We compare the rank ordering of tables 5.5a and 5.5b. Five states are in the top ten of both tables. Generally, however, the ranking in both tables are only weakly correlated. The Spearman rank correlation coefficient is 0.296, positive but not large. This is further evidence that participation and enhancement are distinct.

5.2. Comparison with other findings

We compared our estimates with a previous survey of rural Internet technology development—the United States Department of Agriculture (USDA) estimates for computer and Internet use by U.S. farmers, summarized at the state level.[19] The correlation between participation at rural commercial establishments and farm computer use is 0.41. For enhancement, it is 0.18. While these correlations are positive, only the first one is large. Not surprisingly, we conclude that the USDA survey is an incomplete assessment of nearby commercial Internet use. Our survey and theirs should be positively related, because the level of sophistication of the general population influences adoption at farm and nonfarm establishments. However, the economic costs and benefits from adoption differ between farming and nonfarming establishments. These results warn against inferring much about rural conditions from farm data alone.

As another important lesson in the economic geography of the Internet policies, tables 5.6a and 5.6b include adoption rates for states. As indicated by many previous tables, this level of aggregation hides much variance at

Table 5.5a
Participation in Rural Areas by State

Rank	State	Rural rate	Std error	Observations	Urban rate	Std error	Observations
1	IN	92.9%	0.8%	653	88.9%	0.6%	1745
2	MN	92.9%	0.7%	566	91.0%	0.5%	1628
3	WI	91.9%	0.7%	672	90.9%	0.5%	1728
4	WY	91.6%	1.7%	96	82.1%	2.8%	42
5	NY	91.5%	0.8%	365	89.4%	0.4%	4193
6	NE	91.3%	1.0%	250	91.5%	0.8%	460
7	MI	91.1%	0.9%	532	91.2%	0.5%	2623
8	OH	90.9%	0.8%	735	89.5%	0.4%	3465
9	UT	90.7%	1.3%	124	92.3%	0.7%	627
10	KS	90.6%	1.0%	327	92.8%	0.6%	623
11	SD	90.5%	1.6%	140	88.4%	2.0%	127
12	AR	90.2%	1.1%	371	88.8%	1.0%	481
13	ID	89.9%	1.2%	188	88.0%	1.3%	160
14	IA	89.7%	0.8%	555	88.0%	0.9%	644
15	LA	89.7%	1.3%	228	91.4%	0.7%	992
16	MO	89.4%	1.0%	438	90.1%	0.5%	1505
17	WV	89.3%	1.2%	223	89.3%	1.3%	242
18	IL	89.1%	1.0%	585	89.2%	0.4%	3977
19	AL	89.0%	0.9%	384	90.1%	0.7%	1138
20	VT	89.0%	1.2%	107	78.9%	2.0%	71
21	KY	88.7%	0.8%	574	89.4%	0.7%	798
22	WA	88.7%	1.2%	215	92.1%	0.5%	1408
23	TX	88.5%	0.9%	492	90.1%	0.4%	5073
24	AK	88.4%	1.6%	97	90.1%	2.1%	91
25	NC	88.1%	0.8%	895	89.9%	0.5%	2122
26	SC	87.9%	1.3%	331	87.4%	0.8%	921
27	OK	87.8%	1.5%	238	92.1%	0.7%	683
28	VA	87.4%	1.1%	411	89.2%	0.5%	1603
29	MD	87.2%	2.2%	114	87.8%	0.8%	1352
30	GA	87.1%	0.8%	749	88.1%	0.6%	1859
31	TN	87.1%	1.2%	545	90.3%	0.6%	1463
32	NV	86.6%	2.4%	72	86.0%	1.1%	537
33	NH	86.5%	1.1%	163	88.9%	1.3%	297
34	OR	86.4%	1.4%	224	91.7%	0.6%	855
35	MS	85.7%	1.0%	564	89.6%	1.2%	302
36	CO	84.6%	1.1%	153	90.0%	0.6%	1246
37	PA	84.6%	1.0%	502	89.6%	0.4%	3489
38	ND	83.8%	1.1%	112	89.0%	1.4%	152
39	NM	83.1%	1.9%	131	84.5%	1.2%	261
40	CA	82.0%	1.8%	183	91.4%	0.3%	8379
41	FL	81.9%	1.8%	206	87.9%	0.5%	4289

Table 5.5a
(continued)

Rank	State	Rural rate	Std error	Observations	Urban rate	Std error	Observations
42	MT	81.9%	2.1%	114	72.2%	2.2%	90
43	ME	81.8%	1.2%	202	92.1%	1.5%	217
44	HI	81.2%	1.8%	100	92.4%	1.1%	231
45	AZ	79.1%	2.4%	89	90.0%	0.6%	1300
46	CT	78.9%	1.2%	89	89.7%	0.6%	1136
47	MA	74.0%	3.5%	33	92.6%	0.5%	2221
8	DE	71.5%	4.6%	31	85.5%	1.4%	208
49	RI	67.9%	2.6%	21	92.4%	1.1%	290

the MSA and rural levels. The open question is "How badly do you do if state data is the only thing available?" First, we look at participation. This distribution lacks much variance. The highest state (Massachusetts at 92.4 percent) is hardly higher than the median state (Arkansas at 89.4 percent). Only six states are below 87 percent. Next, we examine enhancement across states. Again, there is not much of a spread. The highest state (Colorado at 16.7 percent) is not much higher than the median (Nebraska at 12.8 percent), and the difference in point estimates is not statistically significant at a 95 percent confidence level. Only three states are less than 10 percent in their point estimates, and none are below 10 percent at traditional significance levels. In general, because urban and rural are not highly correlated, these state-level statistics mask the information in more detailed data. At the same time the rates for participation and enhancement are positively correlated (at 0.40).

5.3. Urban/rural divides in broader perspective

We conclude that enhancement needs to be understood at a fine geographic level, preferably with data relating adoption to MSA and establishments. When this is done, it is apparent that in terms of both participation and enhancement, there are distinct differences between the establishments found in the most populous urban centers and the least dense, even within the same state. We further conclude that concerns about digital divide in commercial establishments are justified, but only if properly qualified. Since participation was not costly, it is surprising and disturbing to find any establishment in any area with low participation.

Table 5.5b
Enhancement in Rural Areas by State

Rank	State	Rural rate	Std error	Observations	Urban rate	Std error	Observations
1	MN	15.5%	1.6%	566	15.5%	0.9%	1628
2	RI	14.9%	6.4%	21	15.5%	2.2%	290
3	SC	14.9%	1.7%	331	10.7%	1.1%	921
4	LA	13.4%	2.3%	228	12.0%	1.2%	992
5	NY	13.0%	1.8%	365	12.7%	0.6%	4193
6	OH	12.5%	1.2%	735	12.4%	0.6%	3465
7	WV	12.5%	2.0%	223	8.6%	1.5%	242
8	WY	12.5%	3.4%	96	18.5%	5.7%	42
9	UT	12.4%	3.0%	124	16.2%	1.6%	627
10	AK	12.2%	3.2%	97	15.2%	3.8%	91
11	DE	12.2%	5.1%	31	14.2%	2.2%	208
12	NV	12.1%	3.8%	72	9.4%	1.3%	537
13	ND	11.8%	3.0%	112	8.7%	2.2%	152
14	CT	11.7%	2.3%	89	14.6%	1.1%	1136
15	WA	11.6%	2.2%	215	13.5%	1.0%	1408
16	WI	11.6%	1.4%	672	13.4%	0.9%	1728
17	IA	11.4%	1.4%	555	15.5%	1.4%	644
18	ID	11.4%	2.4%	188	10.2%	2.5%	160
19	IL	11.4%	1.2%	585	14.3%	0.6%	3977
20	IN	11.4%	1.3%	653	12.2%	0.8%	1745
21	AL	10.9%	1.6%	384	11.9%	1.0%	1138
22	GA	10.8%	1.2%	749	14.0%	0.9%	1859
23	VA	10.3%	1.6%	411	13.8%	0.8%	1603
24	VT	10.2%	2.9%	107	11.3%	3.7%	71
25	OR	10.1%	2.0%	224	14.6%	1.2%	855
26	AR	9.9%	1.6%	371	13.8%	1.6%	481
27	HI	9.6%	3.0%	100	10.1%	2.1%	231
28	KY	9.6%	1.3%	574	13.0%	1.1%	798
29	MO	9.6%	1.5%	438	13.6%	0.8%	1505
30	MS	9.6%	1.3%	564	13.4%	2.0%	302
31	MT	9.4%	2.7%	114	15.3%	3.5%	90
32	TN	9.3%	1.3%	545	12.2%	0.9%	1463
33	TX	9.3%	1.4%	492	14.6%	0.6%	5073
34	OK	9.2%	1.9%	238	15.0%	1.4%	683
35	AZ	9.1%	2.9%	89	11.5%	0.9%	1300
36	CA	9.1%	1.8%	183	13.8%	0.5%	8379
37	CO	9.1%	3.0%	153	16.9%	1.1%	1246
38	NC	8.9%	1.0%	895	12.3%	0.8%	2122
39	KS	8.2%	1.5%	327	13.1%	1.2%	623
40	PA	8.2%	1.3%	502	12.9%	0.6%	3489
41	NE	7.7%	1.7%	250	15.2%	1.7%	460

Table 5.5b
(continued)

Rank	State	Rural rate	Std error	Observations	Urban rate	Std error	Observations
42	NH	7.7%	2.3%	163	11.0%	1.9%	297
43	SD	6.9%	2.3%	140	20.9%	3.6%	127
44	FL	6.8%	1.9%	206	12.8%	0.6%	4289
45	NM	6.4%	2.2%	131	13.4%	2.1%	261
46	MA	5.6%	3.6%	33	14.4%	0.9%	2221
47	MD	5.6%	3.0%	114	15.5%	1.0%	1352
48	ME	5.6%	1.9%	202	11.0%	2.3%	217
49	MI	5.6%	1.4%	532	13.7%	0.8%	2623

To be sure, if these disparities persist, then it is worrisome for business prospects in those locations since nearly every other establishment in the United States takes this technology for granted. Nevertheless, the scope of the problem is limited: Laggard areas do not have large populations.

The dispersion of enhancement provides a different set of insights. This distribution is much more skewed. Yet, such skew is not strong evidence of a digital divide. It is more understandable as an economic matter. First, skew could arise alone from thin technical labor markets in smaller MSAs and rural areas. This would drive up costs of operating facilities employing Internet technology.

Second, this reasoning also suggests that preexisting single-establishment organizations would hesitate to open their own complex Internet facilities until the costs are lower. Either case would lead to more use of enhancement in major urban areas.

6. Conclusions

The diffusion of Internet technology has important consequences for comparative regional economic growth. However, there has been remarkably little statistical evidence documenting the uses and benefits of Internet adoption among commercial organizations. This lack of data has engendered some long-standing misperceptions about Internet use that could potentially cloud decision-making of policymakers. In this chapter, we have developed a framework for understanding commercial Internet

Table 5.6a
Participation among States

Rank	State	Adoption rate	Standard error	Number of observations
1	MA	92.4%	0.4%	2254
2	KS	92.0%	0.5%	950
3	WA	91.9%	0.5%	1624
4	UT	91.8%	0.6%	751
5	CA	91.4%	0.3%	8581
6	MN	91.3%	0.4%	2194
7	OK	91.3%	0.7%	921
8	NE	91.3%	0.7%	710
9	MI	91.2%	0.4%	3159
10	LA	91.2%	0.6%	1220
11	WI	91.0%	0.4%	2400
12	OR	90.6%	0.5%	1079
13	IN	90.6%	0.5%	2398
14	WY	90.6%	1.5%	138
15	SD	90.5%	1.3%	267
16	RI	90.5%	1.0%	311
17	TX	90.0%	0.3%	5572
18	MO	89.8%	0.5%	1943
19	NJ	89.8%	0.5%	2020
20	TN	89.8%	0.6%	2008
21	AL	89.8%	0.6%	1522
22	OH	89.7%	0.4%	4203
23	CO	89.6%	0.6%	1403
24	NY	89.5%	0.3%	4558
25	NC	89.4%	0.4%	3021
26	AK	89.4%	1.4%	188
27	AZ	89.4%	0.6%	1389
28	WV	89.3%	0.9%	465
29	HI	89.3%	0.9%	331
30	PA	89.2%	0.4%	4000
31	IL	89.1%	0.4%	4563
32	AR	89.1%	0.7%	853
33	ID	89.1%	0.9%	348
34	CT	89.0%	0.6%	1199
35	VA	89.0%	0.5%	2015
36	KY	88.9%	0.6%	1372
37	ND	88.8%	0.9%	268
38	IA	88.6%	0.6%	1200
39	NH	88.3%	1.0%	460
40	MD	88.2%	0.5%	1466
41	FL	87.7%	0.5%	4501

Table 5.6a
(continued)

Rank	State	Adoption rate	Standard error	Number of observations
42	GA	87.7%	0.5%	2610
43	SC	87.7%	0.7%	1252
44	ME	87.5%	0.9%	419
45	MS	87.3%	0.8%	866
46	VT	86.6%	1.0%	178
47	NV	86.5%	1.0%	609
48	DC	85.9%	0.5%	285
49	NM	84.1%	1.0%	392
50	DE	84.0%	1.3%	239
51	MT	81.3%	1.5%	204

use and employed a unique data set to clarify the reality of commercial Internet usage.

We demonstrated the importance of distinguishing between different layers of Internet technology. Rapid diffusion in participation did not necessarily imply rapid diffusion in enhancement. This distinction is crucial to understanding the evolution of this technology. The widespread belief that Internet technology diffused rapidly and became table stakes for business was true for participation but not enhancement. Recent concerns that innovation in Internet technology has subsided are misplaced. We speculate that diffusion of enhancement will follow a more traditional path than participation, taking time, innovation, and resources before economic welfare gains are realized. There is still a large possibility that economic gains will manifest themselves in the future.

We showed that Internet use is widely dispersed across geographic regions. It is factually incorrect to characterize regional rivalry in use of the Internet as if use were concentrated. We conclude that research focused on concentration or digital divides—heretofore a central concern of the literature on Internet geography—is a misleading basis for formulating regional economic policy about Internet use in business. To be sure, the concerns about low growth are real for the areas in which adoption lags, but economic policy for laggards has little to do with the majority of areas, which do not lag. Policies for regional development in most

Table 5.6b
Enhancement among States

Rank	State	Adoption rate	Standard error	Number of observations
1	CO	16.7%	1.0%	1403
2	UT	15.6%	1.4%	751
3	MN	15.5%	0.8%	2194
4	RI	15.3%	2.1%	311
5	WY	15.1%	3.0%	138
6	CT	14.5%	1.1%	1199
7	MA	14.3%	0.9%	2254
8	SD	14.2%	2.1%	267
9	TX	14.2%	0.5%	5572
10	DE	14.2%	2.1%	239
11	DC	13.8%	0.9%	285
12	AK	13.8%	2.5%	188
13	OR	13.8%	1.1%	1079
14	IA	13.8%	1.0%	1200
15	NJ	13.8%	0.8%	2020
16	OK	13.7%	1.2%	921
17	IL	13.7%	0.6%	4563
18	CA	13.7%	0.5%	8581
19	WA	13.3%	0.9%	1624
20	VA	13.3%	0.7%	2015
21	MI	13.3%	0.7%	3159
22	GA	13.3%	0.7%	2610
23	WI	13.3%	0.7%	2400
24	MD	13.0%	0.8%	1466
25	MT	12.9%	2.3%	204
26	NE	12.8%	1.3%	710
27	MO	12.8%	0.7%	1943
28	NY	12.7%	0.6%	4558
29	OH	12.5%	0.6%	4203
30	FL	12.5%	0.6%	4501
31	PA	12.4%	0.6%	4000
32	LA	12.2%	1.0%	1220
33	IN	12.1%	0.7%	2398
34	AR	12.0%	1.2%	853
35	KY	11.7%	0.8%	1372
36	AL	11.7%	0.9%	1522
37	KS	11.6%	1.0%	950
38	TN	11.6%	0.8%	2008
39	NM	11.6%	1.7%	392
40	NC	11.5%	0.6%	3021
41	AZ	11.3%	0.9%	1389

Table 5.6b
(continued)

Rank	State	Adoption rate	Standard error	Number of observations
42	MS	11.2%	1.1%	866
43	VT	11.0%	2.3%	178
44	ID	10.9%	1.7%	348
45	NH	10.6%	1.5%	460
46	SC	10.5%	0.9%	1252
47	ND	10.3%	1.8%	268
48	HI	10.0%	1.7%	331
49	ME	9.9%	1.5%	419
50	NV	9.8%	1.2%	609
51	WV	8.8%	1.3%	465

places should devote attention to the factors that are possibly complementary to the use of the Internet for competitive advantage (e.g., such as immobile skilled labor, see Feldman 2002, Kolko 2002). Bottlenecks in complementary factors will determine regional rivalry in the future.

Acknowledgments

We thank Tim Bresnahan, Steve Klepper, Roger Noll, Scott Stern, and Manuel Trajtenberg for comments. We also thank Harte Hanks Market Intelligence for supplying data. We received funding from the Kellogg School of Management and the GM Strategy Center and seed funding from the National Science Foundation and Bureau of Economic Analysis. All opinions and errors are ours alone.

Notes

1. For discussion on how Internet use varies across industries, see Forman, Goldfarb, and Greenstein (2002).

2. Nevertheless, there has been much progress. For information about PC use, see, e.g., National Telecommunications Information Administration (2001), Census (2002), and Goolsbee and Klenow (1999); and for the beginnings in measuring electronic commerce see, e.g., Atrostic, Gates, and Jarmin (2000), Landefeld and Fraumeni (2001), Mesenbourg (2001), or Whinston et al. (2001). We discuss this further below. Several studies have also examined commercial Internet use, but are too small to study dispersion. E.g., Varian et al. (2001), Whinston et al. (2001), Forman (2002), and Kraemer, Dedrick, and Dunkle (2002).

3. See, e.g., Forman (2002), Gertner and Stillman (2001), Carlton and Chevalier (2001).

4. See, e.g., Bresnahan and Trajtenberg (1995), Bresnahan and Greenstein (1997), Helpman (1998).

5. Careful readers will notice that this varies from the definitions employed by Porter (2001). This is due to a difference in research goals. Throughout his article, Porter discusses the determinants of, and shifting boundaries between, investments that provided table stakes and those that complement a firm's strategy and enhance competitive advantage. He argues that these levels vary by industry and firm. This is the proper variance to emphasize when advising managers about their firm's strategic investment. However, when *measuring* this variance for purposes of formulating policy advice it is useful to shift focus. Our measurement goals require both a standardized definition (of something of interest for policy, but consistent with the spirit of strategy research) and a consistent application across industries and locations.

6. This section provides an overview of our methodology. For a more detailed discussion, see Forman, Goldfarb, and Greenstein (2002).

7. Previous studies (Charles, Ives, and Leduc 2002; Census 2002) have shown that Internet participation varies with business size, and that very small establishments rarely make Internet investments for enhancement. Thus, our sampling methodology enables us to track the relevant margin in investments for enhancement, while our participation estimates may overstate participation relative to the population of all business establishments.

8. If we were only interested in the features of the most populated regions of the country, then we could easily rely solely on the most recent data from the latter half of 2000, about 40 percent of the data. However, using only this data would result in very small number of observations for most regions with under one million in population.

9. When two or more MSAs are part of the same urban environment, the census combines them into CMSAs. For example the Dallas-Fort Worth CMSA contains both Dallas and Forth Worth. In table 5.2 we present the CMSA results rather than the individual MSA results when an MSA is part of a CMSA. However, because we will be comparing data between metropolitan areas of different sizes, the only standard of measure we can use is the MSA, as opposed to the CMSA, which applies only to large areas. Thus, in our tables 5.3 and 5.4 we will be discussing rates of adoption in MSAs. This causes for a slight shift in the hierarchy of leaders and laggards. (See, for example, the minor changes in the top ten leaders of enhancement between tables 5.2b and 5.4a.)

10. These are computed using the delta method.

11. For metropolitan areas consisting of more than one PMSA, we use the CMSA rather than PMSA metropolitan area definition.

12. Since these results are simple, we discuss them only in the text. They are not reported in any table.

13. For more details on inter-industry differences in adoption rates, see Forman, Goldfarb and Greenstein (2002).

14. In Forman, Goldfarb, and Greenstein (2002), we show that retailing is a laggard industry. We tried a number of variations on the same type of regression, with similar qualitative results. Hence, we show the simplest result here.

15. Disclosure rules prevent the cell size from becoming too small. Hence, this sample undersamples small MSAs. The smallest cell size for any MSA in this data is six observations, for the next smallest it is eight, for the next it is ten.

16. See, e.g., Kolko (2002), Moss and Townsend (1997), Zooks (2000a, 2000b), Saxenian (1994), Castells (2002).

17. In addition to those already cited, see research on the geography of cyberspace. See, e.g., Cairncross (1997), Kitchin and Dodge (2001), Kotkin (2000), Kolko (2002), Castells (2002) chapter 8, Zooks (2000a, 2000b).

18. New Jersey has only one rural establishment in our data and D.C. has none.

19. The USDA groups several states together, so we only can compare 30 states. We use the data released 30, July 2001, available at http://usda.mannlib.cornell.edu/reports/nassr/other/computer/.

References

Ames, Edward, and Nathan Rosenberg. 1984. Technological change in the machine tool industry, 1840–1910. In *Perspectives on Technology,* second edition, edited by Nathan Rosenberg. New York: Cambridge University Press.

Atrostic, Barbara K., John Gates, and Ron Jarmin. 2000. Measuring the Electronic Economy: Current Status and Next Steps. Working Paper #00-10. Center for Economic Studies, U.S. Census Bureau.

Bresnahan, Timothy, and Shane Greenstein. 1997. Technical Progress in Computing and in the Uses of Computers. *Brookings Papers on Economic Activity, Microeconomics,* 1–78.

Bresnahan, Timothy, and Manuel Trajtenberg. 1995. General Purpose Technologies: "Engines of Growth"? *Journal of Econometrics* 65, no. 1: 83–108.

Cairncross, Frances. 1997. *The Death of Distance.* Cambridge, Mass.: Harvard University Press.

Carlton, Dennis, and Judith Chevalier. 2001. Free Riding and Sales Strategies for the Internet. *Journal of Industrial Economics* 49 (December): 441–461.

Castells, Manuel. 2002. The Internet Galaxy, Reflections on the Internet, Business and Society, Oxford University Press.

Census. 2002. Detailing Tabulations of Manufacturing E-Business Process Use in 2000. Survey Release from United States Department of Commerce. See www.census.gov/epcd/www/ebusiness.htm.

Charles, S., M. Ives, and A. Leduc. 2002. Embracing E-Business: Does Size Matter? Research Report, Statistics Canada.

Cherry, Barbara, Steven S. Wildman, and Alan H. Hammond. 1999. *Making Universal Service Policy: Enhancing the Process Through Multidisciplinary Evaluation.* Mahwah, N.J.: Lawrence Erlbaum Associates.

Compaine, Benjamin. 2001. *The Digital Divide: Facing a Crisis or Creating a Myth?* Cambridge, Mass.: MIT Press.

Feldman, Maryann P. 2002. The Internet Revolution and the Geography of Innovation. Forthcoming in *International Social Sciences Review Journal.*

Forman, Chris. 2002. The Corporate Digital Divide: Determinants of Internet Adoption. Mimeo. Carnegie Mellon University.

Forman, Chris, Avi Goldfarb, and Shane Greenstein. 2002. Digital Dispersion: An Industrial and Geographic Census of Commercial Internet Use. NBER Working Paper #9287.

Gertner, Robert, and Robert Stillman. 2001. Vertical Integration and Internet Strategies in the Apparel Industry. *Journal of Industrial Economics* 49 (December): 415–440.

Goolsbee, Austan, and Peter Klenow. 1999. Evidence on Learning and Network Externalities in the Diffusion of Home Computers. NBER working paper #7329.

Helpman, Elahanan. 1998. *General Purpose Technologies and Economic Growth.* Cambridge, Mass.: The MIT Press.

Kitchen, R., and Dodge, Martin. 2001. "Placing" cyberspace: why geography still matters. *Information Technology, Education and Society* 1, no. 2: 25–46.

Kolko, Jed. 2002. Silicon Mountains, Silicon Molehills, Geographic Concentration and Convergence of Internet Industries in the U.S., *Economics of Information and Policy.*

Kotkin, Joel. 2000. *The New Geography: How the Digital Revolution is Reshaping the American Landscape.* New York: Random House.

Kraemer, Kenneth, Jason Dedrick, and Debora Dunkle. 2002. E-Commerce: A Mile Wide and an Inch Deep. Working Paper, Center for Research on Information Technology and Organizations (CRITO), University of California, Irvine.

Landefeld, Steven, and Barbara Fraumeni. 2001, Measuring the New Economy, *Survey of Current Business,* March., 23–40.

Lipsey, Richard, Cliff Bekar, and Kenneth Carlaw. 1998. What Requires Explanation? In *General Purpose Technologies and Economic Growth,* edited by Elhanan Helpman. Cambridge, Mass.: MIT Press, 15–54.

Mesenbourg, Thomas. 2001. Measuring Electronic Business: Definitions, Underlying Concepts, and Measurement Plans, mimeo released on http://www.census.gov/epcd/www/ebusines.htm.

Moss, Mitchell L., and Anthony M. Townsend. 1997. Tracking the net: using domain names to measure the growth of the Internet in U.S. cities. *Journal of Urban Technology* 4, no. 3:47–60.

National Telecommunications Information Administration, U.S. Department of Commerce. 2001. Falling through the net: Toward Digital Inclusion.

Noll, Roger, Dina Older-Aguilar, Gregory Rosston, and Richard Ross. 2001. "The Digital Divide: Definitions, Measurement, and Policy Issues," Paper presented at the American Economics Association Meetings, January 2002.

Porter, Michael. 2001. "Strategy and the Internet," *Harvard Business Review.* March, 63–78.

Rogers, Everett. 1995. *The Diffusion of Innovations.* Fourth Edition. New York: Free Press.

Saxenian, AnnaLee. 1994. *Regional Advantage: Culture and Competition in Silicon Valley and Route 128.* Cambridge, Mass.: Harvard University Press.

U.S. Department of Agriculture, 2001, Farm Computer Usage and Ownership Report, 07.30.01, http://usda.mannlib.cornell.edu/reports/nassr/other/computer/.

Varian, Hal, Robert E. Litan, Andrew Elder, and Jay Shutter. 2001. The Net Impact Study. www.netimpactstudy.com.

Whinston, Andrew, Anitesh Barua, Jay Shutter, and Brant Wilson. 2001. Measuring the Internet Economy; available from http://www.internetindicators.com/jan_2001.pdf.

Zooks, Matthew. 2000a. "Internet Metrics: Using Hosts and Domain Counts to Map the Internet Globally," *Telecommunications Policy* 24, no. 6/7.

Zooks, Matthew. 2000b. "The Web of Production: the Economic Geography of Commercial Interent Content Production in the United States," *Environment and Planning* 32:411–426.

II

The Evolution of Legal Rights

6

Some Economics of Wireless Communications

Yochai Benkler

I. Introduction

In the first half of the twentieth century there was roughly universal agreement that "spectrum" was scarce, and that if it was to be used efficiently, it had to be regulated by an expert agency. A little over forty years ago, Coase wrote a seminal critique of this system, explaining why "spectrum scarcity" was no more reason for regulation than "wheat scarcity."[1] "Scarcity" was the normal condition of all economic goods, and markets, not regulation, were the preferred mode of allocating scarce resources. In the 1960s and 1970s a number of academic studies of property rights in spectrum elaborated on Coase's work,[2] but these remained largely outside the pale of actual likely policy options. Only in the 1980s did a chairman of FCC voice support for a system of market-base allocation,[3] and Congress eventually permitted the FCC to use auctions instead of comparative hearings in the 1990s.[4] Auctions, however, are but a pale shadow of real market-based allocation. Since the mid-1980s, and with increasing acceptance into the 1990s, arguments emerged within the FCC in favor of introducing a much more serious implementation of market-based allocation.[5] This would call for the definition and auctioning of perpetual, exclusive property rights, which could be divided, aggregated, resold, and reallocated in any form their owners chose to do.

Just as this call for more perfect markets in spectrum allocations began to emerge as a real policy option[6] a very different kind of voice began to be heard on spectrum policy.[7] This position was every bit as radically different from the traditional approach as the perfected property rights approach, but in a radically different way. The argument was that technology had rendered the old dichotomy between government licensing of frequencies and property rights in frequencies obsolete. It was now possible to change our approach, and instead of creating and enforcing property rights in spectrum blocks, we could rely on a market in smart radio equipment that would allow people to communicate without *anyone* having to control "the spectrum." Just as no one "owns the Internet," but intelligent computers communicate with each other using widely accepted sharing protocols, so too could computationally intensive radios. This approach has been called "spectrum commons," because it regards bandwidth as a common resource that all equipment can call on, subject to sharing protocols, rather than as a controlled resource that is always under the control of someone, be it a property owner, a government agency, or both. It is important to understand, however, that resource-focused metaphor has its limitations. As this chapter explains, "spectrum" is not best analyzed as a discrete resource whose optimal utilization is the correct object of policy. The correct object of optimization is wireless network capacity, which may or may not be optimized by focusing on "spectrum utilization." I will therefore mostly refer in this chapter to "open wireless networks" rather than to spectrum commons.

Most of the initial responses to this critique were similar to the responses that greeted the economists' critique forty years ago—incomprehension, disbelief, and mockery,[8] leading Noam to call the standard economists' view "the new orthodoxy."[9] But reality has a way of forcing debates. The most immediate debate-forcing fact is the growth of the market in high-speed unlicensed wireless communications devices, in particular the 802.11x or WiFi family of standards. Particularly when compared to the anemic performance of licensed wireless services in delivering high-speed wireless data services, and the poor performance of other sectors of the telecommunications markets, the success of WiFi forces a serious debate. It now appears, indeed, that serious conversation between the two radical critiques of the licensing regime is beginning to emerge.[10]

What I hope to do in this chapter is (a) provide a concise description of the technological developments that have changed the wireless policy debate; (b) explain how these changes critique the spectrum property rights approach and suggest that open wireless networks will be more efficient; and (c) outline a transition plan that will allow us to facilitate an experiment in both approaches so as to inform ourselves as we make longer term and larger scale policy choices in the coming decade.

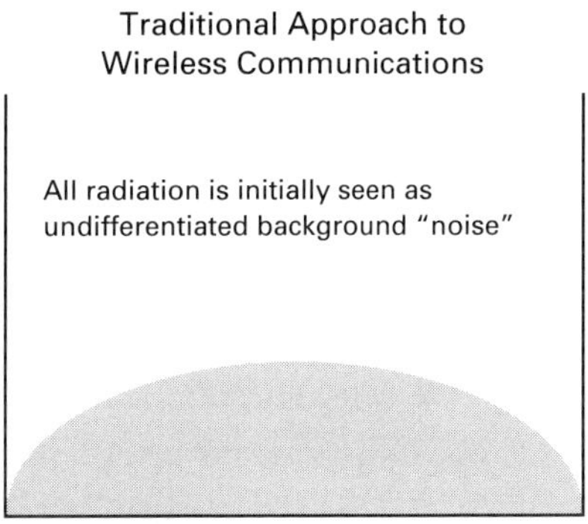

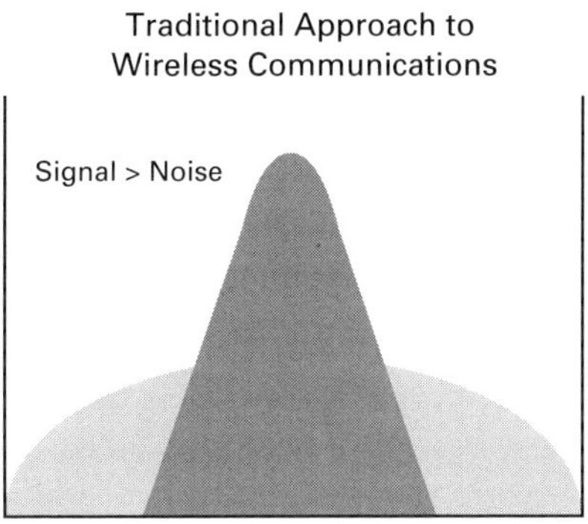

Figure 6.1 (left)
The world in the eyes of a stupid, lone receiver

Figure 6.2 (right)
Receiver treats high-power radiation as signal

II. Technical Background

The traditional model of wireless communications looks at the world through the eyes of a lone, stupid receiver. Stupid, because it is a receiver in whose eyes (or ears) all electromagnetic radiation is equal. It sees the world as a mass of radiation, undifferentiated except by the frequency of its oscillation, so that any given range of frequencies seems undifferentiated, as in figure 6.1.

Lone, because it does not seek or rely in any way on communications with other receivers, it simply waits for some source of radiation that is much more powerful than all other radiation that has a similar frequency, and it treats that radiation as a signal from a "transmitter," which it then translates into human communication—audio, video, or text. The spike in the center of figure 6.2 represents the powerful signal, which is then decoded by the receiver into humanly meaningful communication.

The problem of "interference" occurs when a receiver that is lone and stupid and has such a simple picture of the world encounters more than one source of powerful radiation that it tries to decode and cannot because, as in figure 6.3, neither source is now sufficiently more powerful than *all* other sources of radiation. But "interference" is just a property of the decoding model that the receiver uses, not of nature. The electromagnetic waves do not actually bounce off each other or "fall" to the

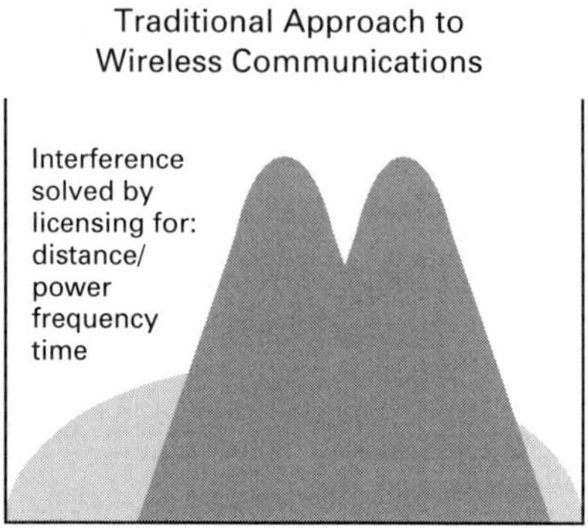

Figure 6.3 (left)
"Interface" in the eyes of a simple receiver

ground before reaching the receiver's antenna. "Interference" describes the condition of a stupid lone receiver faced with multiple sources of radiation that it is trying to decode but, in its simplicity, cannot. To solve this problem, we created and have implemented since 1912 a regulatory system that prohibits everyone from radiating electromagnetic waves at frequencies that we know how to use for communication, and then permits in individual cases someone, somewhere, to radiate within tightly regulated parameters designed to permit the simple receivers to deliver intelligible messages to their human owners.

This model was a reasonably good approximation of the practical characteristics of wireless communications networks given the high cost of computation and the conception of the relationship between terminals and networks that prevailed both in broadcast and in switched telephone networks throughout most of the twentieth century. That is, a computationally intensive machine was not really conceived before Turing in the 1930s, well after the regulatory framework we now have was created, and not really practical as a commercially viable end-user terminal until the early 1990s. The role of terminals in a network—radios and telephones—was largely to be dumb access points to a network whose intelligence resided at the core. The stupid lonely terminal, or receiver, was the correct assumption during this period, and it is what drove the picture of the world upon which both radio regulation and its property-based critique have been based ever since. If in fact all receivers can do to differentiate sources of radiation is to look at their frequency and relative power, and if the ability to listen for frequencies has to be hardwired into the circuits of the receiver, then from the perspective of the receiver there really are "channels" of "bandwidth" that correctly define the way the world is, in the only terms that that machine can perceive the world. It is also then true, as a practical matter, that if more than one person radiates in the "channel" the receiver cannot make head or tail of the message. And when this is the state of commercially available technology for almost a hundred years, we all begin to think of "the airwaves" as being divided into "channels" that can be used for various communications, but only if someone has an exclusive right to transmit. And it is this picture, embedded in our collective minds since our parents or grandparents sat and listened to the magical voices coming from the box in the 1920s, that underlies both current spectrum regulation and its spectrum property alternative.

The traditional model is no longer the most useful model with which to understand the problem of how to permit people to communicate information to each other electronically without being connected by wires. This is so because of one huge practical fact and two fundamental theoretical developments that have intervened since the problem of radio regulation was imprinted on our collective minds. Together they mean that the stupid, lone receiver is the wrong starting point for wireless communications systems design, and hence for the institutional framework designed to support it.

The practical fact is the dramatic decline in the cost of computation. It means that receivers can use computationally intensive approaches for both signal processing and network communications to differentiate among different sources of electromagnetic radiation. No longer are frequency and power the two sole parameters that can be used, nor must any of the differentiating characteristics be hardwired into receivers.

The first theoretical development—Claude Shannon's information theory—is over fifty years old.[11] Shannon was first to define "information" as a measurable entity independent of its transmission medium. Since Shannon, a communication can be described in terms of the probability that a receiver will successfully decode the information content of a transmitter's transmission, rather than the successful delivery of a medium—such as a piece of paper or some specified energy—to a receiver. Among his innovations Shannon developed a formula to represent the information capacity of a noisy communications channel. His capacity theorem implies that there is an inverse correlation between the width of the band of frequencies of electromagnetic radiation that encodes information and the signal to noise ratio—that is, the power of the radiation that encodes the desired communication relative to other sources of radiation with a similar frequency when it reaches the receiver. The implication of this theory is that if a communication is sent using a sufficiently wide band of frequencies, the power of its signal need not be greater than the power of other sources of radiation. This implication was not practically usable for wireless communications until substantial computation became cheap enough to locate in receivers and transmitters, but it is now the basis of most advanced mobile phone standards, as well as of WiFi networks and other wireless systems.

What is crucial to understand about the implication of Shannon's capacity theorem in particular, and his information theory more generally, is the concept of *processing gain.* "Gain" is used in radio technology to refer to a situation where, considering only the power at which the transmitter radiates its signal, the distance between the transmitter and the receiver, and the receiver's required signal-to-interference ratio, the receiver would not be able to tell the difference between signal and noise, but something is done to the receiver or the transmitter, *other than increasing transmission power*, that increases the probability that the receiver will correctly decode the signal *as though* it were more powerful. Antenna gain—the use of a better or more sensitive antenna, is the most intuitively obvious form of gain. You can have bad reception, until you move your antenna, and then you get good reception. The transmitter did not increase power, but your use of the antenna created a perceived gain in signal strength. Processing gain relates to the same idea, but refers to using more complex encoding of the information, and the processing power necessary to decode it, rather than radiation power, to compensate for low transmission power.

A common approach for this is known as direct sequencing spread spectrum. A transmitter will take the message it intends to send, say, "Mary had a little lamb," which in the traditional model would have been sent as the powerful signal described in figure 6.2.

Instead of sending the minimally complex code at the narrowest bandwidth, the transmitter adds more complex encoding, for example, adding

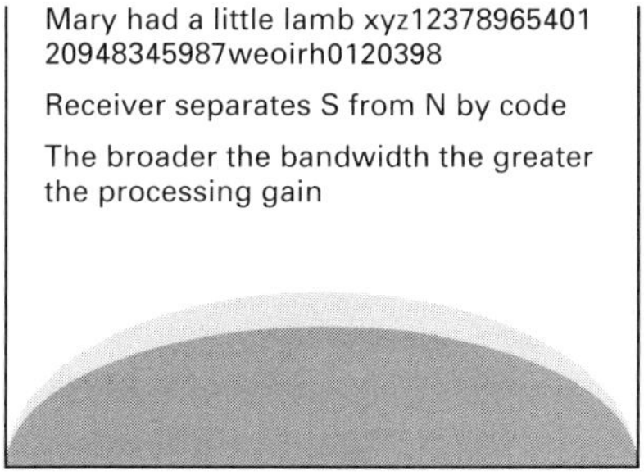

Figure 6.4 (right)
Code-based spread spectrum techniques

xyz123. . . to each packet of data that makes up the message. It then sends this message over a wide band of frequencies, much wider than the minimal frequency bandwidth necessary purely to carry the actual message to a stupid receiver. As figure 6.4 illustrates, because of Shannon's theorem this allows the transmitter to send the message at much lower power than it would have to use were it using a narrow channel—indeed, at such low power that it is no more powerful than other sources of radiation that, in the old model, would have been treated simply as background noise. The receivers, which are in fact computers, listen to very broad ranges of frequencies, and instead of differentiating between sources of radiation by their relative power they identify radiation patterns that coincide with the code that they know is associated with the transmission they are listening for. In our case, whenever a receiver listening for "Mary had a little lamb" perceives radiation that would mean something if correlated to the code xyz123. . ., it treats that radiation as part of the message it is looking for. But it ignores "Mary" from the message "Mary Queen of Scots abc987 . . . ." In effect, the receivers used computation and complex encoding to create a "gain," just like an antenna creates antenna gain, so that the weak signal is comprehended by the smart receiver to the same extent that a stupid receiver would have understood a much stronger signal in a narrow channel. This is called processing gain. Needless to say, the description oversimplifies and the technique I used to illustrate this point is only one of a number of techniques used to attain processing gain. Processing gain challenges the prevailing paradigm because it means there is no necessity that anyone be the *sole* speaker in a given "channel." Many sources can radiate many messages at the same time over wide swaths of frequencies, and there may not be "interference" because the receivers can use techniques that are computationally intensive to differentiate one from the other.

From a policy perspective, the most important thing to understand about processing gain is that it increases as bandwidth and computation available to a wireless network increase. The wider the band, the less power a transmitter-receiver pair needs in order for a receiver to understand the transmitter, but at the cost of more complex computation. Limiting the bandwidth of a signal limits the processing gain a sender-receiver pair can achieve irrespective of how computationally sophisticated the equipment is. As more devices use a band, their low power builds up locally (their effect on unintended receivers rapidly declines as a function of distance from the transmitter), requiring all proximate devices to increase their processing gain. With infinite bandwidth and costless computation, this would not present an efficient limit. With finite bandwidth and costly computation, increased information flow through a network will result in some social cost—either in terms of the cost of computation embedded in the equipment, or in terms of displaced communications—the communications of others who have less sophisticated equipment and cannot achieve the same processing gain. This means that the cost of computation and the permission to use wide swaths of spectrum are the limits on how many users can use a specified band with processing gain. A perfect commons in all frequencies would mean that wireless networks could increase in capacity as a function of the rate of improvement of processors. A licensing or spectrum property regime will limit that growth when, and to the extent that, those who control frequencies release them to open wireless network use more slowly than the rate of growth in computation capabilities of user equipment.

The second theoretical development that works in conjunction with Shannon's theory is tied to the evolution of networked communications that accompanied the development of the Internet, and of work done to improve the efficiency of cellular systems under the rubric of multi-user information theory.[12] This work suggests that there is another source of "gain," independent of processing gain, that every receiver can get from being part of a network of receivers, rather than being a lone receiver. David Reed has described this gain as *cooperation gain*, and has been the most important voice in focusing the public policy debate on the poten-

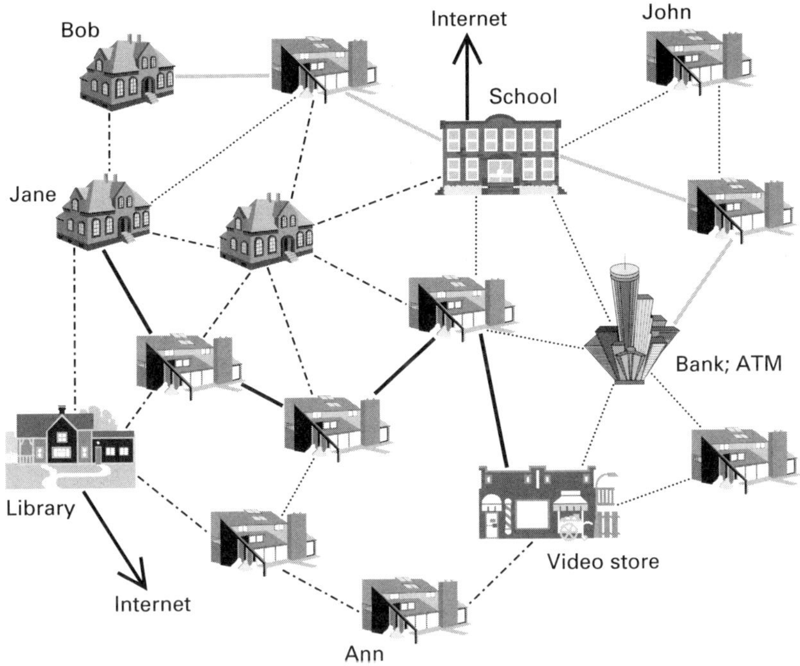

Figure 6.5
Network with repeaters minimizing power

tial of this type of gain to scale capacity proportionately with demand. In multi-user information theory it has been called *diversity gain*. This includes both the value added by repeater networks[13] and the information value that multiple receivers can gain by cooperating to help each other detect signals.

The most intuitive form of cooperation gain is the effect that adopting a repeating, mesh architecture has on the capacity of a wireless communications network.

At the simplest level, take the network depicted in figure 6.5, and imagine that Bob wants to talk to the Bank, while Jane wants to talk to the Video Store, ignoring for a moment processing gain. In the traditional model, they would each have had to radiate with enough power to reach their destination. Because they are closer to each other than to their destination, they would not have been able to do so at the same frequency. In a repeating network, however, neither need radiate at that high power. Instead, each need only reach a neighbor who can further relay the message with several low power hops, none of which is powerful enough to interfere with the parallel path of hops used by the other. Thus, even without processing gain, the two messages could have used the same frequency. This is precisely the rationale of adding cells to a cell phone network in order to increase the number of people who can communicate over the same set of frequencies by "reusing spectrum." The thing to understand is that, just as adding cells to a cell phone network adds capacity to the same band of frequencies, but at the cost of added complexity in network management, so too adding *users* to an open wireless network can *add capacity*, not only demand. But adding cell towers means adding infrastructure to support more users, which is not counterintuitive. The notion that adding users—those who are the source of increased demand for capacity—itself could also add capacity is thoroughly counterintuitive. It shifts the question of network design from one of building enough infrastructure to support *x* number of users, to one concerned with a particular Holy Grail—how to design the equipment and the network so that users add capacity at least proportionately to their added demand. If a network can be designed so that each user can add at least as much capacity as he or she requires from the network, then adding the user to the network is costless except for the cost of the equipment.

Multi-user information theory more generally suggests that there are many techniques for increasing the capacity of a network of users by relying on cooperation among an increasing number of nodes in the network, both as repeaters to lower power output, and as receivers. These include[14] using mobility and relative spatial location of coordinated receivers; allowing receivers to recognize each other's radiation structure, so they can filter it out without complex encoding; and using the antennas of cooperating devices as mutually reinforcing antenna arrays. This body of work shows quite clearly that repeater networks and multi-user detection can be achieved in ad hoc networks, and that cooperation gain can be attained efficiently without a network owner providing centralized coordination.

In combination, these two effects—processing gain and cooperation or diversity gain—convert the fundamental question of "spectrum management"—how to use a finite and fixed resource—into a different fundamental question. The basic point to see is that the bandwidth of the frequencies needed for a communication is not an independent and finite resource, whose amount is fixed prior to the act of communicating, and to which property rights can be affixed so that it is efficiently allocated among communications. Bandwidth is one important parameter in an equation that includes radiation power, processing power of receivers and transmitters, bandwidth, antenna design, and network architecture. Different configurations of these parameters are possible: some will invest more in signal processing, some in network design, some in utilization of specified bandwidth. A policy approach that assumes that bandwidth is "the resource" whose regulation needs to deliver the socially desirable outcome of efficient wireless communications ignores *and burdens* a whole set of strategies to providing the functionality of wireless communication that rely on intensive use of computation and network architecture, rather than on bandwidth intensive usage. The basic economic policy choice we now face is whether wireless communications will be better optimized through the implementation of wireless communications systems designed to scale capacity to meet demand dynamically and locally, or by systems based on licensing or spectrum property rights, designed, at best, more efficiently to allocate capacity that is fixed in the short term and grows slowly.

III. Capacity Growth and Allocation in Wireless Communications Systems

While it is common to talk about optimizing "spectrum use," then, a better definition of what we should optimize is *the capacity of users to communicate information without wires*. This part outlines a framework for specifying and analyzing the social cost of a wireless communication.

The Social Cost of a Wireless Communication

Let *a. . .n* represent a network of devices that enables communications at least among some nodes that are part of this network (some, like base stations, may be dedicated solely to facilitating communication among others). This includes open networks, but is general enough to describe even the receiver-transmitter pair involved in a traditional broadcast transmission or a proprietary cellular communications system. The social cost of a wireless communication between any *a* and *b* that are part of *a. . .n* is defined by three components. First, equipment cost of the network of devices that enables *a* and *b* to communicate, $E_{a \ldots n}$. The equipment cost parameter is intended to be expansive, and to cover all costs including labor and software etc. related to network maintenance necessary to enable the communication. Second, displacement, $\Delta_{a,b}$ where $\Delta_{a,b}$ represents the number of communications between any sender-receiver pair *x*, *y* that the communication between *a*, *b* displaces, and their value to *x*, *y*. Third, overhead O, which refers to transaction and administrative costs. The cost of the communication is, then, $C_{a,b} = E_{a \ldots n} + \Delta_{a,b} + O$.

Equipment At this early stage of the development of equipment markets that are the precursors of open wireless systems it is hard to say anything definitive about the total equipment cost of open wireless networks versus spectrum property based networks. We do, however, have reasons to think that the investment patterns will be different in each of the two systems. Property systems will invest more at the core of the network and less in end user equipment, while open wireless networks will have exactly the opposite capital investment structure.

The end user equipment market is the primary market driving innovation and efficiency in the open wireless network model. Processing gain and cooperation gain increase the capacity of a network, but at a cost of increasing the complexity of the network and the signal processing involved. The cost-benefit tradeoff in open wireless systems is therefore part of the end user equipment cost. Users will generally invest in better equipment up to the point where the value of additional capacity gained from the investment will be less than the incrementally higher cost. It is a dynamic we know well from the computer market, and it is a dynamic we are beginning to see in the WiFi market precisely for wireless communications capabilities, as we begin to see a migration from the cheaper 802.11b equipment to more expensive, higher speed 802.11a equipment. The result is that the value of communicating without wires in an open wireless system is capitalized in the end user equipment, and the sophistication and capacity of a network built of such devices is a function of the demand for computationally intensive end user equipment.

In spectrum property based networks, the efficiency of the system arises from pricing communications over time. It is impossible both to capitalize the value of free communications over the lifetime of the equipment into the *ex ante* price of user equipment, and to price usage *ex post* to achieve efficiency. The prospect of paying *ex post* will lead users to invest less in the computational capabilities of the equipment *ex ante*, leaving the network owner to make up the difference in the intelligence of the network as a whole by investing at the core of the network. These investments can both improve the capacity of the network—for example by adding cell towers to intensify reuse of the same frequencies—and implement pricing, such as by adding local market-exchange servers that would allow the network owner to price efficiently on a dynamic, local basis. Whether these investments, financed in expectation of being covered through usage fees, will be higher or lower in total than the investments to be made by users in open wireless network equipment is not, a priori, clear.

It is important to see, however, that the efficiency with which a spectrum property based system can price bandwidth is limited by its investment in infrastructure equipment. Demand for communication is highly variable, and, as the following section explains, the displacement effect of any given wireless communication is highly localized.[15] In order to price efficiently, a spectrum-property based network must dynamically acquire information about the communications needed and the local conditions under which they must be cleared. Doing so requires deployment of many local market exchanges or pricing points that will collect information about who wants to transmit at a given moment and what their displacement effect will be, so as to price communication for that moment for that locale dynamically. A spectrum property owner will only invest in such equipment up to the point where efficiency gains from investing in the necessary equipment outweigh the cost of the added equipment. At that point, the spectrum owner will price based on more global judgments regarding types of competing uses, rather than on dynamically updated information about actual intended usage and actual local displacement effects.

Displacement The second parameter contributing to the social cost of a communication is its displacement effect—that is, the extent to which the clearance of one communication in its intended time frame displaces the clearance of another in that other communication's intended time frame. While equipment cost is mostly a fixed cost for any specific communication, displacement represents its primary variable cost. In order to see the effects of processing and cooperation gain on displacement, I derive the definition of the economic displacement effect of a transmission from the definition used in multi-user information theory to define the capacity of a sender-receiver pair to transmit information. First, let us define the displacement effect of a communication between sender-receiver pair *a*, *b*, $\Delta_{a,b}$, as $\Sigma\Delta_{x,y}V_{x,y}$, that is, the sum of communications dropped because of the *a*, *b* communication by any other pair, *x*, *y*, each multiplied by its value to its senders and receivers. For purposes of this general analysis, I will assume that any given $\Delta_{x,y}$ has a value of either 0 or 1, that is, it either is dropped or it is not. The value of $\Delta_{a,b}$, will be the total number of communications where the transmission from *a* to *b* causes $\Delta_{x,y}$ to equal 1, multiplied in each case by the value of the communication to its participants. If we wanted a more fine-grained cost-benefit analysis that includes lost speed, we could further refine this definition by treating incremental declines in information throughput rates as independent cases of displaced communication, and treat $\Delta_{x,y}$ as having some value between 0 and 1 based on the number of incremental decreases in throughput.

Here I adapt a multi-user version of Shanon's theorem[16] to define the information that is being lost or communicated as the information in the potentially displaced communication, while separating out the marginal contribution of the communications whose displacement effect we are measuring to the total radiation that the potentially displaced communication must deal with in order to achieve effective communication. Let $P_x(t)$ be the transmit power of node *x*, and $\gamma_{x,y}(t)$ the channel gain between *x* and *y*, such that the received power of the transmission by *x* at *y* is $P_x(t)\gamma_{x,y}(t)$. Let β be the signal to interference ratio needed by *y* for communication, and N_0 the level of electromagnetic radiation treated by *y* as background noise that exists in the channel that *x*, *y*, are using independent of the transmission from *a* to *b*. Let *k* represent any node that is part of *a*. . .*n*, including *a* and *b*, that radiates to facilitate the transmission from *a* to *b*. $P_k(t)\gamma_{k,y}(t)$ is the received power at *y* of the transmission by each *k* as part of the communication *a*, *b*. π represents the processing gain of system *a*. . .*n*, and α the cooperation gain of that system. The value of π is 1 for a system that has no processing gain, and increases as processing gain increases. The value of a is 0 for a system that uses no cooperation gain, and increases as cooperation gain increases.

$\Delta_{x,y} = 1$ when

$$\frac{P_x(t)\gamma_{x,y}(t)}{N_0} \geqq \beta \quad \text{and} \quad \frac{P_x(t)\gamma_{x,y}(t)}{N_0 + \frac{1}{\pi + a}\Sigma_k P_k(t)\gamma_{x,y}(t)} > \beta$$

$\Delta_{x,y} = 0$ when

$$\frac{P_x(t)\gamma_{x,y}(t)}{N_0} < \beta \quad \text{or} \quad \frac{P_x(t)\gamma_{x,y}(t)}{N_0 + \frac{1}{\pi + a}\Sigma_k P_k(t)\gamma_{x,y}(t)} \geqq \beta$$

This is a rather complex formulation of the fairly simple intuition that one communication displaces another when *the marginal contribution* of the former to the total radiation perceived as noise by the receiver of the latter leads that receiver to fail to decode the information. The value of this formulation, nonetheless, is that it separates out the marginal contribution of the communications system involved in transmitting the potentially interfering communication, expresses the effect of processing gain and cooperation gain in determining that marginal contribution, and underscores the externalities imposed by the sensitivity of the displaced communication, expressed by β, in contributing to the perceived social cost of the potentially displacing communication.

Three types of effect on processing gain, cooperation gain, and the signal to interference ratio of devices that might be displaced, β, suggest that the total number of displaced communications is likely to be smaller for communications in open wireless networks than for communications in a spectrum property-based system. First, π increases with bandwidth. A property system that prices bandwidth will, all things being equal, induce lower usage of bandwidth than a system that does not price bandwidth—that is, after all, precisely its purpose. Transaction costs associated with pricing over time contribute further to reducing the bandwidth used. Any wireless communications system that uses less bandwidth than it is computationally capable of using will be able to attain less processing gain than its potential, and hence will displace more communications than it could if it were to use as much bandwidth as it was computationally capable of using in order to attain processing gain.

Second, π and α are a function of the computational ability of the edges of the network—the receivers and the transmitters. Processing gain and cooperation gain increase with computational intensity. As explained in the discussion of equipment investments, the capital investment structure of spectrum property-based systems will generally reduce the computational capacity at the edges, in particular user equipment, be it transmitter or receiver. This is because a property system with a network operator will migrate value into the network as the basis for long term pricing of communications, rather than building all the long-term value of communication into end-user equipment. Assuming that computational capability is the primary source of equipment cost, a system that wants to price usage over time rather than capitalizing the value of free usage over time into the cost of the user equipment will build less computationally intensive user equipment and replace computation at the edges, which is not usage priced, with power and computation in the network, which is susceptible to usage-based pricing. If this in fact describes the likely equipment investment structure of spectrum property systems as a direct consequence that their claim to efficiency requires that they price over time based on usage, then a spectrum property based network will have simpler equipment at its edges, total demand for communications with either system being equal. It will therefore have lower processing and cooperation gain than open wireless networks, unless the spectrum owner invests enough in intelligent network components very close to the end users so as effectively to replicate the spatial diversity and receiver processing capabilities of open wireless networks. That, however, would roughly require replication of the entire investment that an open wireless network would make within the owned infrastructure, and would still require in addition end user equipment and devices intended to price usage. Again, this suggests that an open wireless network will likely displace fewer communications than a spectrum-property based system.

Third, for any receiver, a low β is a positive externality, in that it makes that receiver more impervious to the effects of other transmissions. Communications between any *a*, *b* pair near an *x*, *y* pair are less likely to displace *x*, *y* when *y* has a low β. The *a*, *b* communications therefore impose less social cost than they would have had they displaced *x*, *y*, but the benefit of communications between *a*, *b* made possible by this lower cost are captured by *a*, *b*, not by *y*. Making a receiver more resistant to interference, however, imposes a cost on y. Conversely, all things being equal, having a higher β makes y's receiver cheaper, but causes more communications a, b, to displace it, thereby making the a, b communication appear to have a higher social cost measured in $\Delta_{x,y}$. A high β is therefore a negative externality. Like Coase's famous vibration-sensitive physician, a cheap receiver as much "interferes" with the transmissions of its neighbors as these neighbors "interfere" with the cheap receiver's reception. Receivers designed to be part of open wireless networks need a low β in order to communicate, since they are designed to operate in the presence of many other transceivers sharing the same bandwidth. Each receiver therefore sees at least some of the benefit of the low β as a private gain, and the benefit increases proportionately with the expense of making the receiver require ever-lower signal-to-interference ratios. Receivers designed to be part of spectrum property based systems, on the other hand, will be cheaper and have higher β values. Receivers in such systems need not have a low β because the purpose of the property system is to allow them to be reached by transmitters who need not share the channel with anyone else. The β rate is pure externality to a receiver in a system of property in transmission rights.[17]

Because of these three factors, any a, b, communication cleared through a spectrum property system is likely to have lower values for π and α, and any x, y communication in such a system will likely have higher values for β. If all devices are part of a spectrum property based system, a given a, b communication will cause the largest number of communications to be displaced. In a system where a, b are part of an open wireless network and x, y are not, a, b will have a lower displacement effect. The lowest displacement will occur if all devices involved are part of open wireless networks. If the property system is to be more efficient than the open wireless system, then, it must gain its efficiency from the $V_{x,y}$ element of the displacement parameter. That is, the contribution to the social cost of a property system represented by its displacement factor, $\Sigma\Delta_{x,y}V_{x,y}$, will be lower than the displacement factor of an open network only if the value differential between those communications that each system drops is sufficiently high in favor of the pricing system that it overcomes the higher volume of displacement likely to be caused by the spectrum property based system.

This formulation obviously includes the edge case that all sides to the spectrum policy debate agree on—that is, that where an open network's capacity will be greater than demand, there is no scarcity and the property system imposes a social cost without providing any benefit. Similarly, but less definitively, when the volume of displacement is very small, pricing will improve performance only if the communications that happen to be dropped by the value-insensitive protocol of the open wireless network have an unusually high value.

More generally, it is likely that there will be some range where the total value of the displacement factor for open networks will be smaller than the displacement value of a spectrum property system, and this range will grow as a function of computation, the amount of bandwidth that open systems are permitted to use, and the pricing inefficiencies in spectrum property based networks. The speed of computation growth is given by the innovation rate in the computation markets. Moore's Law has been a reasonable predictor of this for quite a while. The extent to which this growth can be used to improve the efficiency of open wireless networks is partly limited by the total bandwidth that regulation permits equipment manufacturers to use for achieving processing gain.

Overhead: transaction costs and administrative costs—The most important transaction costs associated with open wireless networks are expressed as the network management overhead that devices need to use in order to coordinate their communications. The most important transaction costs associated with spectrum property based markets are those entailed by the need to negotiate clearance of permissions to transmit in a specified bandwidth. Their most important effect will be to cause spectrum-property based systems to price spectrum inefficiently, given the dynamic and local demand for wireless communications capacity. The primary administrative costs of the property system are the definition and judicial enforcement of the property rights. The primary administrative costs of the open wireless system are the standards setting processes and the administrative enforcement of equipment compliance with them.

There are two reasons for the central importance of dynamic adaptation to the efficiency of wireless communications systems. First, wireless communications capacity is mostly determined by local conditions, such as who is trying to communicate in a relatively small geographic space, or whether there are leafy trees between two nodes etc. Second, human demand for communication is highly variable, even over large-scale networks and time frames, and certainly where the relevant demand exists in geographically small areas and for brief time frames.[18]

As I explained in the technical description, open wireless approaches rely on intelligent devices that constantly reconfigure the network architecture, the power used, and the frequency bands used—both in terms of total bandwidth used and in terms of specific frequencies used—to fit the dynamically changing environment and usage patterns. Property rights and pricing mechanisms that are attached to spectrum allocations—that is, to a particular band of frequencies—burden this dynamic adaptation of wireless networks by making the bandwidth parameter "sticky." Bandwidth is one parameter that, unlike computation, architecture, and power, cannot be changed unilaterally and locally by the network of intelligent devices. It requires a transaction.

To illustrate the problem, imagine an automated market exchange point where all proprietors of bandwidth are present to vend their wares, ignoring difficulties with multiple property holders whose systems do not intersect, and ignoring that network equipment costs will limit how much a spectrum property owner will be willing to invest in equipment intended to gather information so as to price efficiently. The optimal transmission path and architecture for any given sender-receiver pair in a network of transceivers is highly local and dynamic. This is particularly so when mobile units are considered. For example, figures 6.6, 6.7, and 6.8 describe three different potential states that a specific sender receiver pair *a*, *b* might be in, depending on the presence or absence of, say, trees and additional nodes available for repeating or collaborative detection.

In each case *M* represents the market exchange point, *x*, *y* represent a sender receiver pair whose displaced communication, if any, represents the social cost of permitting *a* and *b* to communicate as they had intended. I assume for simplicity that there is some mechanism for *a*, *b* to communicate to M that is fixed and does not itself require a negotiation of spectrum rights. In figure 6.6, *a*, *b* is a lonely pair, with no repeating nodes available to cooperate, and with no obstacles to block line of sight. Under these conditions the pair could transmit a given number of bits per second using either high or low frequency spectrum, using a little bit of bandwidth, say, 6MHz, at high power, or a larger amount of bandwidth, say 300MHz, which would give them sufficient processing gain to transmit at such low power that no other sender-receiver pair in the area

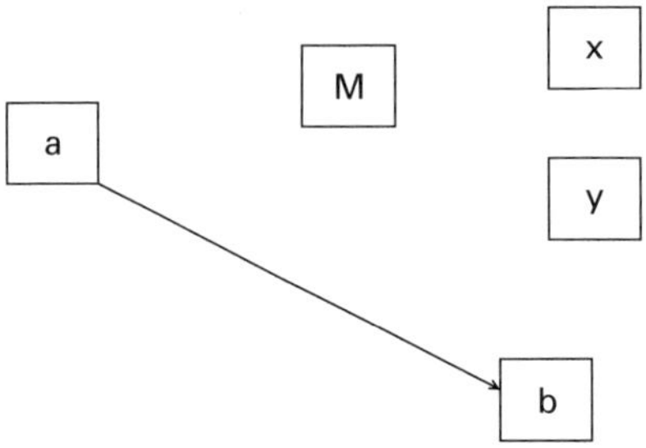

Figures 6.6
a and b have a clear line of sight no other nodes present for repeating

affected by their radiation would be displaced. In other words, given the signal to interference ratio necessary for the pair *x*, *y* to communicate, and the distance from *a*, *b* to *x*, *y*, a transmission in *a*, *b* spread over 300MHz will not affect the ability of *x*, *y* to communicate. In terms of the quasi-formal representation of the displacement parameter, if the signal is spread over 300MHz then $\Delta_{x,y} = 0$, but if it spread over, say, anything less than 50MHz, $\Delta_{x,y} = 1$, while if it is spread between 100–250MHz, $0 < \Delta_{x,y} < 1$. Imagine that all the frequencies are divided into 20MHz to 50MHz blocks. This is a reasonable assumption given that the cost of attaining processing gain is tied to computation and the price of computation drops rapidly over time. We can therefore safely assume that at any given time user equipment will be computationally capable of using more processing gain than it could have used in the past. Hence, even assuming that past aggregations of bandwidth that permitted open wireless operation had a width sufficient to take advantage of all the computation then available, whatever that efficient bandwidth was in the past it will be less than what current computation makes possible at reasonable consumer prices. *a*, *b*, are computationally capable of using 300MHz, but can only communicate over 300MHz if they can get transmission rights from at least six owners, each of whom owns at least 50MHz of bandwidth. As we defined the effect of processing gain achieved by spreading over 300 MHz, the correct price of the transmission right necessary to spread the signal over 300MHz is zero, since spreading the signal to that width will reduce the marginal social cost of the bandwidth used by the communication—its displacement effect—to zero. Yet no single owner would be willing to sell transmission rights over its spectrum for that amount, given nonzero

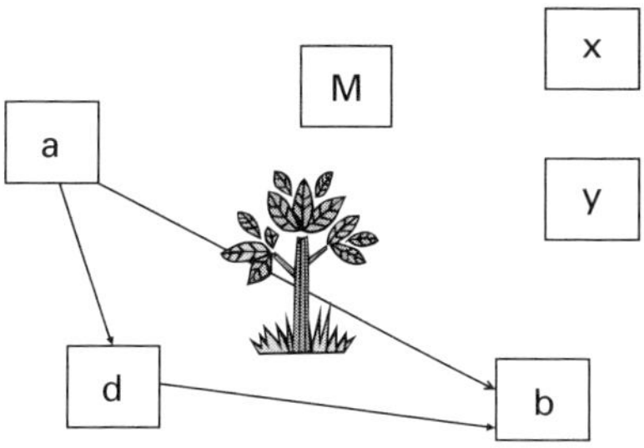

Figures 6.7
a and b have a clear line of sight no other nodes present for repeating

transaction costs associated with fixing the correct price, as well the cost of communications that would be displaced if the signal is spread only to 50MHz, rather than 300MHz, which is all the spectrum owner can secure and monitor unilaterally. All parties would have to negotiate simultaneously as to whether *a*, *b* would spread to 100, 200, or 300 MHz given cumulative transaction costs of deciding which power/bandwidth combination would be less expensive to combine, given the resulting effect, if any, on any pair *x*, *y*. Note that, knowing that they will encounter such transaction cost constraints on their ability to pursue feasible nondisplacing communications, *a* and *b* will both under-invest in high-computation equipment in the amount of lost potential communications over the lifetime of the equipment that they will not be able to achieve because of transaction costs.

Open wireless networks, however, also have transaction costs, specifically the overhead traffic necessary to identify the most efficient transmission path.

If high enough, these will also constrain efficient communication under that approach. While true, it is important to understand that these are transaction costs that *both* open wireless systems and proprietary spectrum systems must incur, if pricing in the latter is to be efficient. Take, for example, a similar situation to the one in figure 6.6, but because *a* and *b* are mobile units, geography and network topology between them change. In figure 6.7, a tree has intervened between them and one more potentially cooperating node has appeared as well. Now *a* and *b* need lower frequency spectrum (which can go through leaves) if they are to communicate directly, but can use higher frequency spectrum with direct line of

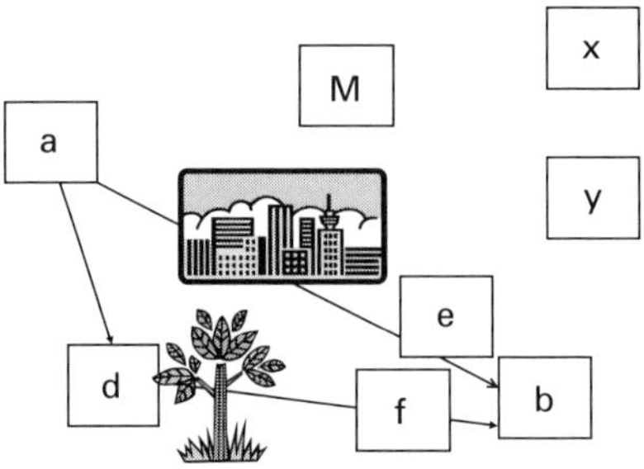

Figures 6.8
a and b have a tree in between then, but clear lines of sight to a cooperating node

sight if they use *d* as a repeater. In figure 6.8, not only has a tree appeared, but so have some buildings, and additional nodes *e* and *f* situated behind a set of buildings whose multi-path effects would let the smart antennas of *e* and *f* achieve multi-user detection in cooperation with *b*. *a* and *b* have to compute which of several strategies would result in the lowest error rate at the minimal power at a desired speed: relying on nodes *e* and *f* as a distributed antenna array for *b*, on *d* and *f* as repeaters for *b* (perhaps alternating high frequency for the first and last hops from *a* to *d* and from *f* to *b*, and low frequency from *d* to *f*), or on direct communication at a low frequency.

Without computing all these potential courses of action, it is impossible to tell what effect the desired communication from a *to* b *would have on any given sender receiver pair* x, y. Because the effect on *x*, *y* represents the marginal social cost of the *a*, *b* communication, *it is impossible efficiently to price the bandwidth* a *and* b *need* before *a* and *b* have all the information they need in order to determine what is the lowest possible displacement they could cause other communications. A spectrum market has higher transaction costs for achieving efficiently priced communications than an open wireless network has for efficient communication at least to the extent of the positive transaction costs incurred after *a*, *b* communicate to the exchange the range of possible communication patterns open to them. The point here is not that property rights burden open wireless networks, as they surely do and as I will discuss in the next few paragraphs. The point is that *any* potential communication from any *a* to any *b*, whether done in an open wireless network or in a proprietary system, will need to map its least displacing configuration, given available network topology and deployed equipment at any given time *t*, as a precondition to efficient pricing. "Spectrum" at that time and place cannot be priced efficiently in the absence of this information, for this is the information that defines the marginal cost of any communication over any wireless network. To the extent that a spectrum property based system cuts its transaction costs by pricing spectrum on less refined information, to that same extent it is pricing inefficiently.

In the most technologically sophisticated version of the property rights argument, Faulhaber and Farber recognize the fact that transaction costs (which they call direct) of a spectrum property system are higher than the transaction costs of an open wireless approach, or a spectrum commons. Indeed, it is to solve this problem that they propose a modified property system, rather than one based on perfect rights of exclusion. Their preferred regime implies into all property rights in spectrum a public "easement" that permits anyone, anywhere, to transmit at any frequency as long as they do not interfere with the owner of the right to transmit in that frequency. This modified property regime is intended to permit very wide band communications that are "below the noise floor" given the operating parameters of the devices that operate with the property owner's permission in the frequencies they use, as well as agile radios that occupy frequencies only when their owner is not using them, and as soon as the owner wants to transmit, hop to a different, unused frequency. While Faulhaber and Farber agree that direct transaction costs are likely to be higher in either property-based approach than in a commons-based approach, they do not attempt to specify the effect of these transaction costs. It is important to emphasize that these transaction costs go precisely to limit the capacity of the spectrum property system to do the one thing that potentially could make it more efficient than open wireless networks—that is, accurately to identify competing values that users place on communications, and to adjust the price of bandwidth accordingly. The important measure of the transaction costs is the extent of the deviation they cause from efficient pricing in the spectrum property based system. The more that transaction costs cause the spectrum property system to use prices that reflect statistical judgments about competing usage patterns, rather than actual real-time bids, the less of a value there is in these systems as compared to an open wireless system that treats all communications equally and will drop high and low value communications with equal probability, but will drop fewer communications in total.

Faulhaber and Farber are also quite cognizant not only of the internal limits that the transaction costs associated with spectrum property rights impose on that system's own efficiency, but also of the externalities that implementing such a system would cause, in terms of the constraints it would place on the technological evolution of efficient open wireless networks. Recall, the problem is that the number of rights holders that a transmitter-receiver pair must negotiate with in order to pull together the right to transmit in a band say, 10GHz wide, is so great that as a practical matter that mode of communication will be unavailable under a pure spectrum property rights approach. They recognize this as a problem akin to what has come to be known as the "anticommons" problem,[19] that is, a particularly virulent version of the Coasian problem of inefficient design or allocation of rights in the presence of high transaction costs. It is a condition where rights are so fragmented relative to the efficient contours of a usable resource that the transaction costs of assembling a usable resource out of these fragments are too high to permit it to occur, causing stasis. Faulhaber and Farber offer two types of solutions to this problem. First, they claim that because of the easement they postulate in their modified property regime, that is, the right of anyone to transmit in any frequency as long as they are not interfering with the licensed user/owner, anyone will be able to transmit the 10GHz wide signal as long as they are "below the noise floor." Second, to the extent that communications that would interfere with other devices are desired, they should be permitted in bands purchased for this purpose by the Federal or even state or local governments, or perhaps by manufacturers of open wireless network equipment, seeking to make the market in their equipment more attractive to consumers.

While their modified system is much better than the pure property system, it is still substantially constraining to open wireless network design, and again it is Coase who helps us understand why. In both his *Federal Communications Commission* piece and in the Nobel-winning article he wrote the following year, *The Problem of Social Cost*, Coase introduces the problem of the physician and the confectioner who are neighbors. The confectioner's equipment makes vibrations that make it difficult for the physician to see patients. Normal legal thinking at the time would treat the confectioner as "causing" damage to the physician by making noise and vibrations. One of Coase's great insights in that article was that the physician is "causing" the damage to the confectioner by being so sensitive, just as much as the opposite is true. Who should be shut down or made to pay cannot therefore be decided on the basis of stating who is "causing harm," but should rather be based on whose activity is more socially valuable. The lesson is directly applicable to the proposition that open wireless networks need not be adversely affected by an exhaustive Big Bang auction of property rights as long as they are permitted to operate without interfering with rights owned under that regime. If, however, we define the operating parameters of open wireless networks based on the sensitivities of the property-based services, we have effectively treated the property-based system as the physician, and the wide band devices and agile radios as the confectioner. But saying that we will allow confectioners so long as their equipment does not vibrate is not to say that we now allow both physicians and confectioners. It is to say that we have chosen to make the world safe for physicians and constrained for confectioners. This may be the right decision or the wrong decision from a social welfare perspective, but it is a decision in favor of one approach, not an accommodation of both.

To be less metaphoric, the level of noncooperating radiation in any given band affects the extent to which a system needs processing and cooperation gain to achieve a certain rate of information delivery through an open wireless network. The more radiation there is, the greater the complexity of the solution to the problem of communicating information through the channel. The greater the complexity of a system, the greater the cost of the equipment needed to implement it. So, holding all other things equal, if you permit *only* open wireless systems to operate in a given range of frequencies, they will be able to achieve a given throughput at lower cost than they could if they need to achieve the same throughput in the presence of high powered communications. So, while the modified property right is much better than the perfect property rights regime in that it does not completely prohibit open wireless systems, it still imposes a burden on the development of those systems. Perhaps the proponents of spectrum property rights are correct, and that burden is socially justified given the relative value of both types of approaches—the proprietary and the open—to wireless communications. But the modified property right does not allow us to eat our cake and have it too. We must still choose how much we will have of each type of wireless communications facility.

As for the suggestion that the Federal or state or local government bodies will buy spectrum to create parks, this is a surprisingly naive proposal from two such sophisticated authors. If one were to think that Congress and the federal government were rational decision makers who operate to optimize the public good with the same efficiency as, say, any large corporation maximizes the benefits to its shareholders, this might not fundamentally be a mistaken approach. But the notion that Congress is equally likely to appropriate \$*x* already in the Treasury as it is to forgo potential revenue by refraining from auctioning the spectrum, particularly under the politically palatable heading of reserving it for a public trust, is surprising. As a matter of treating the government as a rational agent responding to real costs, forgoing *x* millions of dollars by refraining from auctioning the spectrum is identical to spending that amount of money after the spectrum is sold. As a matter of practical politics, they are not similar in the least. I suspect that the reason for this aspect of the Faulhaber and Farber proposal has more to do with the integrity of the transition policy—that is, with the big bang auction that is intended to launch the property system. But this is a transition policy that would result in substantially lower public investment in space for open wireless networks than might a differently framed public debate, and the transition policy should not be allowed to preempt the outcome of such a controversial policy question.

As for administrative costs, or what Faulhaber and Farber call indirect transaction costs, they suggest that the open wireless approach has the highest indirect costs, because uncertainty as to what equipment is "interfering" or complying with the open protocols and what is not will be confusing and difficult, and hence costly, for courts to sort out, and will lead to much litigation. They claim that the pure property regime will have the lowest indirect costs, because courts are most adept at solving property rights disputes. And they see their own preferred modified property regime as having higher administrative costs than pure property, because the boundary between the easement and the property rights will lead to difficult litigation, but lower than the administrative costs of open systems, because courts, familiar with property disputes, will find a property framework easier to design and enforce than an open system.

This view of the administrative costs takes a somewhat more rosy view of property litigation, and a more dim view of administrative equipment certification and private standard setting, than I would. All one need do is look at the decades-long history of some of the cases that I teach every year in my first year property course to see that courts do indeed resolve property disputes, but to say that they will do so efficiently because of their experience with real property is somewhat optimistic. It is important in this regard to see that disputes about use of open wireless networks will occur not with regard to property-type claims, but with regard to equipment compliance with standards. Here, standards can be set for whole industries by open processes like the IEEE or W3C standard setting processes. The FCC can then certify equipment as it does now on a Part 15 model. In all these cases, the administrative costs are incurred, but once incurred apply industry wide, and can be enforced against noncomplying equipment fairly simply by engineering tests of the equipment. This is by no means a costless exercise for dispute resolution, but it is vastly cheaper and more certain than relying, say, on the owner of property rights in 724–744MHz in Des Moines, Iowa, to sue the owner of 745–747MHz in one neighborhood there for using a particular antenna configuration, with the owners of 748–768MHz and 712–23MHz as third party intervenors, and then on courts of appeal to resolve conflicts between how the Iowa court and another court, say adjudicating similar claims in the 2GHz band in Memphis Tennessee, decided the case by applying the *sic utere tuo ut alicnum non laedaes* maxim rather than *damnum absque injuria*.

More generally, others who have written in favor of property rights in spectrum have treated "property" as being antiregulation, and commons as being regulation by the back door.[20] The dichotomy between the two in this regard is, however, overstated. In order to have efficient property rights, it is necessary to define, enforce, and update the definition of the content of property rights.[21] These are all functions that require thoughtful institutional design, initially through Congress, later through enforcement agencies or courts. None of this is new or surprising to anyone who teaches a first year property course, and must take the students through the centuries of choices made by judges and legislatures between barons and King, modernizing landowners and their overbearing dead ancestors, or developers and the neighbors who wanted a quite residential community, not a gas station next door. Lacking the benefit of centuries of gradual development, property rights in spectrum are unlikely to involve less explicit regulatory choices, and Faulhaber & Farber correctly identify the need for well designed governmental planning in the initial creation of the property rights and well functioning dispute resolution to fine tune the rights when realty teaches us the limitations of the original design.[22] Similarly, in order to have efficient commons, some set of rules about usage may well be necessary. Property rights can be defined or interpreted in an inefficient and corrupt manner, as can a commons oriented regulatory processes. The trick in setting up either arrangement will be to make sure that they are designed so as not to allow the recreation of command and control regulation through the backdoor.

Pricing, Block Allocations, QoS, and the Big Bang

The transaction costs analysis suggests three additional observations with regard to the policy implications of the potential value of pricing. Recall that the efficiency with which open wireless networks can provide wireless communications capacity does not necessarily mean that there will never be situations where pricing of bandwidth can improve the efficiency of communication. It is possible that when demand exceeds capacity of a given network of devices, as deployed in a given locale at a given time, introducing pricing will improve allocation of whatever capacity is attainable by the network of devices in place. Three points need to be made with regard to this observation, however. First, the introduction of pricing does not itself support the creation of property rights in blocks of spectrum, as compared to a single fluid market exchange in spectrum on the model proposed by Eli Noam.[23] Second, even if some quality of service (QoS) assurance is attainable through the introduction of pricing, that still does not mean that the game is worth the candle—that is, that the cost and implications of introducing a pricing system for assuring QoS is worth the social cost of setting up the pricing system. The experience of wired networks suggests otherwise. Whether it is or is not is a question that can only be determined empirically over time, as we get better information about wireless network usage and capacity given the presence of open wireless networks. Third, whatever the possible merits of pricing, they do not merit, based on our present knowledge, a "big bang" auction of all spectrum, but at most the dedication of some bands to provide pricing to handle peak utilization periods.

First, the dynamic, local, and highly variable nature of demand for wireless communication suggests that block allocation will be systematically inefficient. Similar to demand for electric power distribution, designing capacity to meet highly variable demand will be more efficient if demand can be averaged over all users throughout the year rather than if it is averaged among the contingent distributions of customers of different firms.[24] One does not want transaction costs involved in shifting users from, say, 724–726 MHz to 964–966MHz to be higher than shifting those same users to 728–730MHz, as they might be if there is one owner for 720–730MHz and a different one for 960–980MHz. If transaction costs are higher in this way, then there will be situations where a communication would have cleared given over-utilization of the 720–730MHz band but under-utilization of the 960–980MHz band had these bands been part of a single transactional unit, but will not clear because these bands are separated into two transactional units. This inefficiency of block allocation is central to the efficiencies of the Noam-style market, where all spectrum is available all the time for both spot market and forward contract purchases, so that the local and dynamic variability in demand can be averaged over the entire usable spectrum as opposed to over smaller ranges of bands. To the extent that the presence of rights in discrete blocks of spectrum add stickiness to the efficiency of the market clearance of bandwidth, to that same extent rights in blocks of spectrum will be less efficient than a single dynamic market in all usable frequencies.

Second, the case of demand occasionally exceeding capacity in a system that throughout many moments has an excess of capacity is very similar to the problems of quality of service presented by wired networks, and for which models of pricing bits have been proposed.[25] Pricing-based QoS solutions in wired networks have not, however, been adopted, and there are some reasons to think that they are unnecessary in the foreseeable future for wireless networks. Partly this is due to the fact that computation, storage and caching capabilities have grown so quickly that adding capacity to more than meet demand has been a more efficient solution in the wired world than accepting that capacity cannot meet demand

and allocating slow-growing capacity to meet it. In wireless, it is likely that the declining price of computation and the growing market in wireless communications devices will, for any useful time horizon, make it cheaper to increase supply by improving the end user devices than to introduce a pricing system to allocate slower growing capacity. There is perhaps a more systematic problem with pricing bandwidth as a means of assuring QoS. At all times when demand is not high, pricing the allocation of spectrum introduces a pure transaction cost of maintaining a system that will be available to clear excess demand in those more rare events when demand exceeds capacity. It is only in those peak moments that pricing could in principle improve the efficiency of communications. The aggregate cost-benefit analysis of any pricing system must compute the total transaction costs attached to all communications, relative to the benefit attained in the moments where demand exceeds capacity. While there is no *a priori* reason to think that pricing will not be beneficial, whether or not it will in fact be beneficial would largely depend on traffic patterns in a system whose characteristics may change dramatically over the time between now and when capacity will begin to grow slowly enough to justify pricing.

Finally, while it is possible that some pricing of spectrum will improve efficiency of some systems sometimes, that possibility does not support a "big bang auction" to create property in all spectrum, always, everywhere, now. In public highways, for example, it is likely that creating a pricing system by using toll roads or paid carpool lanes in specific locations with predictable congestion patterns will improve efficient traffic flows. This may indeed recommend introduction of pricing in some predictably congestion-prone roads. But it would be odd to derive from that likely geographically and temporally focused improvement that we would be better off introducing property rights, toll-booths, and electronic payment systems for use in all city streets and sidewalks, dirt roads, or highways at nighttime, on the off chance that sometimes these too may become congested and pricing could then be useful to help improve their efficient utilization. It is, in other words, possible that benefits could be attained by allowing some "spectrum" to be treated as a reservoir of bands usable for pricing to serve QoS needs. But that is no basis to have a big bang auction of all usable frequencies, nationwide, before we know how the market in open wireless network equipment develops, and before we know how much spectrum, if at all, could usefully be priced, sometimes, in some locales. At most, the theoretical value of pricing suggests that it would be plausible to adopt a policy of increasing the flexibility permitted to current licensees to use their presently owned bands for resale when utilization is low, or perhaps for dedicating some bands to be run on the Noam pricing model.

Capacity, Growth, and Efficiency: Conclusion

The economic comparison between the efficiencies of property rights in spectrum allocations and those of open wireless networks can be restated in the following main points:

- The choice is between a market in infrastructure rights and a market in equipment, not between a market approach and a nonmarket approach
- Evaluating the social cost of a communication in either system requires evaluating the equipment cost involved in enabling the communication, the displacement effect a cleared communication has on other communications that are not cleared because of it, and the overhead involved in clearing the communication in terms of transaction costs and administrative costs.
- It is difficult to predict the total cost of equipment necessary for spectrum-property based communications relative to the cost of open wireless network equipment. It is likely that investment in a spectrum property model will be more centralized at the core of the network, with cheaper end user devices, and investment in an open wireless model will be more decentralized and located in the hands of users, representing a capitalization of the value of communications over the useful lifetime of the equipment either in the hands of the network owner (with spectrum property) or in the hands of users, in the absence of a network owner.
- Open wireless systems are likely to be able to grow capacity more rapidly than spectrum property based systems, because the free availability of bandwidth and the higher computational intensity of end user equipment will allow such systems to use and improve processing and cooperation gain in pace with the price/power growth in processing, while property based systems will be limited by the lower computational complexity of end user devices, the relative stickiness of proprietary bandwidth, and the likely higher signal-to-noise ratio required by receivers.
- The relative advantage of pricing bandwidth will occur, if at all, only at peak utilization moments, and is akin to pricing based QoS approaches in wired networks. Attaining that advantage may not be

worth investing in deploying these approaches at all, as it has not in the unregulated world of wired networks.

- Transaction and administrative costs of markets in spectrum are likely to be higher than those associated with communications in open wireless networks
  - Direct transaction costs will limit the ability of spectrum property based systems to price efficiently
    - Given that spectrum property based systems grow capacity more slowly than open wireless systems, the limitations on their capacity to price efficiently may be fatal to their justifiability
  - Administrative costs of litigation in a property system are likely to be higher than the administrative costs of equipment certification processes, at least if the latter are designed to be no more burdensome than current equipment certification programs, and particularly if those are streamlined for standards set in open private standard setting processes.

IV. Innovation, Welfare, and Security

In addition to the fundamental point about the limitations of property in spectrum allocations as mechanisms for optimizing the capacity of users to communicate without wires, there are other, more general points to make regarding the likely advantages and limitations of open wireless systems. These fall into the categories of innovation, welfare enhancement, and security.

Innovation In addition to reasons we have to think that property in spectrum will not improve capacity utilization over time, we have reasons to believe that an open wireless system will have better characteristics where innovation is concerned. The property-in-spectrum model relies on the owners of spectrum to innovate in order to increase the value of their spectrum. The open wireless approach, on the other hand, relies on the openness of the system and on the notion that the smartest people usually work for someone else. That is, the principle of innovation underlying the Internet, as Lessig described so forcefully in *The Future of Ideas*, is the idea that the network itself is simple and open. Everyone then gets to innovate as they wish, and can use the network as long as they can translate their new applications into simple modules that can be transmitted using TCP/IP, the open protocol underlying Internet communication. This is fundamentally different from innovation in the Bell System—an immensely innovative company in its own right—where innovation occurs primarily in Bell Labs, because only they have permission to implement. Innovations from the outside are permitted if and only if they fit the Bell revenue model. In wireless systems design too, owners of spectrum are likely to invest in innovation so as to increase the value of "their" spectrum. But they will likely prevent the implementation of innovative communications technology by most outsiders unless it fits their revenue model and they can appropriate it. With a commons approach toward spectrum, however, anyone can innovate. Anyone can develop a device, and if it works better, users will adopt it. Anyone can set up an Internet service, anywhere, and if it offers better service—faster or more robust, cleaner of commercial advertising or better at delivering targeted advertising—that person can offer the service without asking permission of an "owner" of the system, as one would need today for cable or licensed wireless Internet access. This freedom to innovate and implement has proven enormously important to growth and innovation in the space of computers and the Internet. Adopting an open spectrum policy would structure the environment for innovation in the wireless communications systems design along the same lines, rather than on the old, centralized innovation model.

Welfare optimization While much of Part IV was devoted to describing the comparative welfare implications of each approach, there is a separate element of welfare optimization that merits separated note. A particular type of constraint on the ability of spectrum property based systems to price efficiently has to do with the difference in their investment structure. As Part IV explains, open wireless systems are built of end user equipment designed to optimize end user capacity to communicate, while owned networks rely on greater investment at the core of the network in terms of designing capacity optimization and pricing. A consequence of this differential investment pattern is that open wireless networks are likely to adapt more rapidly to changing consumer preferences than proprietary networks.

Posed with the need to invest in infrastructure and in a system to collect information about preferences and to minimize transaction costs associated with satisfying them, proprietary network owners must make decisions regarding for what they will optimize the network and pricing schemes. If SMS messaging is the big thing today, and the network provider believes that mobile gaming is the killer app of tomorrow, then the provider will design the network to serve the present and expected future applications best. If it turns out that some portion of the population, immediately or thereafter, wants to use the system to compare live feeds of traffic from automobile-mounted webcams, and the system does not price or service that use well, the operator will have to recognize that use, compare it to others, and optimize equipment within the network to service and price it. The lag between the redesign of the network and the contracts and the changing needs of consumers is a source of welfare loss.

Open wireless systems, on the other hand, are built by equipment manufacturers that capture the future communications value embedded in the equipment at the moment of sale. To do so, they are likely to design flexible devices that can adapt to give their owners whatever utility the owner wishes over time. That is precisely the value embedded in general purpose computers, and it is precisely this agility and built-in capacity to be repurposed by the user as the user's preferences change over time that has driven the value of the computer market. Wireless equipment manufacturers too will try to embody as great a future value as possible in the equipment, in order to enhance its value to users. Innovation and changing possibility sets lead consumers to have rapidly evolving preferences. A system that allows users dynamically to repurpose the network will enhance welfare as compared to a system that requires a centralized decision to shift optimized uses to fit demand.

Security In the context of communications networks in general, and wireless networks in particular, security usually arises in the context of three questions: how easy it is to cause the network to collapse, how easy it is to infiltrate and eavesdrop on the network, and how readily available it is for security forces to use in emergencies. The Internet and the encryption debates of the 1990s have shown us that there are real tradeoffs between closed proprietary and open networks in these regards. While it is hard to specify which approach will be better under all conditions, open networks have important characteristics that could make them more, rather than less secure than closed networks. First, because open networks rely on densely deployed, self-configuring, decentralized mesh networks, physically bringing down the network is extremely difficult. On September 11, 2001, for example, traditional wired and wireless telephone networks were overloaded, New York City's public radio station was down, but email, instant messaging, and IP-based voice communications applications like NetMeeting were functioning. High-speed data connections were available downtown for the first few weeks only by using WiFi networks.[26] The basic characteristic of the Internet's robustness—its redundancy and decentralized architecture—is replicated in open wireless networks at the physical layer of the communications infrastructure. Second, communications that rely on processing gain and encryption are much harder to tap and infiltrate than communications that use high power communications. They are widely deployed by the military, which, of course, cannot assume that its enemies will comply with FCC regulations as to spectrum use, and so must design its systems for environments where no one has exclusive control over spectrum. Third, both of these characteristics also suggest that widespread deployment of redundant, robust communications networks that rely on encryption will actually provide a more robust system for public security communications in time of emergency than approaches that rely on proprietary or regulated control over specified blocks of spectrum, which depend on infrastructure that can be bombed or communications protocols that can be jammed. The physical infrastructure of an open wireless network will be more robustly and ubiquitously available and the platform it will offer will be less susceptible to jamming. All that needs to be implemented, if anything need be done, is to build into the network protocols an ability to recognize an emergency signal and give it precedence to overcome the potential for latency.

V. Policy Recommendations

The conclusion of my analysis suggests that there are strong reasons to think that permitting the operation of open wireless networks will be a superior approach toward wireless communications than creating an exhaustive system of property rights in spectrum. Nonetheless, the reasons to think that an equipment market based on open wireless policies will be better than an infrastructure market based on property rights in "spectrum allocations" are not *a priori* determinative. This leaves us, as a polity, in a position of doubt, not knowing which of two policy alternatives is preferable, yet convinced that one, or the other, or some combination of the two is likely to be dramatically better than the present regulatory system. Under these conditions, it seems that the most prudent course would be to begin to transition away from the present system by setting up a sufficiently robust experiment with both approaches, so that experience over the next few years will provide useful information about the longer-term choice. The elements of such a framework would include;

- Creating a commons of sufficient magnitude and stability to allow a credible investment effort by toolmakers—equipment manufacturers and software companies—in building the tools that can take advantage of an ownerless wireless infrastructure
- Implementing some flexible property rights on a more limited and experimental basis than proponents of the Big Bang approach propose
- Building revisability into both systems through recovery options designed to permit policy to abandon or scale back either alternative, should experience suggest that one is decisively superior, designed so as to minimize the effect of potential future abandonment on the efficiency of current pricing of spectrum rights or on investment incentives in the equipment market

Expanding the Commons

An expansion of the commons should include three primary components:

- Commons-only designations. Like the U-NII band in the 5 GHz range, some bands will be designated for use solely by unlicensed device.
    - Unlike the U-NII Band, the regulatory constraints on devices operated in these commons-only designated bands should be based solely on the needs of unlicensed devices, and not aimed at protecting incumbent services. Incumbent services sharing these bands should be cleared.
    - Some bands should permit unlicensed operation below 2GHz, so as to improve their capacity to pass through walls and be deployed relatively simply in most urban environments. One plausible space for such a dedication is the 700MHz band of UHF channels that was slated for auctioning, but for which there was so little demand given the present state of capital markets that the auctions were called off.
    - These bands will likely be subject to an equipment certification process for equipment intended to operate in the unlicensed bands.

To avoid turning the equipment certification model into back-door command and control regulation, the process should include fast track approval for all equipment complying with standards set in open standard setting processes, and should impose only the minimal nonharmfulness or coexistence standards.

- Public Trust. Some of these bands could be donated into a nongovernmental body, roughly akin to conservation trusts, whose charter would require it to permit operation of any and all devices that meet minimal sharing standards to be set in an open professional standards setting process, along the lines of the W3C or the IEEE. The trust would be funded by fees paid by members on the W3C model, not from spectrum usage fees The trust's certification and standards decisions would be relatively insulated from regulation by either regulatory agencies or judicial review by treating the trust's control over "its" frequencies as equivalent to the decisions of a private licensee in the most flexible licensing frameworks, or of a spectrum property owner should any bands be transitioned to that model

Underlay and interweaving rights

- Separate from any specific band designations, we should introduce a general privilege to transmit consistent with the basic principle that noninterfering uses should be permitted wherever possible.
- The specific requirement here would be to revisit the FCC's UWB Order (Ultrawideband) and the SDR (software defined radio) process, so as to, in each case, expand to the extent possible the permission for wireless devices to use any frequency they wish as long as they comply with one of two basic constraints
  - Devices operate at a level of power so low that it does not appreciably affect the information flow rate of licensed incumbent devices deployed on the day of the approval. All licensed devices introduced thereafter will not be protected if designed to be less robust to interference from underlay devices than the incumbent devices were.
  - Devices implement some mechanism for agile radio use, so as to sense the presence of radiation in a band, and to recognize radiation from the licensed owner of the band they are using. The devices must automatically vacate the frequency upon sensing an attempted use by the licensed owner of the frequency. This would assure that these devices only use frequencies when the licensed owner is not attempting to use them. Because "spectrum" is perfectly renewable and reusable with no degradation, such use imposes no cost on the licensed owner, but offers socially valuable communications capacity.

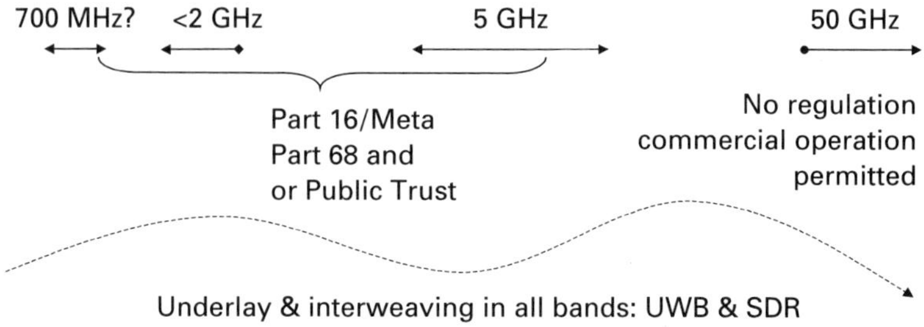

Figure 6.9
Interlocking commons for open wireless networks

The overall system would look roughly as described in figure 6.9.

Revisability: Experimenting with Spectrum Rights

In a similar fashion, we should work to identify a series of ranges of frequencies that have roughly similar propagation characteristics, and that could be subject to greater flexibility. The spectrum needed for this experiment will be easier to locate, because the experiment will represent a windfall to the incumbent licensees, wherever it is located. The big bang auction design is intended to create incentives for incumbent licensees to participate. To do so, it gives licensees a right to all the money paid at the auction, and gives them a choice between, on the one hand, not placing their rights up for auction and retaining the precise contours of their license, and on the other hand placing their licensed spectrum up for auction but retaining a right to refuse to sell if the bids do not meet their reservation price. This option makes the licensees strictly better off by being designated as eligible to participate.

The primary institutional design question here, then, is how to experiment with the spectrum property idea without imposing too great a difficulty on reversing course in a few years, if our experience with the two systems strongly suggests that the preferable solution is to have less property in spectrum and more open wireless networks. The concern, of course, is that should property rights be created in too much spectrum, their incumbents will prove very difficult to clear to make way for open wireless networks. A parallel right to redesignation should be implemented for the spectrum commons bands should the opposite conclusion emerge from experience.

The institutional design should include two constraints. First, no more frequencies should be designated for the spectrum market experiment than necessary to make it viable. Certainly, this should be no more than the bandwidth set aside specifically for open wireless networks, given that this approach is most effective at allocating narrow bands, whereas open wireless networks rely on wide bands as a baseline requirement.

Second, the property rights should include a recovery reservation, such that, should our understanding of the relative value of the approaches over time develop to favor much broader permission for open wireless networks, the cost of implementing the change will not be prohibitive. The trick will be to design the recovery system in such a way so as not to burden too much the present efficient pricing of the spectrum auctioned. The primary vehicle here could be to create a preset low cost buyback option in the government, that would allow the government the option to redesignate the frequencies to open wireless network use upon payment of a reduced recovery rate. The "redesignation" option, rather than a more generally defined repurchase option, is intended to prevent the government from simply speculating in spectrum, exercising the option and then selling back into a proprietary system. The exercise date must be set sufficiently far into the future that present discount rates in the wireless communications industry would make the discounted value of the option very low. 10 years may be a good baseline, but the precise term should be based on investment practices in the industry regarding when projected returns are no longer usefully considered in making an investment decision. The terms of the option would be set and known before the auction, so that no expectations are violated after the purchase of the rights. To prevent inefficient pricing over time as the exercise date grew near, Congress could create a periodic review process, whereby every three years, for example, Congress could decide to extend the option exercise period to the original full period, to cancel the option, or to do nothing, and keep the option date unchanged. It would choose the first option if information was still lacking on the relative performance of the two approaches to wireless communications policy, the second if the spectrum property approach appeared to be better, and the third if open wireless networks seemed to be preferable. A similar redesignation caveat should be included in the instruments permitting various forms of open wireless communications equipment to function, adjusted to the discount rates in the equipment manufacturing industry, which is the primary industry whose investment incentives will be affected by the option.

VI. Conclusion

Current wireless communications policy operates on technical assumptions largely unchanged since Marconi's time. While there is relatively widespread agreement that, at least from an efficiency perspective, the licensing regime that still regulates use of almost the entire usable range of frequencies is obsolete and should be abandoned, there is quite substantial disagreement over what its replacement should be. In particular, there are two primary alternative approaches. The first involves the creation of more or less perfect property rights in spectrum allocations, so as to allow bandwidth to be allocated based on market pricing of these exclusive transmission rights. The second involves the removal of current prohibitions on wireless communications equipment that prevent the emergence of open wireless networks built entirely of end user equipment.

The tradeoff between spectrum property markets and open wireless networks is primarily a tradeoff between the rate of increase in the capacity of the network and the efficiency with which a given capacity is allocated among competing uses. Spectrum property based markets are likely to grow capacity more slowly than open wireless networks. Because they will price usage, however, they are in theory likely, at least at peak utilization moments, to allocate the capacity they have more efficiently than would an open wireless network. Open wireless networks, however, are likely to increase capacity more rapidly, and if unconstrained by band use regulation, could increase capacity at the rate of growth of computation. Some research suggests that they may even be able to increase capacity proportionately with the increase of the number of users. Our experience in wired networks, both the public Internet and proprietary corporate networks, has been that strategies that have relied on rapid growth of capacity have been widely adopted, while strategies that assume slow growing capacity and seek efficiency through pricing to achieve quality of service assurances have not. It seems odd, in the name of the efficiency of markets, to adopt by regulation a system of property rights in spectrum that makes exactly the opposite choice than the one that users and corporations have made in the actual market environment, when presented with a parallel choice in the context of unregulated wired networks.

At present, however, the lack of clear empirical evidence in favor of one or the other of the two radical critiques of the prevailing licensing regime cautions against any kind of "big bang" approach that will preempt future policy making. What we need is a relatively large-scale experiment in both markets. On the one hand, we should move to deregulate wireless communications equipment capable of functioning on the open wireless networks model. This move should be substantial enough to give equipment manufacturers a credible playing field for which to invest in equipment design, production, and marketing for ownerless networks. In parallel, it may be useful to permit some experimentation with spectrum property allocations, carefully designed so as to preserve longer term flexibility and avoid being locked in to the spectrum property model should open wireless networks prove to be the better approach.

Acknowledgments

Many of the insights in this article are the product of many conversations and email exchanges over the past 18 months with participants in the Open Spectrum Project, an ad hoc multidisciplinary group of academics, engineers from the industry, and policy experts. In particular my work has been influenced by Dewayne Hendricks, Andy Lippman, Larry Lessig, David Reed, Jerry Saltzer, and Tim Shepard.

Notes

This chapter is an abridged version of a paper I gave at the Telecommunications Policy Research Conference, September 2002. The full paper is published as *Some Economics of Wireless Communications*, 16 Harv. J. L. & Tech. (Fall 2002).

1. Ronald Coase, *The Federal Communications Commission*, 2 J of Law & Econ. 1 (1959).

2. William K, Jones, *Use and Regulation of the Radio Spectrum: Report on a Conference*, 1968 Wash. L.Q.71 (1968); Arthur S. De Vany, Ross D. Eckert, Charles J. Meyers, Donald J. O'Hara, Richard C. Scott, *A Property System for Market Allocation of the Electromagnetic Spectrum: A Legal-Economic-Engineering Study*, 21 Stan. L. Rev. 1499 (1969); Harvey J. Levin, The Invisible Resource: Use and Regulation of the Radio Spectrum (1971); and Jora R. Minasian, *Property Rights in Radiation: An Alternative Approach to Radio Frequency Allocation*, 18 J. Law & Econ. 221 (1975).

3. See Mark S. Fowler and Daniel Brenner, *A Marketplace Approach to Broadcast Regulation* , 60 Tex. L. Rev. 207 (1982) (Fowler was Chairman of the FCC under President Reagan). By the 1990s this position had become the mainstream, cutting across the political spectrum, as indicated by the remarks of FCC

Chairman under President Clinton as well. Reed E. Hundt, *Remarks To Citizens For A Sound Economy Spectrum Policy and Auctions: What's Right, What's Left* (Washington, D.C. June 18, 1997) (Chairman of the FCC stating in his introduction that "for the first time ever the FCC truly follows a market-based approach to the allocation and use of spectrum.")

4. See Pub. Law No. 103–66, 6001–02, 107 Stat. 379, 379–401 (1993).

5. See Evan R. Kwerel and Alex D. Felker, 1985, *Using Auctions to Select FCC Licensees*, OPP Working Paper No. 16 (May 1985); Evan R Kwerel and John R. Williams, *Changing Channels: Voluntary Reallocation of UHF Television Spectrum*, OPP Working Paper No. 27 (November 1992); Gregory L. Rosston, Jeffrey S. Steinberg, *Using Market-Based Spectrum Policy to Promote the Public Interest* (OPP Working Paper 1997)

6. See, e.g., Spectrum Bill Discussion Draft: The Electromagnetic Spectrum Management Policy Reform And Privatization Act, CR, 104th Cong., S4928–4936 (May 9, 1996) (Senator Pressler)

7. The policy implications of computationally intensive radios using wide bands were first raised, to my knowledge, by George Gilder in *The New Rule of the Wireless*, Forbes ASAP, March 29th, 1993, and by Paul Baran, *Visions of the 21st Century Communications: Is the Shortage of Radio Spectrum for Broadband Networks of the Future a Self Made Problem?* Keynote Talk Transcript, 8th Annual Conference on Next Generation Networks Washington, D.C., November 9, 1994, available at http://www.eff.org/pub/GII_NII/Wireless_cellular_ radio/false_scarcity_baran_cngn94.transcript. Both statements focused on the potential abundance of spectrum, and how it renders "spectrum management" obsolete. Eli Noam was the first to point out that, even if one did not buy the idea that computationally intensive radios eliminated scarcity, they still rendered spectrum *property rights* obsolete, and enabled instead a fluid, dynamic, real time market in spectrum clearance rights. See Eli Noam, *Taking the Next Step Beyond Spectrum Auctions: Open Spectrum Access*, 33 IEEE Comm. Mag. 66 (1995); later elaborated in Eli Noam, *Spectrum Auction: Yesterday's Heresy, Today's Orthodoxy, Tomorrow's Anachronism. Taking the Next Step to Open Spectrum Access*, 41 J. Law & Econ. 765, 778–80 (1998). The argument that equipment markets based on a spectrum commons, or free access to frequencies, could replace the role planned for markets in spectrum property rights with computationally intensive equipment and sophisticated network sharing protocols, and would likely be more efficient even assuming that scarcity persists, is Yochai Benkler, *Overcoming Agoraphobia: Building the Commons of the Digitally Networked Environment*, 11 Harv. J. L & Tech. 287 (Winter 1997–98). Noam 1998, *supra*, and Yochai Benkler and Lawrence Lessig, *Net Gains: Is CBS Unconstitutional*, The New Republic (December 1998) suggested that the obsolescence of the controlled spectrum approach raises concerns as to whether the present licensing regime was unconstitutional as a matter of contemporary First Amendment law. Lawrence Lessig, CODE (1999) and Lessig, THE FUTURE OF IDEAS (2001) developed the argument that relied on the parallel structure of innovation in the original Internet end-to-end design architecture and of open wireless networks, offering a strong rationale based on the innovation dynamic in support of

the economic value of open wireless networks. David Reed, *Why Spectrum is Not Property, The Case for an Entirely New Regime of Wireless Communications Policy* (February 2001) available at http://www.reed.com/dprframeweb/dprframe.asp?section=paper&fn=openspec.html and David Reed, *Comments for FCC Spectrum Task Force on Spectrum Policy*, filed July 10, 2002, and available at http://gullfoss2.fcc.gov/prod/ecfs/retrieve.cgi?native_or_pdf=pdf&id_document=6513202407 crystallized the technical underpinnings and limitations of the idea that spectrum can be regarded as property. Comments to the Task Force generally were the first substantial set of public comments in favor of a spectrum commons. Kevin Werbach, *Open Spectrum: The Paradise of the Commons*, Release 1.0 (November 2001) provided a crystallizing overview of the state of this critique and how it relates to the implementation of WiFi.

8. See Thomas Hazlett, *Spectrum Flash Dance: Eli Noam's Proposal for "Open Access" to Radio Waves*, 41 J.L. & Econ. 805 (1998); Thomas Hazlett, *The Wireless Craze, the Unlimited Bandwidth Myth, the Spectrum Auction Faux Pas, and the Punchline to Ronald Coase's "Big Jokes: An Essay on Airwave Allocation Policy*, 14 Harv. J. L. & Tech. (Spring 2001).

9. See Noam, 1998, *supra*.

10. See, e.g., Gerald Faulhaber and David Farber, *Spectrum Management: Property Rights, Markets, and the Commons* (this volume).

11. Claude E. Shannon, *A Mathematical Theory of Communication*, 27 Bell System Technical 379–423, 623–56 (1948).

12. See bibliography in the full version, 16 Harv. J L & Tech.

13. The first model for practical implementation of this approach was Timothy Shepard, *Decentralized Channel Management in Scalable Multi-hop Spread Spectrum Packet Radio Networks*, PhD Dissertation, MIT 1995). Information theoretical work includes Knopp and Humblet; Gupta and Kumar; Grossglauser and Tse, cited in 16 Harv. J. L & Tech.

14. See citations in Benkler, 16 Harv. J. L & Tech.

15. Even in open areas the power of a radio signal fades as a function of the square of the distance, and where there are buildings, trees etc, it fades even more rapidly, typically as a function of distance to the third or fourth power.

16. In particular I modify here Equation 1 from Grossglauser and Tse, cited in Benkler, 16 Harv. J L & Tech.

17. Indeed, it was precisely the need to provide for the growing market of relatively cheap receiver sets that drove the development of the radio industry in the 1920s and formed the basis of the band-licensing model that has been with us ever since. See Benkler, *Overcoming Agoraphobia*, *supra*.

18. Because of fading and the fact that "spectrum" is perfectly renewable from one moment to the next, the relevant locale and time frame for gauging demand and displacement are geographically proximate and temporally dynamic. With packet based communications, the relevant time frame is on the order of milliseconds.

19. Heller, *supra*.

20. Hazlett, *Wireless Craze*, *supra.*

21. See, e.g., DeVany et al., *supra.*

22. Faulhaber and Farber 7–8 ("In the case of spectrum, spillovers in the form of out-of-band power in adjacent frequencies are important, and can generally be controlled by the careful definition of property rights. In today's regime, spectrum licensees operate under a set of technical restrictions regarding power and place of emission, and possibly direction and time of emission. In a property rights regime, these restrictions would be codified in the property rights of the frequency owner, who would then be subject to civil penalties should he or she violate these restrictions. In fact, such restrictions are often codified in property rights and laws. My right to use my automobile is restricted by speed limits; my right to use my real property is restricted by noise and nuisance statutes of my state, county and local municipality. Property rights in spectrum would be similarly constrained, and in fact we already know what the constraints are: they are largely defined by the technical restrictions in current licenses.")

23. Noam 1998, *supra.*

24. In the context of wired networks, the benefits of aggregating users to lower the cost of provisioning for bursty peak utilization and its relationship to industry structure is discussed in David Clark, William Lehr, and Ian Liu, *Provisioning for Bursty Internet Traffic: Implications for Industry and Internet Structure* MIT ITC Workshop on Internet Quality of Service. (Nov. 1999).

25. An economists' version of this proposal is Jefferey Jackie Mason and Hal Varian, *Economic FAQs About the Internet*, 8 J. Econ. Persp. 75 (1994). For a technologist's view see S. Shenker, D. Clark, D. Estrin, S. Herzog, Pricing in computer networks: reshaping the research agenda, 20(3) Telecomms. Policy (1996)

26. Peter Meyers, In Crisis Zone, A Wireless Patch, NYT Oct. 4, 2001, Sec. G, p. 8., col 1.

7

Spectrum Management: Property Rights, Markets, and the Commons

Gerald R. Faulhaber and David J. Farber

Introduction

Since 1927, the electromagnetic spectrum has been allocated to uses and users by the Federal government, covering broadcast radio, microwave communications systems, broadcast television, satellites, dispatch, police and national defense needs, among many others. Assignees receive a license to broadcast certain material (say, taxi dispatch) at a specified frequency and a specified power level (and perhaps direction). For many purposes, this license is time-limited, but with a presumption of renewal; in fact, radio licenses are almost always renewed. Licensees can only use the spectrum for the specified purpose and may not sell or lease it to others.

Economists since Ronald Coase (1959) have argued strongly and persuasively that allocating a scarce resource by administrative fiat makes little sense; establishing a market for spectrum, in which owners could buy, sell, subdivide and aggregate spectrum parcels would lead to a much more efficient allocation of this scarce resource. The Federal Communications Commission (FCC) has gradually been allocating more spectrum for flexible use and since 1993 has been using auctions to award most new wireless licenses. However, this experiment in bringing market forces to bear to allocate radio spectrum has been applied to only about 10 percent of the most valuable spectrum. Economists continue to press for "marketizing" spectrum as the surest means to use this important national resource efficiently (White (2001)).

Meanwhile, substantial strides have been made in radio technology, including wideband radio (such as spread spectrum and ultra wideband (UWB)), "agile" radio (one of several applications of software defined radio (SDR)) and mesh networks (including ad hoc networks and other forms of peer-to-peer infrastructure architectures). The developers of these technologies note that the products based on these technologies undermine the current system of administrative allocation of exclusive-use licenses, and call for an "open range," or commons, approach to the spectrum that would do away with exclusive use. "Removing the fences," in this view, will lead to more efficient use of the spectrum.

While both economists and radio engineers believe the present system of spectrum allocation is inefficient and wasteful, they appear to have diametrically opposed views of what should replace it. Economists seek to unleash the power of the market to achieve efficient outcomes; engineers seek to unleash the power of the commons to achieve efficient outcomes. Which is right?

We argue in this chapter that this is a false dichotomy. We propose a legal regime rooted in property rights that can simultaneously support both private markets and a commons that can accommodate the rapid diffusion of the new radio technologies, leading to a far more efficient allocation of this important and limited national resource.

I. Early Radio History: From Innovation to Government Allocation[1]

At its earliest inception, radio was seen as useful primarily for marine communications: ship-to-shore telephony. The failure to heed disaster calls from the Titanic in 1912 and the failure to fully realize the naval benefits of wireless in World War I created a public sentiment to improve the maritime uses of wireless communications, leading to the U.S. Navy's efforts to cartelize the industry in 1919–1921.

Broadcast radio seems to have arisen spontaneously in 1920–21, when the first broadcast stations in New York and Pittsburgh went on the air, reaching thousands of hobbyists with crystal radios. The popularity of broadcast radio spread very quickly, and its commercial possibilities were realized almost immediately. However, the problem of interference was recognized early. If two (or more) broadcasters in the same city chose to transmit at the same time on the same (or very close) frequency, then each interfered with the other's signals and radio listeners were treated to cacophony. This was good for no one, and in the early years, a *de facto* property right standard of "priority in use" arose; quite simply, the first user "owned" the frequency, and subsequent users had to broadcast elsewhere. This property right was supported by the Department of Commerce and by 1926 was recognized by several courts.

In 1926, Herbert Hoover, Secretary of the Commerce Department, ordered that the Department stop supporting priority in use claims following two conflicting court decisions. The result was rather chaotic; in major radio markets, interference became the norm as new firms attempted to poach on the frequencies of popular radio stations. In the resulting outcry, Congress passed the Radio Act of 1927, which established the Federal Radio Agency (FRA) with the responsibility of stewardship of the spectrum and the sole right to determine what various frequencies could be used for and who could use them. In the ensuing years, virtually every country in the world emulated the U.S. by establishing a national agency solely in charge of allocating spectrum to uses and assigning it to users. All national agencies gather every three years at the World Radiocommunications Conference to discuss and resolve radio spectrum problems across administrative boundaries.

In the United States, the Communications Act of 1934 created the Federal Communications Commission (FCC), vesting in it the FRA's spectrum allocation authority[2] (and abolishing the FRA). Since its inception, the FCC has interpreted its authority as the nation's spectrum manager rather broadly. Until quite recently, it imposed the Fairness Doctrine on broadcast networks and stations, by which broadcasters were required to cover controversial issues and provide airtime for contrasting viewpoints. Currently, the FCC also has the authority to review all corporate mergers and acquisitions that result in the transfer of radio licenses; the standard governing this review is a rather general "public interest" standard.

The standard procedure (until quite recently) was that an individual or firm wishing to use spectrum for a specific purpose license for a particular frequency in a particular location applied to the FCC for a license that covered only that purpose, frequency and place. After public notice, anyone else could also apply for the same frequency and location; should there be more than one applicant, a comparative hearing was held to determine which applicant was "more suitable" to discharge the public interest obligations of license-holding. Numerous critics have charged that this process could be politically influenced; one of the more notorious cases concerns the radio licenses obtained by Lyndon Johnson in the 1940s while he was a Congressman, which licenses became the foundation of his personal fortune (see Caro (1991)). Applicants were issued licenses for specified purposes; a license for taxi dispatch could not be used for ham radio, for example. Further, the license was limited to ten years, although issued with the presumption of renewal. Recently, renewal has become as easy as sending the FCC a postcard, but in the past license renewals could be and were challenged.

The award of the license did not grant the licensee any property rights in the spectrum beyond that of the license. The licensee could not use it for any purpose other than that specified in the license. If the licensee were purchased, or merged with another firm, the transfer of the license had to be approved by the FCC.

More recently, the FCC and Congress have retreated from the comparative hearings model. After a brief foray into licensing analog cellular licenses by lottery, Congress gave the FCC authority to conduct auctions for licenses for commercial services, excluding broadcasting. (Currently, all mutually exclusive FCC licenses except those used for satellite and public safety services are subject to auction). A number of auctions have since been held, raising over $14 billion for the U.S. Treasury.[3] Again, the auction winners, like other licensees, do not actually own the spectrum, but merely the license to operate mobile or fixed service (excluding broadcasting). The FCC (nor NTIA) does not assert ownership of the spectrum, but does retain all rights to control it, including the issuance, conditioning and revocation of licenses; however, a recent ruling by a bankruptcy court in the NextWave case[4] ruled that a wireless license is considered an asset of the firm and the FCC has no primacy over other creditors in reclaiming this particular asset. This would suggest that the FCC's residual control of all wireless licenses is not absolute. Additionally, wireless licenses granted to satellite systems have been explicitly excluded from the auction process.[5]

The results of this process are not difficult to predict. Holders of spectrum are unwilling to give it up, even when they are unable to make use of it. For example, the FCC's experience in the 1950s with UHF television assigned 330 Mhz of spectrum to this use.[6] The experience was not successful, and this band is extremely underused. However, license holders are unable to use the spectrum for any other purpose (such as wireless telephony) and are unwilling to give it back (see footnote 39). Thus, this prime spectrum provides little value to consumers, while other uses (such as wireless telephony) claim to be in a "spectrum drought." The political nature of spectrum allocation is illustrated by Congress' direction to the FCC[7] to allocate spectrum to the broadcast industry for DTV (digital television), which has allocated channels 2–51 for this purpose.[8] The broadcast industry appears to be stoutly resisting the deployment of DTV and yet it is unwilling to give up the spectrum Congress gave it for this purpose. Again, valuable spectrum provides little value to consumers while other uses are starved for spectrum.[9]

There are several efforts underway at the FCC to improve this highly inefficient use of the spectrum. "Flexible use" is a policy initiative in which wireless license holders are permitted to use their spectrum for products not specified in their original license. For example, if flexible use were applied to the UHF channels, then UHF license holders could use their spectrum for wireless telephony (or any other use).[10] Nextel is an entrepreneur that has already taken full advantage of flexible use, offering cellphone service using spectrum from the taxi dispatch band. "Band managers" would permit the licensing of spectrum to firms who could then lease this spectrum to others on commercial terms.[11] The FCC is also engaged in band clearing, in which current license holders are offered spectrum in other bands to give up their current allocation that could be more constructively deployed in other uses. Currently, the UHF channels 52–69 are targeted for band clearing.

Despite the recent moves toward more market-based spectrum allocation, the dominant mode of managing the spectrum is administrative fiat. Perhaps the closest analogy to the U.S.'s current approach is that of GOSPLAN, the central planning agency in the former Soviet Union. GOSPLAN drew up plans for every sector of the Soviet economy, determined how much of each scarce input was required for each industry and each factory, and then issued orders to each factory as to how much it was to produce and to whom it was to be shipped. GOSPLAN was subject to intense lobbying by powerful factory bosses regarding quotas and shipments, and allocations were politically mediated. While the FCC only controls the electromagnetic spectrum, it has operated in a very similar manner, and subject to the same political pressures. It should be no surprise that both GOSPLAN and the FCC processes have had similar results: woeful inefficiencies and wasted resources (see, for example, Kwerel and Felker (1985) and Kwerel and Williams (1992)).

The basics of the system we use today were established when the most important use of the spectrum was broadcasting and the range of usable spectrum was about 1 percent of what it is today. Few would argue that this system is optimal today, but many may lose if the system were changed. The system is so embedded in how we use the spectrum that change is practically unthinkable. Current licensees received scarce spectrum years ago at zero cost from the government under the expectation that it would be theirs forever. These licensees include not only TV broadcasters and telephone companies using microwave relay systems, but police and fire departments, Department of Defense, taxi dispatchers and paging companies. While zero-cost transfers represent a windfall gain to many licensees, to many others it is a component of their public service obligation that they could not otherwise afford. Is this a system that is admittedly highly inefficient yet with so many stakeholders that it cannot be changed?

II. The Economists' Critique

Ronald Coase The seminal contribution of economists to the issue of spectrum allocation was made by Ronald Coase (1959). Coase was awarded the Nobel Prize in Economics in 1991, and in his Nobel autobiography, wrote of this work:

> I made a study of the Federal Communications Commission which regulated the broadcasting industry in the United States, including the allocation of the radio frequency spectrum. I wrote an article, published in 1959, which discussed the procedures followed by the Commission and suggested that it would be better if use of the spectrum was determined by the pricing system and was awarded to the highest bidder. (Coase 1991)

To an economist, this critique is as natural for the FCC's method of allocating a scarce resource as it was for the Soviet Union's method of running its economy. The market is a far more powerful and efficient allocator of resources than administrators and bureaucrats can ever be, no matter how knowledgeable and well intentioned. Efficient markets can realize their magic because they are highly decentralized processors of information. Prices are determined by buyers and sellers interacting in the market, to ensure that demand and supply are equated. The ability of the market price to capture all the information regarding supply and demand is far greater than that of a centralized planner no matter how sophisticated their planning and allocation tools.

Coase's critique seems, in retrospect, blindingly obvious. For almost all activities in the U.S. economy we rely on markets to allocate resources, and markets work somewhere between pretty well and extremely well. Why is spectrum allocated using this wildly inefficient, Soviet-style means of administrative fiat? Coase's solution was to create sufficient property rights in spectrum so that it could be sold to private owners who would then be free to buy, sell and lease spectrum. In legal terms, ownership of spectrum would be ownership in fee simple.[12] Spectrum could be aggregated or subdivided, according to the needs of customers as expressed through the market. As a result, all frequencies would move to their highest valued use. For example, owners of inefficiently used UHF channels would have both the ability and incentive to sell or lease their spectrum to wireless telephony firms, or even become such firms themselves.[13] The price at which such transactions occur would reflect the demand and supply for spectrum; since certain frequencies are particularly useful for certain in-demand applications, these frequencies might well command a price premium relative to other frequencies, as the market dictates.[14]

Fundamental to the efficiency of markets is scarcity. If resources are not scarce, if consumers can pick their food off trees that are never exhausted and if there is infinite bandwidth, then there is simply no need to have markets, which have costs to organize, administer and maintain. Early hunter-gatherer cultures existed in such a world of plenty; unfortunately, as populations expand, the previously plentiful becomes scarce and people must find a way to allocate these scarce resources. In our own time, we have seen the oceans undergo the same transformation, as fisheries historically treated as an international commons became overfished and stocks have had to be allocated. Over the long haul, costly trial and error has demonstrated that when resources are scarce, markets are the most efficient way to allocate these resources. Grand experiments with government (rather than market) allocation of economic resources have ended badly, to say the least.[15]

Markets have also shown themselves to be particularly friendly to innovation, as owners of assets strive to make their property more valuable through the use of new technology. Restricted licensing of spectrum, however, has the opposite effect. Since a licensee can only use his or her frequencies for their designated purpose, the incentives to innovate for a licensee are mitigated. An existing license holder may have incentives to innovate to increase the capacity of its frequency band if it can thereby serve more customers. For example, current licensees of satellite bands may have incentive to convert these bands to terrestrial digital cellular to make more efficient use of this spectrum. But since they are barred from different uses, innovation is limited only to existing authorized uses so that licensees' incentives to innovate are less than they otherwise would be.

As with any social change, transiting from a government-assigned licensing regime to a market regime almost always involves costs to incumbents who have large stakes in the existing system. As mentioned in the previous section, there are many beneficiaries of the current system and they can be expected to resist strongly any solution that involves taking back their long-held assets. We address this question in "Transitioning to Markets: A Modest Proposal," below. For the remainder of this section, we analyze a market-based system ignoring for the moment the problems of actually getting there.

As many college freshmen learn in Econ 1, not all markets work perfectly, and there is an extensive theory of "market failure."[16] One such "failure" that can arise from unrestricted use of property is a "spillover," in which one property owner's use creates costs (or benefits) to others. For example, a factory may produce pollution that is costly to others; alternatively, the owner of an apple orchard creates a positive spillover for the beekeeper next door (and vice versa). In the case of spectrum, spillovers in the form of out-of-band power in adjacent frequencies are important, and can generally be controlled by the careful definition of property rights. In today's regime, wireless licensees operate under a set of technical restrictions regarding power and place of emission, and possibly direction and time of emission. In a property rights regime, these restrictions would be codified in the property rights of the frequency owner, who would then be subject to civil penalties should he or she violate these restrictions. In fact, such restrictions are often codified in property rights and laws. My right to use my automobile is restricted by speed limits; my right to use my real property is restricted by noise and nuisance statutes of my state, county and local municipality. Property rights in spectrum would be similarly constrained, and in fact we already know what the constraints are: they are largely defined by the technical restrictions in current licenses. These licenses may also include both use restrictions and equipment restrictions that would *not* be included in property rights. The spillover of interference in adjacent bands can thus be eliminated by suitably constraining each owner's property right to use his or her frequency, exactly as we do today. Therefore, the spillovers associated with out-of-band out-of-area frequency emissions can be fully controlled through the appropriate and careful definition of the owner's property rights; emitters who violated these restrictions could be sued by those who suffered from the resultant spillovers for damages and perhaps penalties.

Interference From the economic perspective, radio interference is the spillover that is the primary rationale for government control of the spectrum. It is the interference spillover that requires limitations on the property rights of ownership in a market regime. While we focus on the property rights of the transmitters of radio energy, the problem of interference involves both transmitters and receivers. Restrictions on transmitters include in-band power restrictions, so one transmitter doesn't interfere with a transmitter at a distant location, and out-of-band power restrictions, to control emissions in frequency bands in use by others. But these constraints are based on the ability of the intended *receivers* to filter out spurious signals. For example, early TV receivers had little ability to reject power spills from adjacent TV broadcast bands. As a consequence, "guard bands" of spectrum were designated between each usable bands so that out-of-band power leakage would not impinge on nearby signals. The use of guard bands is wasteful of spectrum today, but was necessary given the technology of the time. Because they employed unsophisticated tuners, early TV sets were relatively inexpensive. Today the ability to discriminate and filter out of-band power leakage is very inexpensive to build into TV sets. However, the wasted spectrum is still there, "protecting" TV sets, so television set manufacturers have no incentive to install more sophisticated tuners. The inefficiency of spectrum use is locked in because of receivers, not transmitters, require the use of guard bands.[17]

Today's technical rules on interference are likely to become tomorrow's property rights in spectrum. They are based on a balancing of the current technology of both transmitters and receivers. As the technology has evolved, the current licensing system has not been particularly successful at reclaiming valuable spectrum by changing the rules. An important question for any property rights regime is how well it permits property rights to evolve with technology.

Markets and Property Rights Coase's critique of the FCC concluded that spectrum should be allocated by a market, not administrative fiat. However, markets can function without explicit assignment of property rights. Indeed, the current model of FCC auctions for spectrum use is just that—-a market without property rights. The government conducts the auction for licenses to use the spectrum with only limited ability of the licensees to deploy new services through their licensed spectrum and to transfer or sell it. More complete models of markets without property rights have been suggested by Noam (1998).[18]

Theoretically, the benefits of the market could be realized via a government-conducted scheme. As a practical matter, such benefits have been limited by the lack of a secondary market and by continued political actions to interfere with the operation of the market, such as the aforementioned problems with NextWave, et al. Even with a well-intentioned and relatively skillful FCC, outcomes of politically controlled market mechanisms fall short of a well-functioning private market. But private markets depend upon clearly defined property rights; in order to realize the full power of the market to bring about more efficient allocations of spectrum, private markets and therefore property rights are essential.

Enforcement All property rights must be enforceable if they are to be meaningful. Today's licensees must be able to enforce their licenses, and if ownership of spectrum is permitted, owners must have a way to enforce their property rights.

Typically, property rights are enforced by the rights-holder lodging a complaint against an alleged infringer. This might be a simple call to the police that a stranger is trespassing on my land and refuses to leave. It could be a patent holder filing suit in court against another party accused of infringing on his or her patent. Under the current system, a licensee complains to the FCC who may then investigate the complaint and, if appropriate, punish the infringer. In an ownership regime, the rights-holder brings a civil suit against the infringer.[19] In certain cases, such as patent law, special courts are available for adjudicating such cases because of the specialized knowledge required. In a spectrum ownership regime, the FCC could retain an enforcement role, or this role could be subsumed by special "spectrum" courts, or by the general court system. Thus, there are a variety of enforcement models available for an ownership regime. Which venue is most appropriate depends upon the transaction costs of each. The general court system has the great benefit that it is ubiquitous and available locally anywhere in the country. However, if special expertise is required to litigate spectrum claims because of technical complexity, then special courts or the FCC may be needed, albeit more costly. Such agencies are subject to "capture" by their constituent firms, which raises their cost to the economy. If property rights are sufficiently simple and clear, then the general courts may be the preferred venue. Should the property rights be less than simple and clear, however, general courts may be more prone to error and inconsistency, thus raising their costs to the economy.

Assumptions Underlying Fee Simple Ownership Since the earliest days of broadcast, the use of spectrum by licensees has properties that are facilitated by a fee simple property rights regime (and facilitated, less efficiently, by the current licensing regime). These properties are:

High Power Within the relevant geographic region, emission is at a high enough power that more than one emitter at the same (or similar) frequency will cause damaging interference to the signal of at least one emitter. In many cases, broadcasters emit 24 hours a day, 7 days a week, and noninterfering frequency sharing has not been easy or obvious.

Dedicated Frequencies Most broadcasters emit at a particular frequency (or a limited set of frequencies) so that simple receivers can easily locate them.

Under these assumptions, dedicating certain frequencies to high-powered licensees/owners is an efficient response to the interference problem. The difference between a fee simple property rights regime and the current licensing system is that a market-based regime is far more flexible than the rigid bureaucratic processes of regulation. It is therefore a far more powerful mechanism to achieve an efficient allocation of the scarce resource of spectrum, as it harnesses the self-interest of owners rather than relying on bureaucratic processes. However, technology has not been standing still, and new technologies have begun to undermine these assumptions of high power and dedicated frequencies.

The Engineer's Critique

Since 1938, the FCC has used its "Part 15" rules to permit the unlicensed use of certain "intentional emitters," such as garage door openers and cordless phones.[20] Such unlicensed emitters have been constrained to operate only within certain frequency bands and at relatively low power. These limits are enforced by requiring the manufacturers of emitting devices to certify their products as having been tested and found to be within the FCC's frequency and power limits. Manufacturers are required to submit their devices to the FCC or an FCC-approved testing lab. The FCC may sample the product for compliance. Certification is required for imported as well as domestically produced electronic products. While there are opportunities for cheating the system, the consensus within the industry and the FCC[21] is that type certification has generally worked well at controlling interference, and industry cooperation on device design to control interference has been successful.

The openness of Part 15 spectrum has also promoted innovation in spectrum use. Within the FCC constraints, engineers and scientists have developed systems for spread spectrum technology into cordless phones, wireless broadband networks into neighborhoods (such as Metricom's Ricochet service), short-range wireless LANs and wireless home networks (such as "Wi-Fi"). Not surprisingly, radio engineers have lauded the openness of Part 15 spectrum as a boon to innovation.

Further, many have noted that Part 15 spectrum has property rights akin to that of a commons: an asset available for the use of all, with common restrictions governing use restrictions for all.[22] If innovation has been so forthcoming in a commons environment of unlicensed use, then why not extend the commons environment to the entire spectrum? Advocates of this approach compare the level of innovation that has occurred under this commons model with the much more disappointing level of innovation under the current licensing regime, which they sometimes refer to as a private property regime (which it clearly isn't). There does not appear to be an evidentiary base for this assertion, however.

Engineers point to two recent developments that would seem to make use of the commons model especially well: ultra-wide band (UWB) radio and software-defined radio (SDR). These two applications show great commercial promise, and appear on the surface to be incompatible with both the existing licensing model as well as a property rights market-based model. We discuss each in turn:

Wideband This form of radio emissions can be used for a variety of purposes, including ground penetration, through-the-wall imaging, and short-range "radar" for vehicles. It can also be used for two-way communications. The most successful wideband application today is spread spectrum, used in many cordless phones. This technology allows a signal to be "spread" across a range of frequencies, trading off power for bandwidth. Ultra-wideband (UWB) operates similarly but in a more extreme form. The signal to be transmitted is captured in small time intervals (about 1 microsecond) and the signal is converted to a set of very short pulses (about 1 picosecond) and these pulses are broadcasted over a very wide bandwidth (greater than 1 Ghz); the broadcaster emits this picosecond pulse in a time slot every microsecond at very low power; the receiver (which must be synchronized) picks up the low power signal over this wide bandwidth, and converts it back to (a very good approximation of) the original signal.

UWB radios essentially trades off lots of power for lots of bandwidth. The power of the emission is extremely low;[23] for most purposes, it is part of the background radio noise, and non-UWB receivers that are designed to reject noise would not recognize the signal, so there is no interference with high-powered broadcasters. The useful range of UWB at these power levels is rather short, at most a hundred meters at currently authorized power levels. Interference with other UWB emitters is unlikely; emitters more than, say, five miles apart can use the same transmit time slot without interference with each other, and there are many time slots. Additionally, UWB is fault-tolerant, in that the frequency pattern transmitted in the picosecond burst can suffer some degradation and the original signal can still be recovered.

On the other hand, the bandwidth of the UWB signal spans a large fraction of the total frequency available to all, and appears (if undetected) at many frequencies for which licensees hold exclusive use. In a property rights market regime, UWB signals would also appear in frequencies owned by others, even if not detectable.[24]

Perhaps the clearest analogy is the right of an aircraft to pass over my home. As the property owner, I do not have the right to forbid aircraft to do so, nor may I charge them a fee to do so. However, aircraft regulations require that aircraft not fly lower than 1000 ft. over any obstacle within 2000' so as not to create a noise or safety nuisance.[25] The property rights of aircraft owners and pilots are restricted so as not to interfere (by noise or safety) with my property right to enjoy my home.[26] In a similar vein, the FCC's recent ruling on UWB limits the power of emissions across the frequency band so as not to interfere with licensees' rights to use their frequencies.

Agile Radio This is a form of software defined radio (SDR), a term that covers a rather broad category of devices and includes any device in which the received radio signal is processed by software. "Agile" radios are devices in which a radio can determine if a specific frequency band is currently in use, emit in that band if not, and switch to another band in microseconds if another user begins to emit in that band. Both transmitter and receiver must be agile for this system to function. For example, in principle an agile radio transmitter could use an empty ham radio band (or government military band) to communicate with an agile radio receiver; should a ham operator (or military user) start using that band,[27] the transmitter would shift to another band within microseconds (the receiver presumably shifting as well, according to a prearranged script) and the agile radio communication could continue while the ham operator used of original band. Provided the agile radio switches its emissions to another band, it need not interfere with the ham band. As long as there are sufficient frequency bands so that the agile radio pair can always find an unused band, agile radio achieves a more efficient use of bandwidth without interference with existing licensees (or owners, in a property rights market regime).

Agile radio creates this increased efficiency by *dynamic* allocation of spectrum, rather than the current *static* allocation approach, common to both the current licensing regime and a property rights regime. For many purposes, static allocation is the efficient solution; AM-FM and TV broadcasting of continuous content to the existing huge base of relatively simple receivers will be a very important spectrum use for years to come, and static allocation works perfectly for this application. But dynamic allocation for certain uses can improve the efficiency of spectrum allocation, perhaps dramatically. In light of the inefficiencies of the current licensing regime, this would appear to be an important improvement.

Mesh Networks Mesh networking is a wireless architecture that can use different forms of radio transmission, including UWB, agile radio, even cellular. A mesh network of (say) computers[28] in a neighborhood could communicate (possibly at high bandwidth) with a Neighborhood Access Point (NAP) that could connect directly into the Internet (or possibly the telephone network).[29] Computers out of the immediate range of the NAP could connect to the NAP using other computers as relay points, thus extending its range through the use of single or multiple relay "hops" via the other computers in the network. Apart from the few NAPs required to seed the network, there is no infrastructure such as cables or fiber optics needed for mesh networks. The wireless devices themselves form the network, much as the Internet currently operates.

Mesh networks use much less power than conventional systems which need every computer to reach a central antenna. Mesh networked computers need only reach the computer next door, and thus need less power. The architecture takes full advantage of the relay capabilities of the mesh devices to lower power requirements and therefore minimize interference problems. Because of this, mesh networks actually increase their capacity as the geographic density of users increases; in other networks (such as cellular), increasing density actually decreases available capacity because of interference.

IV. New Technology and Property Rights

While the new technology opens up new opportunities for efficient use of spectrum, using either of these technologies appears to violate the license rights of current licensees. It also appears to be incompatible with a property rights market regime as well. Proponents of these technologies claim that they should be deployed in the context of a commons model, in which all can use the spectrum whenever they want, as long as we adopt simple rules to keep out of each other's way. In this view, property rights are the problem, not the solution; "building fences" of property rights violates the commons principle.

It is understandable that the developers of these new technologies hold the view that these innovations are likely to deploy most quickly and effectively in a commons regime. After all, much of the research was conducted within the Part 15 unlicensed spectrum, which is a commons regime. Further, the new technologies appear to use spectrum in new ways that don't easily fit into the legacy business model of high-powered dedicated frequency broadcasting. Why adopt a legacy-driven property rights model when the new technologies promise an end to scarcity? In this view, the commons model is best suited to the new technologies.[30]

Central to the choice between a property right regime and a commons regime are (i) scarcity and (ii) transaction costs. If a resource is scarce in that many people contend for its use, then a commons regime will be afflicted with the "tragedy of the commons," in which the resource is overused; in spectrum terms, we experience interference. In the face of scarcity, a property rights regime will function to ration the scarce resource; the resource will have a positive price and contention for it is resolved in the market. However, if the resource isn't scarce, then a commons regime works quite well without incurring the cost of a property rights regime. Further, if a property rights regime is imposed where scarcity is not present, the price of the resource at the margin falls to zero.[31]

The structure and magnitude of transaction costs determine the boundary between efficient regimes. If transactions costs of a property rights regime are quite high, then the costs of the tragedy of the commons must be quite high indeed to justify using a market regime. If the costs of a property rights regime are relatively low, then it is likely more efficient than a commons regime even at low levels of contention costs.[32]

In order to focus on these central issues, we first examine two property rights regimes that appear to release both the power of the market and the power of the new sharing technologies to improve the efficiency of spectrum use.

Fee Simple Ownership with Noninterference Easement[33] In this regime, individuals and corporations would be able to buy, sell and lease specific frequencies in specific locations subject to power (and other technical) limitations, and would possess the right to emit at any time *without interference*. Other emitters could use this spectrum, but only on condition that they not *meaningfully* interfere with the owner's right to clear broadcast. Thus, UWB emitters that maintained power levels below the noise threshold would be noninterferers. Agile radio emitters that vacated a frequency within (say) one microsecond after the frequency owner began broadcasting would be noninterferers. Conversely, either a UWB emitter exceeding its power ceiling or an agile radio emitter taking too long to vacate is an interfering user and becomes subject to penalties.

In this regime, spectrum would be owned but subject to an easement that any and all users that did not meaningfully interfere with the owner's right to the spectrum could not be excluded from using the spectrum. In effect, this easement creates a commons at all frequencies and in all locations of a special type: noninterfering uses only.

Enforcement under this regime would require that UWB and agile radio emitters transmit a unique identifier (similar to identifiers built into computer network interface cards) and frequency owners could monitor and record violations. Penalties could be assessed much as traffic violations are handled; it is likely that third-party collection agencies would arise to handle these violations on behalf of owners. Such monitoring would result in costs to owners. Fines for violations could recompense owners for these expenses.

Pure Fee Simple Ownership In this regime, individuals would be able to buy, sell and lease specific frequencies in specific locations subject to power (and other technical) restrictions, and would possess the right of *exclusive use*. Other emitters could use this spectrum, but only upon payment of a fee to the owner. Sharing fees could cover a range of options, from a long-term lease for the entire band to agile radio noninterfering use. The prices would vary, depending on the nature of the lease arrangement, with noninterfering uses such as agile radio most likely priced the lowest. Agile radio users could negotiate long-term use of a band ("forward contract") or negotiate band use at the moment of use ("spot market"). We would expect agile radio users would negotiate with various band owners in both markets. Prices in the two markets would generally differ.

In the case of spot markets for spectrum, transactions costs are likely to be significant, as owners would have to monitor *all* uses, not just interfering uses.[34] Just as agile radio transmitters would be required to broadcast a unique identifier, owners would have to broadcast their price for use, and this would likely be most efficiently accomplished by ensuring that all equipment was fully compatible. Agile radio transmitters could thus "shop" for the least expensive frequencies.[35] It is likely that third-party collection agencies could manage the flow of lease revenues from users to owners, which may well involve thousands of lessees making very small payments each to thousands of lessors. However, there exist institutions that can handle this problem at minimum transactions cost, even without the magic of computers. A similar situation arises in the payment of royalties owed to musicians every time a song is played on the radio or in a jukebox. There are two associations, BMI and ASCAP, that monitor radio playlists and jukebox records, bill the responsible parties and send the receipts to the owners of the music. A similar arrangement is likely to be successful for band use micropayments as well. However, such a real-time spot market system will only arise if the transaction costs of owners is less than the value of the spectrum to lessors.

This regime would generally have higher direct transactions costs than the easement regime, and may be somewhat less encouraging of innovative noninterfering uses. The magnitude of indirect transactions cost is less clear; litigation regarding the use of the easement may well be extensive and costly.

These two property rights regimes focus on the *emitters* of radio energy; how about the *receivers*? The introduction of new technology in one band may only be possible if receivers in adjacent bands can accommodate the new technology, which may require a costly upgrade. For example, if legacy receivers had inexpensive tuners that picked up emissions in neighboring bands, then technologies that uses those bands would only be noninterfering if the legacy tuners were upgraded to filter out their emissions. We noted above that in the case of many receivers in adjacent bands, this could be infeasible if the new service providers had to convince owners of legacy receivers in adjacent bands to upgrade. However, the current radio industries have been successful using voluntary standard setting among manufacturers. This model focuses on the manufacturers of receivers rather than end-customers, in particular on the chip manufacturers whose products constitute the core of both receivers and transmitters. If the industry can agree that (say) the introduction of agile radio is likely to result in more business for all participants, but at the cost of increasing filtering capabilities for receivers in adjacent bands, then chip manufacturers may agree to establish enhanced standards for new receivers (for these adjacent bands) effective immediately. If the average life of such a receiver is (say) three years, then the agile radio service providers could begin using their technology in bands adjacent to the interferes after (say) twice the average receiver life, or six years, assuming that most receivers in the field at that point incorporate the enhanced standards.

The use of voluntary industry standards appears to have worked successfully in computer hardware and software, which are of course governed by the market. As new bus architectures have been developed in the PC market, software developers and peripheral manufacturers produce to the new standard while maintaining backward compatibility for some period of time. Eventually, compatibility of complementary products with sufficiently old systems is dropped, and the technology moves on. We believe this model is likely to work in the wireless world as well in a property-rights-with-noninterference-easement market regime.

Military and Public Emergency Spectrum Use This unique use places unique demands on spectrum management. During an earthquake or defense contingency (local or national), there is no time to ask permission or negotiate with other parties; military and public emergency personnel need to have immediate preemption capabilities for spectrum capacity substantially larger than their everyday administrative needs. Under the current system of allocating spectrum, this requires that the maximum amount of spectrum be allocated to these uses, even though it is hardly ever used. Using agile radio technologies, this spectrum can be made available to others for routine use, with the contractual proviso that military and public emergency users have an absolute and immediate preemption right to the spectrum. There is a strong precedent for this; all private broadcast and cable systems can be immediately preempted by civil defense authorities who can commandeer their spectrum as part of the nation's Emergency Alert System,[36] which has a history of over half a century.

Transactions Cost and the Tragedy of the Anticommons There are two forms of transaction costs of concern: (i) direct transaction costs of spectrum buyers and sellers; (ii) indirect transaction costs of dispute resolution. Disputes regarding interference will arise in either a commons regime or the two property rights regime; it is likely that courts will be called upon to resolve such disputes, and it is likely that courts will be more efficient in dealing with the familiar territory of property rights. However, the property rights with easement may require extensive litigation prior to establishing clear easement rights. We thus view pure ownership as having the lowest indirect transaction costs, ownership with noninterfering easement as next lowest indirect transaction costs, and commons as the highest indirect transaction costs.

Direct transaction costs show the opposite ordering. A commons regime has almost no direct transaction costs as no one is paying anyone.[37] The ownership regimes will incur costs for normal transactions among parties leasing or selling spectrum, which are unlikely to be significant.[38] However, transactions between owners and users of the newer technologies may have higher direct costs if buyers and sellers prefer a spot market. In this case, equipment capable of identifying and negotiating electronically within microseconds would need to be deployed. The technology and cost of this equipment is likely to be commensurate with the technology and cost of the advanced devices themselves. The capabilities of an agile radio, for example, are similar to the capabilities of devices required to identify and negotiate with multiple customers at very high speeds. Whether or not a spot market would be preferred over longer term contracts is not clear. The cost of the enabling devices for spot markets may well affect their popularity.

A more serious problem is that of the tragedy of the anticommons, a phrase coined by Heller (1998). If property has *too many* owners, each of which must agree before the property can be put to effective use, then each owner may attempt to "hold up" the other owners for a greater share of the rewards to effective use, thus barring the deployment of the property. Heller and Eisenberg (1998) applied this to patents in biomedical research, and in Heller (1999) he outlined a general theory of the boundaries of private property. Benkler (1997) uses the idea of the anticommons in the context of radio spectrum to argue that the transactions cost of a property rights regime may be prohibitive for the new technologies if legacy owners assert ownership rights.

The argument is perhaps clearest in the case of UWB. Suppose that the spectrum is exhaustively sold, so that an individual or firm owns each frequency band in each locality. Now consider a UWB transmitter, which requires the use of hundreds of these frequency bands (albeit at very low power) to transmit its signal. If the UWB transmitter is required to negotiate a contract with every single owner, and cannot broadcast until every single owner agrees, then the transaction costs are indeed quite high and the transmitter unlikely to be successful. The problem is much less severe for agile radio; if only half the owners agree to transmit short signal bursts from one agile radio to another, this is more than enough. Not every owner must agree, and therefore there is no "hold up" problem. Note also that this problem does not arise at all in the ownership with easement regime.

The tragedy of the anticommons ensures that the direct transaction costs for the pure ownership regime may be particularly high for UWB. For this reason, we favor the ownership with easement regime over the pure ownership regime.

Ownership and the Commons Establishing property rights in spectrum is often portrayed as eliminating the commons (Benkler (1997), Reed (2002), Ikeda (2002)); this is not the case. Commons (and more generally sharing) can exist within an ownership regime; our recommended ownership regime with an easement for noninterfering uses establishes such a commons via the easement. Should it be necessary to have a commons for potentially interfering uses, the most obvious avenue is for the Federal government can purchase a block of spectrum (which it then owns) and open the band to general use under terms and conditions similar to Part 15 (for example). In fact, any state or local government can do the same thing, establishing a "park" in which users are completely free to use the spectrum without permission provided they follow the rules laid down by the owner of the "park." This is perfectly analogous to public lands, such as National and State Parks, National and State Forests, and municipal parks. Further, private foundations could establish such "parks;" for example, there are many horticultural parks open to the public that are maintained by private foundations. Local neighborhood cooperatives could achieve the same end, possibly requiring a one-time or monthly fee for use. Similarly, private firms could establish such "parks," charging a one-time or monthly fee for use. We would expect that manufacturers of mesh network devices, for example, may choose to "prime the pump" by establishing spectrum parks in various localities to increase their equipment sales. We might also see existing wireless providers of cellular service extend their functionality; for example, Sprint is reported to be expanding into Wi-Fi networks for its customers (Charney (2002)).

Any or all of these mechanisms would permit mesh networks to flourish. The authors cited above have alleged that an ownership regime is fundamentally incompatible with the deployment of mesh networks. In the paragraph above, we count at least six ways in which mesh networks can flourish in the ownership regime with noninterfering easement. While we agree with these authors that mesh networking is an exciting new technology that may well shape the future of communications, we have demonstrated that their assertion regarding mesh networking's incompatibility with an ownership regime is incorrect.

Scarcity, Markets, and New Technology Both economists and engineers agree that the current licensing regime has led to grossly inefficient use of the spectrum resource. If the ownership with easement regime is universally adopted, the alleged "spectrum drought" will almost surely turn into a "spectrum flood," as large amounts of underused spectrum come into the market. Current inefficient uses such as UHF TV[39] will come to market quickly once a market regime is in place, with more than enough bandwidth to satisfy immediate demands. Based on this presumption, we conclude that *in the short run,* excess demand will likely turn into excess supply, except in certain especially useful frequency bands. In this situation, *the price of spectrum at the margin is likely to be zero* (or very close to it).[40] This short-run excess supply occurs as a result of markets eliminating current inefficient uses. While this may not be good news to cellular carriers who have spent billions on bandwidth made scarce by government regulation, it is good news to the consuming public and we should welcome it. Under either regime, the artificial scarcity created by the current licensing regime is eliminated.

We do not expect this short-run excess supply to last. New uses of radio spectrum should come on stream fairly quickly, promising to fill this newly available spectrum. But we also expect the new technologies of UWB, agile radio, and mesh networks to come on stream in parallel, and these technologies will again result in excess supply of spectrum, certainly for the medium term. In the long term, we expect that new uses for radio spectrum will use the spectrum fully, and the demand and supply of this important resource will come into balance. The demand for spectrum is likely to grow very rapidly; in the not-too-distant future, this new "unlimited bandwidth"[41] would become limited indeed, as demand grew to meet the available supply. The nature of the market changes, and spectrum bandwidth now becomes a scarce resource; not now, but in the future.

In a long run world of spectrum scarcity (real this time, not the artificial scarcity of government allocation), prices are no longer zero and the commons model breaks down. Agile radios will find the next frequency they hop to is busy, as is the next, and the next, and so forth. As the airwaves congest, the best solution will be the market, as it is for virtually every other economic good or service. In the long run, therefore, the commons portion of the spectrum (including the noninterfering easement) will be highly congested, and many users will migrate toward owned spectrum to ensure access and quality. In a world of real spectrum scarcity, owners will invest in metering gear and charge users a positive price, ensuring that the spectrum is allocated, in real time and otherwise, to its highest valued use.

Pure Commons Regime How would a pure commons regime work? Unfortunately, high power dedicated spectrum uses are likely to be a fixture of any system for a long time, and such uses fare poorly in a commons model as there is no guarantee of noninterference from other high power dedicated frequency users. If a commons regime were to be adopted, this would reproduce the radio world of the early 1920s. If *all* users were forced to undertake a costly upgrade to agile radio (or UWB, if feasible), then a commons regime may be workable in the short run, as long as scarcity is not an issue. However, as new devices and new uses proliferate, spectrum scarcity will become reality. There still is a limited amount to go around, and at some point it will get used up. This is especially true of "sweet spot" spectrum that is particularly good for certain popular services such as cell phones. In this long run view, a commons regime is quite limiting, and another regime change to markets will be required.

Is it likely that in the long run spectrum will indeed become scarce? While today's massive underutilization of spectrum suggests that markets and new technology may increase available spectrum by orders of magnitude, we have no doubt that clever engineers and aggressive marketers will find ways to fill that spectrum with new and useful gadgets that we all must have. We believe the long run answer is clear: ways will be found to use all the spectrum we can make available, and eventually it will become scarce.[42]

Conclusion A market-based ownership with noninterfering easement regime is compatible with the deployment of UWB, agile radio and mesh networks. In the short run, we believe this regime is likely to free up so much spectrum that this resource will be in excess supply. In the long run, as this resource becomes better used and spectrum becomes scarce, we expect that owned spectrum becomes more attractive as a superior method to manage scarcity.

V. Transition to a Market-based Regime

Our chapter thus far has compared the "end-states" of two regimes: the current licensing regime, the ownership with noninterfering easement regime, and a commons regime, without discussing how the ownership regime could actually be obtained in the context of spectrum politics. We argue above that the market-based regime has more attractive economic properties than either the commons regime or the current licensing regime, especially the real-time leasing regime in the long term.

Any transition plan from the current regime to a market-based regime inevitably will create winners and losers. Losers, of course, will oppose the transition, and winners may favor it but seek even greater gains. In other words, the process is essentially political and the transition must be structured to ensure that all or most stakeholders are not harmed.[43] We thus take the world as it is (warts and all) and seek a politically viable transition plan to a more efficient regime.

Defining Property Rights Constructing the bundle of rights that constitute property in spectrum must be done with great care, and must precede any attempt to institute markets. In particular, the scope of property must be economically viable in order to avoid the tragedy of the anticommons. But it must not be so large as to encourage market dominance. DeVany et al. (1969) discussed in detail how to define property rights in their seminal article, and is an excellent starting point for this exercise. White (2001) is also useful in this regard. Generally, these authors recommend that technical constraints regarding time, area (including power limitations) and frequency should constitute the property bundle. Additionally, our recommended option of a noninterfering easement requires a careful definition of what constitutes interference. Perhaps more important is to put in place an efficient dispute resolution system, such as arbitration with technical expertise. Such a dispute resolution system could establish case precedent to correct any mistakes of the original property rights distribution (such as power levels for UWB).

Determining this bundle of rights promises to be a daunting task with technical, economic and political components to that task. The measure of difficulty can be assessed by noting the intensity of the recent debate at the FCC regarding appropriate power and interference levels for ultra wideband deployment. Getting the bundle perfect is not necessary, as mistakes can be remedied by private contracts later. However, getting the bundle approximately correct is important so that post-market bargaining is more efficient.

Broadly speaking, current licenses constitute economically viable bundles, and the technical requirements of these licenses would be an excellent starting point for a property rights bundle. However, current licenses also have use restrictions and in some cases actual equipment restrictions. Such restrictions should not be incorporated into the property rights bundle. In some cases, the current license is tightly tied to a particular use; for example, point-to-point microwave licenses are geographically restricted so that they can be used for little else except microwave, thus limiting their marketability. Such anomalies may need correcting before adopting a market-based regime.

Getting to Market We present this proposal in broadest conceptual outline, without pretense that the technical details have been worked through. We do not claim authorship of this proposal; this transition plan has been put forward by Kwerel and Williams (2002) of the FCC. We endorse this plan as a starting point for a "win-win" transition to the market-based technology-friendly regime we believe we need.

The main features of this transition plan are: (i) it moves from a government allocation scheme to a market-based regime; (ii) it is wholly voluntary on the part of current license holders; (iii) incentives are provided so that current licensees will place their current license asset into the market; and (iv) it eliminates all use restrictions and keeps all technical restrictions as limits on the eventual owners' property rights.

The process:[44]

1. The FCC and NTIA announce that in one year's time, an auction will be held for all spectrum use rights technically available for broadcast, including all government-held spectrum for defense, police, fire and other public safety uses, and "white space" spectrum held by the FCC.
2. Each licensee may choose to place its spectrum in this auction; it need not do so, but if it does not, then for a period of five years it is prohibited from taking advantage of buying, selling or leasing spectrum use rights and will continue to be limited to its licensed use.
3. A licensee may place its spectrum use rights into the auction simply by notifying the FCC of its decision.
4. The auction is held; any party can bid on any spectrum band it wishes, including part of an existing wireless license.[45] If its bid is accepted, *the current licensee receives the full bid payment.* The successful bidder acquires ownership in fee simple with a noninterfering easement with no restrictions on use but all restrictions relating to interference.
5. No current licensee is required to accept a bid for spectrum it has placed in the auction; it has the "right of first refusal," and may keep the spectrum use right regardless of the bid.[46] If the licensee accepts the bid, *then the entire bid is paid to the existing licensee.*
6. If the current licensee decides to keep all or part of the frequency band of his license, it becomes his property (under the ownership with noninterfering easement regime previously discussed); all use restrictions are lifted, all technical restrictions remain. The owner is now free to buy more spectrum, sell all or part of his or her existing spectrum or lease its spectrum for any length of time.
7. After this "big bang" auction, we expect an active secondary market in spectrum to arise, in which owners of spectrum can trade freely. The FCC (and NTIA) would exit the spectrum management business altogether (except possibly for certain enforcement duties).

The purpose of holding the auction of all spectrum at the same time is to ensure liquidity; there is enough spectrum available that bidders can be assured of getting what they want and selling what they want. Additionally, the single auction becomes a salient event, capturing the attention of top corporate managers. This ensures that top management becomes aware that they may be able to capitalize their wireless license asset to improve shareholder value. Spectrum managers further down in the organization may have no such incentive, preferring simply to hold on to their jobs as experts in FCC regulations. With top-level corporate attention, it is more likely that spectrum would end up in the auction.[47]

Government Role The role of the Federal government in this "big bang" auction is twofold: (i) to conduct the auction, and (ii) to participate in the auction as a buyer or seller to own blocks of spectrum for (a) governmental purposes, such as defense, and (b) public spectrum, or commons, for use by anyone. We envision the FCC conducting the auction; it has more operational expertise in this function than any other agency in the world. We envision an operating arm of the Federal government (perhaps the Department of Commerce) deciding how much spectrum is needed for governmental purposes and for public commons purposes, as directed by Congress. After the auction, the government can go to the secondary market if it needs more or less spectrum for its purposes. Thus, the extent of public spectrum held as a commons is a political decision made in the broader context of a property rights-based regime.[48]

Most important, there would seem to be few if any losers from participating in this process. Current holders of wireless licenses would be afforded the opportunity to capitalize some or all of their assets; if they chose not to do so, they now own these assets and can use, sell or lease them as they wish in the future. Those who are not current licensees but who require spectrum for their business plans now have the opportunity to buy it on the open market. No one is forced to put their spectrum at auction; but if they choose not to do so, they cannot take advantage of the new regime for five years. Everyone is better off participating in this process rather than not.[49]

We note the similarity of our proposal to that of Lessig (2001), who also proposes a mixed system of property and of commons. We arrive at our solution from a property base, while Lessig appears to arrive at his from a regulatory base. Nevertheless, we arrive at similar recommendations from very different bases, suggesting a common ground between market advocates and commons advocates.

Conclusion

In this chapter, we considered property rights regimes and a commons regime in spectrum as alternatives to the current licensing regime, which appears to lead to substantial inefficiencies in spectrum allocation. We noted that economists have favored a market-based regime while engineers have favored a commons-based regime to promote new technologies. We show that there is a property rights market-based regimes that unleash the power of the market and unleash the power of the new technologies to efficiently allocate spectrum that is likely to meet our needs for the near-term future. The presumed dichotomy between the market-based and the commons-based views has been resolved, so that both objectives can be realized. We also outline a transition process to achieve the desired regime outcome that is a "win-win" for all stakeholders, and could be politically feasible. The change to a property rights regime is likely to lower the cost of spectrum substantially, in many cases to zero. Both a commons model and a market model can co-exist it would seem, at least until spectrum becomes truly scarce.

Acknowledgments

Many colleagues were gracious enough to contribute their comments and ideas to this chapter; we especially would like to acknowledge the extensive efforts of Lawrence Lessig, Thomas Hazlett, and Evan Kwerel and John Williams. The ideas expressed in this chapter do not represent the Federal Communications Commission, the University of Pennsylvania, or any known institution other than the authors.

Notes

1. The historical material presented here is drawn from Hazlett (1998), to whom the authors are indebted for his work in spectrum economics spanning over a decade, and from Benkler (1997), who presents a somewhat different view of the early history of radio.

2. The Commerce Department retained control over all spectrum used by the Federal government. This authority is now vested in the National Telecommunications and Information Agency (NTIA) within Commerce.

3. *http://www.fcc.gov/auctions/summary*

4. NextWave Personal Communications Inc. v. FCC, 254 F.3d 130 (D.C. Cir. 2001). The FCC has appealed this ruling to the Supreme Court; the issue remains unsettled as of this writing.

5. ORBIT Act, Public Law 106–180, 114 Stat. 48 (2000).

6. By way of comparison, the FCC auctioned a total of 120 Mhz (in each metro area) for PCS use.

7. Balanced Budget Act of 1997. U.S. Public Law 105–33, 111Stat 258,105th Cong.,1st sess., 5 August 1997

8. 3 FCC Rcd 7418 (1998)

9. Hazlett (2001) presents a thorough and carefully documented history of FCC spectrum decisions, illustrating the systematic inefficiencies of the administrative process with extensive case studies.

10. Kwerel and Williams (1992).

11. FCC, 2000 *Second Report and Order,* Service Rules for the 746–764 and 776–794 MHz Bands, and Revisions to Part 27 of the Commission's Rules, WT Docket No. 99–168, FCC 00–90 (rel. March 9, 2000).

12. Fee simple is the most common type of ownership (usually applied to real estate, more generally any ownership) that allows the owner to have unlimited control over a property. *Black's Law Dictionary* (6th ed., St. Paul, Minn.: West Publishing Co. at p. 615, 1990) defines fee simple as follows: "A fee simple estate is one in which the owner is entitled to the entire property, with unconditional power of disposition during one's life, and descending to one's heirs and legal representatives upon one's death intestate. Such estate is unlimited as to duration, disposition, and descendibility."

13. Ownership generally confers two social benefits: (i) the owner has an incentive to deploy his or her assets in a way that maximizes the value of that asset, including selling or leasing it, which ensures that the asset is employed in its most valued use; (ii) the owner has a stewardship incentive to improve the asset (or not let it depreciate) if that increases its net value, such as improving land (in some cases, net value may be increased by permitting the property to depreciate). Spectrum ownership would satisfy the first but not the second condition, as it is neither improvable nor depreciable. While ownership permits spectrum assets to move to their highest valued use, the lack of a stewardship function may lead spectrum owners to be viewed as mere *rentiers* or "middlemen," an economic function historically held in low regard by the general public.

14. In some cases, a use may be highly valued publicly but not be amenable to private production. For example, PBS is a public broadcasting network that produces TV shows that might otherwise be produced but have some public benefit and so receives both governmental and charitable support. There are, of course, other examples of worthy endeavors that require governmental or charitable sup-

port, such as live opera. In a market model, PBS (or a similar service) would buy its spectrum with government/charitable funds if the sponsoring organizations believed this to be the best use of their funds for the public benefit. If they believed some other use superior, then PBS may not survive. But this is a decision best taken by this venture's sponsoring organizations.

15. The government must provide the essential infrastructure of laws, regulations, and courts to ensure that markets can perform their job of allocating resources well. But government provision of the market infrastructure is different than government substituting for the market.

16. Such failures include public goods (such as national defense and the justice system), information asymmetries (such as consumers' lack of knowledge about drug efficacy), natural monopolies (such as electric power distribution), and spillovers (such as pollution or network effects). Of these possible market failures, *only* spillovers appear to be present in the case of spectrum (although the *use* of spectrum may have public good aspects, such as Part 15 spectrum).

17. In fact, all modern TV sets have digital filters, simply because they are now cheaper and produce a better picture quality than the older filters.

18. Noam's plan appears to involve a government-operated central monitor and market-making computer to clear all transactions in real time, that is, a spot market for spectrum use.

19. Public enforcement, such as the police, is usually only available if there is an immediate threat to life or property.

20. Part 15 rules were originally adopted to cover "wireless phonograph," a device whose time has not yet arrived. It was later used to govern "unintentional emitters," such as televisions and personal computers, whose operation caused the emission of electromagnetic radiation. The rules limited both the power and the frequency of the emissions of such devices

21. John Reed, Senior Engineer, Technical Rules Branch, FCC, personal conversation 4/10/02.

22. We oversimplify; restricted sharing is permitted in certain other bands, in which low power devices are permitted to emit radiation in licensed bands.

23. With the exception of ground-penetrating radar (GPR), which is quite powerful and would be an interfering use if not pointed into the ground.

24. Note that UWB radio could broadcast at much higher power and have a greatly extended range; however, that would lift emissions out of the noise and become an interfering use. Even now, certain existing low power uses such as Global Positioning System (GPS) receivers claim UWB can cause interference with their systems if operated at somewhat higher power levels than recently approved by the FCC.

25. Title 14, Code of Federal Regulations, Section 91.119 of the General Operating and Flight Rules

26. Note that the current property right regime for real property could well be modified to permit homeowners to restrict aircraft overflight rights or set a price for each overflight, perhaps dependent upon altitude. There would clearly be a

cost to such a system (see our discussion below regarding the tragedy of the anti-commons), but only justifiable if airspace were a scarce resource, subject to congestion. Currently, airspace is regulated for safety and congestion concerns by the FAA (in the United States) so a price system based on overflight rights is neither necessary nor particularly efficient.

27. Current technologies that use "listen before talk" may not completely avoid interference with agile radio. Some form of "get permission before talk" may be necessary.

28. Mesh network architecture can be used not only for computers but also for voice and indeed any radio transmission; it can also be used with a mix of transmission technologies, such as agile, UWB, cellular, CB radio, etc.

29. A current example of a mesh network is Metricom's Ricochet network (now emerging from bankruptcy) which had many thousands of users in multiple cities at its peak. Metricom was based on ideas and patents of Paul Baran (see *http://www.ricochet.net*).

30. A number of technical and legal scholars have made this argument persuasively, including Lessig (2001), Benkler (1997), Jackson (1999), Ikeda (2002), and Reed (2002).

31. In the case of a property rights regime for spectrum, this does not mean that *all* spectrum would carry a zero price; there may be legacy uses of certain frequencies in certain locations that would continue to carry a high price. But it does mean that should spectrum not be scarce, then *some* spectrum would be available at a near-zero price.

32. For an early but complete discussion of the role of property rights and their emergence, see Harold Demsetz (1967)

33. We use the term "easement" somewhat freely, to indicate a restriction on ownership that specified others may use the property for specified purposes under specified conditions.

34. Obviously, such measuring and metering devices do not exist today, as there is no use for them in the current licensing regime. The technology to create such devices is well within today's state of the art; if produced in volume are likely to be low cost. However, they do represent a transaction cost to operating a market system.

35. This plan is quite similar to that suggested by Eli Noam, *op.cit.* As noted above, Noam's plan involves a government-operated central monitor and market-making computer to clear all transactions. We envision each owner implementing such a system (if economically feasible). See also a critique of this plan by Thomas Hazlett (1998).

36. See *http://www.fcc.gov/eb/easfact.html* for a description of the Emergency Alert System.

37. This may not be true; if the government is the controller of the commons, it may assess a fee to all users to cover administrative expenses, including dispute resolution costs.

38. Such transactions occur in all other sectors of the economy: the owner of a

factory in New Jersey (or of 20 Mhz of spectrum in New York City) may sell this asset to another party as a normal commercial transaction.

39. At present, UHF stations are broadcasting and virtually no one is watching; the reason is the FCC's "must carry" rule: any local station doing over-the-air broadcasting must be carried by local cable television. Therefore, any station broadcasting, even though no one is watching the over-the-air broadcast, get carried on cable TV, where lots of households are watching. We would propose that the FCC grandfather the "must carry" rule for all currently broadcasting stations; without requiring them to continue this unnecessary activity.

40. Not *all* spectrum will be priced near zero; for example, FM radio station frequencies and cellular wireless frequencies will continue to command a premium. Our assertion is that *some* spectrum will be available at low cost.

41. The pre–1996 Internet community was particularly fond of the "unlimited bandwidth" vision of the Internet. Everything could be free, it was argued, because the bandwidth of the Internet was virtually unlimited. Post–1996, the phenomenal growth of Internet traffic quickly dispelled the notion of unlimited bandwidth; new applications engendered new demand that quickly exhausted what had appeared to be unlimited supply, and then some. Similarly, we have great faith in electronic engineers and entrepreneurs to create a demand for spectrum that will fill every nook and cranny of it.

42. A more subtle point is that technological advances can increase the efficiency with which we use spectrum. But if more spectrum is available at zero cost, then it doesn't pay to invest in using spectrum more efficiently. Only as spectrum becomes scarce (as it is now, artificially, and as it will be in the future, for real) does it pay to invest in more efficient use.

43. Inevitably, that means perceived inequities that have been built into current system will not be "corrected." Some may view certain current licensees as undeserving of reward, either because they received their licenses through questionable political dealings or from corporate power. We believe that moving toward a more efficient regime of spectrum allocation is far more important than correcting for perceived inequities in the current allocation of licenses.

44. Our characterization of the Kwerel-Williams plan is somewhat more aggressive than that discussed in their paper; we both admire and borrow the concept but supply some details ourselves, for which we beg the authors' indulgence.

45. It would be preferable for bidders to be permitted "combinatorial" bids, in which they may bid on a combination of existing licenses. See *http://wireless.fcc.gov/auctions/31/releases/milgrom_reply.pdf* for a description of combinatorial bidding in the context of licenses for wireless communications.

46. It would appear that there is no economic reason to hold spectrum back from the auction, as the current incumbent always has the right to refuse all bids. This is correct; incumbents are better off placing their spectrum into the auction than not. The holdback option gives all incumbents a pure "no change" option, and can help focus managers and shareowners on the benefits of using the auction process to value their asset and possibly monetize it.

47. A similar situation obtains in the public sector. A police chief has little incentive to put his or her excess public safety band in an auction; however, his mayor and city council might consider a partial sell-off of police bandwidth a good budgetary tradeoff.

48. Our proposal is perfectly analogous to land use. All land in the United States is owned, and the Federal government is the largest owner of land in the country. Some of this land is owned for government business and much is owned as a public resource. How much land is committed to each use is a political decision, implemented through real property markets.

49. However, the process may result in some parties being made worse off, compared to the existing regime. For example, if we are correct that the price of spectrum use rights at the margin will decline, then parties with large investments in current licenses will see the price of their asset decline.

References

Benkler, Yochai. 1997., *Overcoming Agoraphobia: Building the Commons of the Digitally Networked Environment*, 11 Harv. J. L & Tech. 287 (Winter 1997–1998).

Caro, Robert A. 1991. *The Means of Ascent (The Years of Lyndon Johnson, Vol. 2)*, New York: Knopf.

Charney, Ben. 2002. "Sprint PCS roaming into Wi-Fi," *ZDNet News*, Oct 31, at *http://zdnet.com.com/2100–1105–964118.html*

Coase, Ronald. 1959. "The Federal Communications Commission," *Journal of Political Economy*, 2 (2) *J. Law. & Econ*, 1.

Coase, Ronald. 1991. "Autobiography of Ronald Coase, Nobel Laureate" at *http://www.nobel.se/economics/laureates/1991/coase-autobio.html*

Demsetz, Harold. 1967. "Toward a Theory of Property Rights," *Amer Econ Rev*, 57 (2) (May), 347–359.

DeVany, Arthur, Ros Eckert, Charles Meyers, Donald O'Hara, and Richard Scott. 1969. "A Property System for Market Allocation of the Electromagnetic Spectrum: A Legal-Economic-Engineering Study," 21 *Stanford Law Review* 1499–1561 (March).

Federal Communications Commission, 2001. "Part 15 Rules," at *http://ftp.fcc.gov/oet/info/rules/part15/part15_dec18_01.pdf*

Hazlett, Thomas. 1998. "Assigning Property Rights to Radio Spectrum Users: Why Did FCC License Auctions Take 67 Years?" *J. Law & Econ*, 41 (2), (October) 529.

Hazlett, Thomas. 1998. "Spectrum Flash Dance: Eli Noam's Proposal for 'Open Access' to Radio Waves," *J. Law & Econ*, 41 (2) (October), 805.

Hazlett, Thomas. 2001. "The Wireless Craze, the Unlimited Bandwidth Myth, the Spectrum Auction Faux Pas, and the Punchline to Ronald Coase's "Big Jokes":

An Essay on Airwave Allocation Policy," *Harvard J. Law & Tech*, 14 (2), (spring), 335–545.

Heller, Michael. 1998. "The Tragedy of the Anticommons: Property in the Transition from Marx to Markets," 111 *Harvard Law Review*, 621, 622–625.

Heller, Michael. 1999. "The Boundaries of Private Property," 108 *Yale Law Journal*, 1163–1223.

Heller, Michael, and Rebecca Eisenberg. 1998. "Can Patents Deter Innovation? The Anticommons in Biomedical Research," 280 *Science* 698.

Ikeda, Nobuo, 2002. "The Spectrum as Commons: Digital Wireless Technologies and the Radio Policy," RIETI Discussion Paper Series 02-E–002 (March).

Kwerel, Evan R., and Alex D. Felker. 1985. "Using Auctions to Select FCC Licensees," *OPP Working Paper no. 16* (May).

Kwerel, Evan R., and John R. Williams. 1992. "Changing Channels: Voluntary Reallocation of UHF Television Spectrum," *OPP Working Paper no. 27* (November).

Kwerel, Evan R., and John R. Williams, 2002. "A Proposal for a Rapid Transition to Market Allocation of Radio Spectrum," *FCC OPP Working Paper no. 38*, (November) at
http://hraunfoss.fcc.gov/edocs_public/attachmatch/DOC–228552A1.pdf

Jackson, Charles. 1999. "Was a Common Law Solution to Chaos in the Radio Waves Reasonable in 1927?" Working Paper at
http://www.jacksons.net/working%20papers/Was%20a%20Common%20Law%20Solution%20to%20Chaos%20in%20the%20Radio%20Waves%20Reasonable%20in%201927.doc

Lessig, Lawrence. 2001. *The Future of Ideas: The Fate of the Commons in a Connected World.* Random House: New York.

Noam, Eli. 1998. "Spectrum Auctions: Yesterday's Heresy, Today's Orthodoxy, Tomorrow's Anachronism," *J. Law & Econ*, 41 (October), 765.

Reed, David. 2002. "Comments for FCC Spectrum Task Force on Spectrum Policy," filed July 10 at
http://gullfoss2.fcc.gov/prod/ecfs/retrieve.cgi?native_or_pdf=pdf&id_document=6513202407

White, Lawrence. 2001. "'Propertyzing' the Electromagnetic Spectrum: Why It's Important, and How to Begin," in *Communications Deregulation and FCC Reform: Finishing the Job,* edited by J. A. Eisenach and R. J. May. Kluwer.

8

"Functionality" as the Distinction Between Patent and Copyright Subject Matter

Dennis S. Karjala

In *State Street Bank & Trust Co. v. Signature Financial. Group, Inc.*,[1] the Court of Appeals for the Federal Circuit rejected the long-accepted exclusion of "business methods" as patent subject matter. Shortly thereafter that same court held that physical transformation of material from one state to another was not a requirement for patentability.[2] The test, rather, is whether an algorithm or process is "applied in a practical manner to produce a useful result."[3] The scholarly commentary on these developments has ranged largely from highly critical to neutral.[4]

This article argues that digital information technology has largely forced these developments in patent law. For one thing, information as such as well as useful and creative ways of organizing and processing information, are increasingly the end products of industry. If patent is to be limited to physical artifacts and their manipulation by industrial processes, it is likely to become increasingly irrelevant as we move more and more to an information-as-product economy. Even more important, the already much eroded distinction between patent and copyright subject matter would almost completely disappear for digital technologies, resulting in expanded copyright coverage of many works that are technological in nature. If patent is bad for any of these things, copyright's long term and broad scope of protection make copyright almost surely worse. Yet, if patent protection is categorically denied on subject matter grounds, courts are likely to try to fill the apparent gap in protection with copyright. The article thus concludes that accepting applied and useful, although nonphysical, processes as patent subject matter is the only way to maintain the critical distinction from copyright subject matter and that we should look to the other requirements for patentability, such as objectively verifiable usefulness, novelty, and nonobviousness, for limiting untoward effects.

I. Patent and Copyright

The patent and copyright regimes have a common overall goal. Each seeks to promote the general welfare by protecting the fruits of intellectual creativity from activities that would undermine the author's or inventor's ability to reap a fair return from investments of time, money, or talent. They go about their daily business, however, in very different ways.[5] Copyright arises automatically upon fixation of copyright subject matter in a tangible medium, the scope of copyright protection is defined by the vague idea/expression dichotomy, copyright infringement is determined by the equally vague "substantial similarity" standard, and the term of copyright protection endures for 70 after the death of the author. Patents, on the other hand, issue only upon formal application and after examination by a skilled examiner for "novelty," "utility," and "nonobviousness," the scope of patent protection is defined and narrowly limited by the claims, and the term of patent protection is only 20 years. As long as the distinctions between the respective subject matters covered by these regimes were well defined, creators might have had an abstract preference for one regime over the other but there was little they could do about it. Nonfunctional works of art, literature, and music were covered by copyright, and functional works of technology were covered by patent.

The advent of digital technologies changed all that. Patent's initial foot-dragging with respect to computer programs (functional subject matter)[6] allowed copyright to establish a firm beachhead in this crucial field of technology.[7] Now patent is a major player in the field of software protection[8] and has recently expanded its subject matter coverage to include "business methods."[9] The former expansion is, or should be, relatively uncontroversial, at least in principle,[10] but the latter has come under heavy criticism[11] and has engendered a number of suggestions for damage control.[12] Under the authority of *Baker v. Selden*,[13] courts over the years have denied copyright protection to a variety of schemes and plans for accomplishing real-world tasks.[14] However, especially in recent years, copyright has quietly been expanding its role in the protection of functional works beyond computer programs, with courts protecting taxonomies and other systems for presenting or organizing information, techniques for producing factual information, and even business and teaching methods.

I suggest that the clamor over business methods as patent subject matter basically misses the real issue. The principle of hardware/software equivalence—that is, the principle that for every general purpose computer running under the control of computer software there is an equivalent device consisting solely of hardware that is indistinguishable[15]—basically eliminates the subject-matter issue for the programmed machine itself. Supposes, for example, that someone has an electronic machine composed solely of hardware (physical wires, transistors and other circuit elements, solder joints, and so forth) that accomplishes a particular task (such as receiving and playing radio signals or calculating sums). No one would deny that this machine constitutes patent subject matter.[16] The principle of hardware/software equivalence says that the function of this machine may be exactly duplicated solely with software and a general purpose computer (and perhaps some input/output devices). If the pure hardware device is patent subject matter, which it is, there can be no basis for denying similar status to a machine built with software that does exactly the same thing. We are thus forced to inquire more carefully into the process the programmed machine is implementing via the software to determine whether a patent is available.[17] Abstract ideas, laws of nature, and natural processes, as well as the art, literature, and music constituting the subject matter of traditional copyright,[18] remain outside the bounds of patent. We must, therefore, distinguish between the technological software that implements any of these patent-unprotected subject matters and the subject matters themselves.[19] Retaining the exclusion of business methods from patent subject matter would not eliminate the need for this distinction.[20] While that alone is not a ground for deeming business methods to be patent subject matter, a valid ground may arise from consideration of what would likely happen under copyright if they remain categorically excluded from patent coverage.

Patent's initial failure to protect computer software to any degree, coupled with congressional reliance on copyright to protect against piracy of code, led initially to a great expansion of copyright protection for computer programs. Courts not only held that code was protected by copyright but also the "structure, sequence, and organization" of the program as well as functional aspects of the user interfaces.[21] Since 1992 courts seem to have cut back somewhat on the heavily overprotective scheme they had devised for computer programs,[22] but judicial misunderstanding of the scope and purpose of a software copyright remains widespread.[23] Most business methods for which patents are sought are implemented by computer. Thus, copyright protects the implementing computer program. If the process implemented by the computer, through the program, is seen by copyright courts as "creative" (even though the program code itself may be perfectly routine), they are likely to try to correct with copyright what they view as wrongful appropriation of the fruits of another's creativity.[24] This could easily lead to the same intellectual property mess for business methodologies that we had for computer software.

Basically, if patent is bad for business methods, copyright is worse. It is therefore better to deal with this subject matter under patent and try to limit the damage by insisting on application of the traditional patent norms of objectively verifiable and specific claims, a meaningful utility requirement, narrow notions of equivalence in determining infringement, and a real nonobviousness hurdle. The same reasoning that leads to this conclusion for business methods, however, also applies more generally to any functional process. Part II of this article argues for functionality and incremental improvability as the touchstone for denying copyright protection. A longer version of this article[25] gives extended consideration to a number of situations in which copyright courts have either protected or come dangerously close to protecting subject matter that should, under this functionality analysis, lie outside of copyright. These fact situations involve accounting systems and similar business methods, maps and compilations, methods of gathering and presenting information, taxonomies, model codes and statutes, teaching methodologies, standardized tests, and various miscellaneous "systems." Part III of this article addresses the question of whether patent must inevitably be the quid pro quo for denying copyright protection or whether we can safely attempt to deny protection, on categorical subject-matter grounds, under both regimes.

II. Patent and Copyright Subject Matter—Functionality and Incremental Improvability

Why do we have two very different statutes aimed at protecting the fruits of intellectual creativity? I have addressed this question on numerous occasions in attempting to define the appropriate scope, beyond code, of the copyright in a computer program.[26] In that context, I have argued that patent protects creative but functional invention, while copyright protects creative but nonfunctional authorship.[27] For these purposes, I have defined "functionality" to be distinct from merely "useful"[28] and have relied on the Copyright Act's definition of a "useful article" as a starting point for focusing on the differences between patent and copyright subject matter: "[A] 'useful article' is an article having an intrinsic utilitarian function that is not merely to portray the appearance of the article or to convey information. . . ."[29] Thus, maps and dictionaries, while often "useful," are not "useful articles" under this definition (and are therefore not "functional" under my definition) because their sole utility is to convey information. This disarmingly simple definition captures much of the distinction between the functional subject matter of traditional patent law and the informational subject matter (however useful) of traditional copyright law. We have made this distinction because the social desirability of allowing later creators of functional works to build on and improve what has come before necessitates a more clearly defined property right in technological advances, a shorter term of protection, and (at least in theory) a more significant (nonobvious) step forward.[30] Whatever the correctness of copyright's broad scope of protection for traditional literary and artistic works[31] and its extremely long term of protection, they are wholly inappropriate to useful arts that progress by incremental contributions from a wide variety of sources.

Traditional copyright subject matter was indeed nonfunctional under this definition, because it was useful only in entertaining (by presenting an appearance or sound) or informing human beings.[32] However, this approach of treating functional works as patent subject matter must make an exception for computer program code (at least application program code).[33] While code is clearly functional under my definition, Congress mandated copyright protection for code in adopting the recommendations of the National Commission on New Technological Uses of Copyright Works (CONTU).[34] An exception for program code is also necessary as a policy matter, because of its vulnerability to fast and inexpensive electronic copying.[35] Program code is therefore a sui generis exception to section 102(b)[36] and to *Baker v. Selden*,[37] which is the foundation of the general rule that patent, and not copyright, has the job of protecting function.[38] Excepting functional program code from the functionality test for distinguishing patent and copyright subject matter, however, does not mean that functionality should play no future role in separating these two classes of intellectual property.[39] Indeed, if functionality were to be abandoned, we would be forced to inquire what sense, if any, it would make to continue with two such different modes of protection.

Functionality is thus the basic determinant of the patent/copyright boundary: Anything that is a "useful article" under the Copyright Act's definition[40] should be at least prima facie patent, and not copyright, subject matter. Exceptions might be made for certain classes of works for which misappropriation by means of fast, cheap, and easy copying appears to be a particular danger.[41] The Copyright Act's "useful article" definition, however, does not capture the whole of the distinction between patent and copyright subject matter. Copyright protection does not extend, for example, to a "system" or "process,"[42] although applied processes have long been patent subject matter.[43] In particular, many systems (such as accounting systems) or processes (such as how to bake a cake) are conceptual algorithms that inform human beings how to do something but are not self-executing. They are therefore not "functional" under direct application of the Copyright Act's definition of "useful article."[44] We must therefore probe more deeply to find the dividing line between patent and copyright for such intellectual creations.

In seeking a more general principle for determining the appropriate 21st-century boundary between patent and copyright, it is instructive to inquire into the historical reasons for the differences between the patent and copyright systems. Why is it that patents are more narrowly defined than copyrights, are harder to get, persist for a shorter time, and are easier to defend against in cases of nonliteral copying, even though both patent and copyright are aimed at the same overall goal of promoting and protecting the fruits of intellectual creativity?[45] We can begin to get a handle on this question by looking at the differences between traditional nonfunctional copyright subject matter and functional patent subject matter. Functional works—works that are used by human beings to do something other than to inform or entertain themselves or others—are usually amenable to objectively measurable incremental improvement. Attaching a motor to a drive shaft and some wheels may be a fundamental advance in the art of transport, but once available that basic vehicle can be improved in an incremental series of smaller steps by the addition of brakes, headlights, a roof, windows, bumpers and other crash protection devices, and so forth. These incremental improvements can themselves each usually be improved, again most often in incremental fashion. If the original vehicle were protected by copyright, there is a good chance that the improved vehicle (with the brakes, headlights, etc.) would infringe under copyright's "substantial similarity" test for infringement. Moreover, the copyright would persist for 70 years after the death of the "author" of the original vehicle, quite possibly well over 100 years.[46] The inappropriateness of copyright in this case is self-evident. Broad protection for such a long time would give the original vehicle inventor the sole right to make improvements for a century, inhibiting much more improvement in the useful art of vehicle transportation than it would engender.[47]

The policy basis for the radically different natures of the patent and copyright regimes is thus the social desirability of allowing all those skilled in the relevant art to try their hands at incremental improvement of functional works.[48] The same can be said of the industrial processes that, while not directly functional under the definition offered above, have long been a part of patent subject matter. A rubber-curing process, for example, can be improved by using a computer to make the complex real-time calculations necessary for determining exactly when to open the mold.[49] Allowing for incremental improvement in functional works, as defined above, is important for socially desirable technological advance. Moreover, "improvement" in functional works is usually objectively measurable, not in the sense that the improved version is desirable (that is for the market to determine under either the patent or copyright system) but rather that it does what it does, for example, faster, more accurately, more quietly, or using fewer resources, perhaps with a concomitant array of disadvantages over the prior art (takes up more space, costs more to produce, is somewhat more dangerous or difficult to use, and so forth). Whether the improved version actually embodies a particular array of characteristics will be a matter of general agreement, even if not everyone (or indeed anyone) actually chooses to buy or use it.[50]

Traditional copyright subject matter, on the other hand, is not subject to improvement in the same way; or, at least, allowing incremental "improvement" of traditional works of art, literature, and music is less pressing than for functional works.[51] It is less pressing for informational works like compilations and dictionaries, not because they cannot be incrementally improved but because the disincentive to initial creation would outweigh the benefits of incremental improvements like error corrections. For copyright-protected works like novels or films, it is even less important to encourage incremental improvement by all comers. First, no one can say what an "improvement" is in these cases, because the appeal is to people's aesthetic taste. Second, "improvement" of these works does not build continuously. Few, if any, book lovers would read what is essentially the same novel ten times to see what "improvements" were made by ten successive improvers. This does not necessarily justify the broad scope of copyright protection, which prevents rewriting even entirely new sequels involving the characters created in a protected work.[52] It says merely that disallowing incremental improvement of art, literature, and music does not as strongly conflict with underlying economic policy goals as would similar limitations with respect to functional subject matter.

Systems and processes that are "applied in a practical manner to produce a useful result"[53] are thus those whose usefulness for their intended purposes generally improves in an objectively articulable, incremental manner.[54] Such systems and processes are "functional" and should seek intellectual property protection in the patent, and not the copyright, regime, even though they are not directly functional under the "useful article" definition. The policy basis for channeling these works to the patent system is incremental improvability, in the sense that such systems and processes can be made to achieve their results faster, more cheaply, more efficiently, more accurately, or in more user-friendly way by allowing general tinkering with their component steps, without unduly undermining incentives for their initial creation or their ongoing development. Copyright protection would give a very long-term monopoly in making those improvements to the first person who describes the system or process in an underlying copyright-protected work. The undesirability of this result is the reason for section 102(b). My suggestion here is thus an amplification of what these terms in section 102(b) should actually mean in practice. I am further suggesting that courts interpreting section 102(b) in application to such works are more likely to get the correct result—exclusion from protection as copyright subject matter—if the functional system or process in question is considered patent subject matter.

In drawing the subject matter boundary between patent and copyright for processes relating to information, we must distinguish between improvements in the quality of information content and improvements in the means for gathering, presenting, or using information. Information with no other application other than its appeal to the judgment of a human audience is, and for the foreseeable future will remain, the core of copyright, as reflected in the Copyright Act's definition of a "useful article,"[55] even where for reasons unrelated to the patent/copyright boundary we deliberately exclude certain information, such as "facts,"[56] from copyright protection. Therefore, "improvements" in information content as such should be considered to lie outside of patent coverage, no less than traditional works or art, music, and literature.[57]

On the other hand, some methods or ways of gathering, organizing, presenting, or using information can be improved incrementally by reducing costs, enhancing accuracy, augmenting ease of use or modification, or, more generally, making the use of available information more effective in some objective sense. Like most incremental improvements of existing technology, we may assume that most improvements in information-handling methodologies will not be patentable. The thesis presented here is that, if any such methodology *is* unpatentable, it should not be on the ground that the method is not patent subject matter. Rather, denial of patentability should be `grounded in anticipation, lack of objectively verifiable utility (in the sense that the claimant cannot demonstrate that the methodology does what she claims), failure to enable or distinctly claim, or obviousness. Concomitantly, categorizing these methodologies as patent subject matter should result in their *exclusion* from copyright protection under section 102(b) of the Copyright Act. The only "arrangements" of information that should be treated as copyright subject matter are those whose appeal lies wholly in the aesthetic appearance of the arrangement or are chosen out of the wholly subjective judgment of the arranger. The longer version of this article develops these ideas more fully.

"Functionality," it must be admitted, is not wholly successful at describing the historical distinctions we have made between patent and copyright subject matter,[58] although I believe it does describe the distinctions as well as any single concept can. Its importance lies not in its descriptive power but rather in the normative analytical generality that allows it to be applied to a number of troublesome works, without forcing many changes in established principles. The full version of this article applies the "incremental improvability" concept to business methods, legal forms, methods of presenting information, taxonomies, and other functional systems and methodologies to argue that patent, strictly construed, is generally a socially more desirable means of protecting such subject matters than copyright. We skip this detailed analysis here and consider next the more general question of whether functional works, as defined herein, should *always* be considered patent subject matter or whether, on the other hand, there remains a basis for categorical exclusions from patent notwithstanding functionality.

III. Must We Choose Between Patent and Copyright for Functional Processes?

Detailed analysis of a number of information-delivery methodologies and systems, which is undertaken in the longer version of this article, demonstrates that courts are often willing to protect functional methodologies as nonliteral elements of the copyright-protected works from which the methodologies are discernible. For the situations analyzed, such as accounting systems, taxonomies, and methods for presenting information, the suggested remedy is usually to treat the methods or systems in question as patent subject matter. This then concedes patentability if the patent hurdles of anticipation, obviousness, demonstrable utility, and clear specification can be surmounted, subject to the limitation that abstract ideas and laws of nature, as well as traditional information works of art, music, and literature,[59] are still excluded. There may well be cases, however, where society will reap the benefit of the advance even without offering the incentive of patent law.[60] United States intellectual property law is incentive based. That is, we recognize patent and copyright rights for the purpose of encouraging the creation and public distribution of desirable works; reward to the author or inventor is a simply a means of achieving that end.[61] The question, then, is whether there are not some classes of functional processes or methodologies for which categorical denial of *both* patent and copyright protection is justified, even for works within those classes that otherwise meet the conditions specified by one of those two regimes.

On the one hand, detailed analysis of the copyright cases demonstrates the inappropriateness of copyright for functional works, as defined via the Copyright Act's definition of a "useful article," including systems and processes that are practically applied to achieve a useful result as informed by the concept of incremental improvability.[62] Denying copyright protection for functional processes and methodologies, regardless of their degree of creativity, is much more important than insuring that a given process or methodology in politics, religion, or business gets a fair hearing under patent law's utility, anticipation, and obviousness tests. If we could be confident of our ability to educate the courts to restrain their "restitutionary impulses"[63] so that they would simply deny copyright protection to functional processes, regardless of their field of use, much of the battle would be won. In principle, moreover, there should be little reason for a copyright court to care whether denial of patent protection is based on the substantive standards of patent law (verifiable utility, nonanticipation, nonobviousness, definite claims, understandable specification) or on a categorical decision that patents are simply not available for that class of work (a subject-matter limitation). Copyright courts should recognize that these methodologies are not copyright-protectable, even if they are elements contained in or implemented by works that *are* copyright protected, and should leave such works to their fate under the patent and trade secret regimes.

On the other hand, our confidence that we can educate the courts to deny copyright protection to functional processes must be tempered by our empirical observation of the strength of the restitutionary impulse in copyright courts. The numerous examples contained in the full version of this article show its vibrancy well over 100 years after *Baker v. Selden* should have quelled it. The real question therefore boils down to whether recognition of all functional processes and methodologies as patent subject matter (subject always to the substantive limitations concerning what patent subject matter is indeed patentable) will reduce this restitutionary impulse in copyright courts and, if so, whether this benefit would be outweighed by creating an even bigger mess in the patent arena than we manage to clean up on the copyright side.

Conclusion

Digital technology increasingly presents the problem of deciding the appropriate level and type of intellectual property protection for various methodologies of gathering, organizing, and presenting information.

Business methods implemented by computer program are only one example. Until we radically change the intellectual property system, one of the two major paradigms—patent or copyright—will be the choice. Given the copyright protectability of computer program code, and the implicit congressional exception of code from section 102(b) of the Copyright Act and from the functionality doctrine of *Baker v. Selden*, it is easy for courts to forget that section 102(b) and *Baker v. Selden* are both alive and well with respect to all other types of subject matter. *Information* is the subject matter of copyright—works that have no function other than to inform, entertain, or present an appearance to human beings. *Function* is the subject matter of patent—works that do have a function beyond informing, entertaining, or presenting an appearance to human beings, including methodologies for gathering, organizing, and presenting information accurately and efficiently. Many of these methodologies should not be patentable, but on grounds of anticipation, nonutility, or obviousness rather than on the ground that they do not constitute patent subject matter. Categorical exclusion as patent subject matter increases the likelihood that courts will treat such methods as a "creative" element of their implementing computer program, protected by copyright as long as other methods can be imagined that accomplish the same general result. Whatever the objections to protection of such methodologies under patent law, the broad scope and long term of copyright makes copyright protection even worse.

Notes

1. 149 F.3d 1368, 1375 (Fed. Cir. 1998).

2. AT&T Corp. v. Excel Communications, Inc., 172 F.3d 1352, 1358 (Fed. Cir. 1999).

3. Id. at 1360.

4. See sources cited infra notes 9–12.

5. See Dennis S. Karjala, *A Coherent Theory for the Copyright Protection of Computer Software and Recent Judicial Interpretations*, 66 U. CINCINNATI L. REV. 53, 57 (1997).

6. The sordid tale of patent's slow recognition of computer software's technological nature and the relevance of patent law to portions of that technology has been told many times. Examples include Vincent Chiappetta, *Patentability of Computer Software Instruction as an "Article of Manufacture:" Software as Such as the Right Stuff*, 17 J. MARSHALL J. COMPUTER & INFO. L. 89, 97–126 (1998);

Julie E. Cohen & Mark A. Lemley, *Patent Scope and Innovation in the Software Industry*, 89 CALIF. L. REV. 1, 8–11 (2001).

7. Computer programs differ from most earlier works of technology in that they can be copied for competitive commercial redistribution essentially without start-up costs for the competitor, thereby severely reducing or even eliminating the original creator's lead time. E.g., Dennis S. Karjala, *Copyright, Computer Software, and the New Protectionism*, 28 JURIMETRICS J. 33, 41 (1987) [hereinafter referred to as *New Protectionism*]. Moreover, patent is insufficient to protect all computer programs from incentive-eroding piracy, because many computer programs will not contain any patentable technology. E.g., *A Coherent Theory*, supra note 5, at 67. Therefore, even if patent had recognized its role at an earlier stage, some form of additional protection against slavish copying of object code, at a minimum, would likely have been deemed necessary.

8. Julie E. Cohen & Mark A. Lemley, supra note 6, at 11.

9. State Street Bank & Trust Co. v. Signature Fin. Group, Inc., 149 F.3d 1368, 1375 (Fed. Cir. 1998). Prior to *State Street Bank* courts had generally required the instantiation of business-related inventions in some physical structure, such as a railway ticket physically detachable from its base. Cincinnati Traction Co. v. Pope, 210 F. 443, 446 (6th Cir. 1913); see John R. Thomas, *The Post-Industrial System*, 10 FORDHAM INTELL. PROP., MEDIA & ENT. L.J. 3, 12–13 (1999).

10. Software, as such, is the technology for making computers operate, and it is difficult to see now just why patent had so much trouble reaching this conclusion. In any event, software as technology does not cause any patent-subject-matter concerns today. Vincent Chiappetta, *Defining the Proper Scope of Internet Patents: If We Don't Know Where We Want to Go, We're Unlikely to Get There*, 7 MICH. TELECOMM. & TECH. L. REV. 289, 296 (2001). The key problem is distinguishing between software as computer-use technology and software that routinely implements a perhaps novel algorithm or composes a novel piece of music. Id. at 296 n.21; Dennis S. Karjala, *The Relative Roles of Patent and Copyright in the Protection of Computer Programs*, 17 JOHN MARSHALL J. COMP. & INFO. L. 41, 57–63 (1998)[hereinafter referred to as *Relative Roles*]; see also R. Carl Moy, *Subjecting Rembrandt to the Rule of Law: Rule-Based Solutions for Determining the Patentability of Business Methods*, 28 WILLIAM MITCHELL L. REV. 1047, 1071–77 (2002)(discussing the artificiality of the method/apparatus distinction and noting that methods recognized as nonstatutory but claimed as apparatus can be weeded out via the novelty or nonobviousness criteria).

11. E.g., Rochelle C. Dreyfuss, *Are Business Method Patents Bad for Business?*, 16 SANTA CLARA COMPUTER & HIGH TECH. L.J. 263 (2000); Alan L. Durham,*"Useful Arts" in the Information Age*, 1999 B.Y.U L. REV. 1419 (arguing for a distinction between technological computer programming and the non-technological ideas that are programmed, with only the former belonging to the category of "useful arts" invention); Julia Alpert Gladstone, *Why Patenting Information Technology and Business Methods Is Not Sound Policy: Lessons from History and Prophecies for the Future*, 25 HAMLINE L. REV. 217 (2002) (arguing that business method patents do not stimulate invention and reinforce present structures of power, wealth, and opportunity); Malla Pollack, *The*

Multiple Unconstitutionality of Business Method Patents: Common Sense, Congressional Consideration, and Constitutional History, 28 RUTGERS COMP. & TECH. L. J. 61 (2002) (arguing that business method patents are both unnecessary and unconstitutional); Leo J. Raskind, *The* State Street Bank *Decision: The Bad Business of Unlimited Patent Protection for Methods of Doing Business* 10 FORD. L. REV. 61 (1999); Jason Taketa, Note, *The Future of Business Method Software Patents in the International Intellectual Property System*, 75 S. Cal. L. Rev. 943, 974 (2002) (lack of international uniformity advises using a system of default liabilities rather than the patent system to protect business methods); John R. Thomas, supra note 9; *but* see David T. Dutcher, *Patents on Methods of Doing Business*, 79 DENVER U. L. REV. 173 (2001) (reviewing the history and presenting the arguments pro and con for business method patents) .

12. E.g., Margo A. Bagley, *Internet Business Model Patents: Obvious by Analogy*, 7 MICH. TELECOMM. TECH. L. REV. 253 (2001) (recommending expansion of what is considered "analogous prior art" and narrowing the doctrine of equivalents to deal with business-method patents); Vincent Chiappetta, supra note 10, at 348–60 (recommending sui generis legislation to handle methods in the "competitive arts" but suggesting various means of implementing current law and practice in application to patent disputes resolved under current law); John Kasdan, *Obviousness and New Technologies*, 10 FORD. L. REV. 159 (1999) (recommending Patent and Trademark Office employment of more, and more qualified, examiners for business patents and legislative provision for PTO waiver of the presumption of validity); Robert P. Merges, *As Many as Six Impossible Patents Before Breakfast: Property Rights for Business Concepts and Patent System Reform*, 14 BERKELEY TECH. L.J. 577 (1999) (suggesting steps for limiting any negative effects of business-method patents); R. Carl Moy, supra note 10, at 1062 (reform requires addressing the fundamental nature of business method patents as well as the decisional criteria used to determine patent subject matter); Maureen A. O'Rourke, *Toward a Doctrine of Fair Use in Patent Law*, 100 COLUM. L. REV. 1177 (2000) (recommending adoption of a fair use doctrine tailored to the needs of patent law); Richard H. Stern, *Scope-of-Protection Problems With Patents and Copyrights on Methods of Doing Business*, 10 FORD. L. REV. 105, 153 & n.179 (1999) (recommending adapting copyright's scènes á faire doctrine for patent, so that a patent claim that preempts a business will be deemed to fall outside the coverage of patent subject matter); *cf.* Julie E. Cohen & Mark A. Lemley, supra note 6 (recommending the judicial adoption of interpretative canons to limit the scope of software claims and for assessing equivalence); but see Gregory S. Fine, Note, *To Issue or Not to Issue: Analysis of the Business Method Patent Controversy on the Internet*, 42 B.C. L. REV. 1195 (2001) (arguing that the criticisms aimed at business-method patents ignore the realities of the internet). For a good review of the current dispute over business method patents, see R. Carl Moy, supra note 10, at 1053–61.

13. 101 U.S. 99 (1879).

14. E.g., Chamberlin v. Uris Sales Corp., 150 F.2d 512, 513 (2d Cir. 1945)(rulebook for "Acey-Ducy" card game not infringed by description of same game with different language); Affiliated Enter., Inc. v. Gantz, 86 F.2d 597, 598 (10th Cir.

1936) (pamphlet describing system for enticing customers to entertainment events by a free lottery does not protect the system); Affiliated Enter., Inc. v. Gruber, 86 F.2d 958 (1st Cir. 1936) (same as *Gantz*, stating that however good a plan or system is, it cannot be copyrighted and, if not patented, becomes the property of the public upon disclosure); Brief English Systems, Inc. v. Owen, 48 F.2d 555, 556 (2d Cir. 1931) (system of shorthand writing by condensing words into fewer letters not protected by copyright); Taylor v. Commissioner of Internal Revenue, 51 F.2d 915, 917 (3d Cir. 1931) (plan for acquiring supremacy of classified advertising by newspapers is not protected by the copyright in its disclosure and so is not depreciable property under the income tax laws); Burk v. Johnson, 146 F. 209, 213 (8th Cir. 1906) (copyright in "Articles of Association" gave no exclusive right to organize and operate under the plan disclosed); Kepner-Tregoe, Inc. v. Carabio, 203 U.S.P.Q. 124, 130 (E.D. Mi. 1979) (copyright in teaching materials does not extend to the problem-solving techniques they describe); Briggs v. New Hampshire Trotting & Breeding Ass'n, Inc., 191 F. Supp. 234, 236 (D.N.H. 1960) (horse race betting system unprotected as a game and as being too elementary and ordinary); Aldrich v. Remington Rand, Inc., 52 F. Supp. 732, 734 (N.D. Tex. 1942) (forms included in brochure describing system to facilitate collection of taxes and necessary to use the system are not protected); Seltzer v. Sunbrock, 22 F. Supp. 621, 630 (S.D. Cal. 1938) (system for conducting races on roller skates is not copyright protectable); see generally Pamela Samuelson, *Computer Programs, User Interfaces, and section 102(b) of the Copyright Act of 1976: A Critique of* Lotus v. Paperback, 6 HIGH TECH. L.J. 209, 226–27 & n.73 (1992).

15. See, e.g., Overhead Door Corp. v. Chamberlain group, Inc., 194 F.3d 1261, 1269 (Fed. Cir. 1999) (noting that any software process can be transformed into an equivalent hardware process and vice versa); Pamela Samuelson, Randall Davis, Mitchell D. Kapor, & J.H. Reichman, *A Manifesto Concerning the Legal Protection of Computer Programs*, 94 COLUM. L. REV. 2308, 2319 (1994) (noting that hardware and software are interchangeable); Margo A. Bagley, supra note 12, at 276 (noting that any software process can be transformed into an equivalent hardware process and vice versa).

16. 35 U.S.C. 101 ("Whoever invents . . . any new and useful . . . machine . . . may obtain a patent therefore, subject to . . . this title").

17. Robert P. Merges, supra note 12, at 586 ("When these softwareembedded concepts are characterized as novel computer programs, there is little to separate them from any other computer program").

18. The exclusion of art, literature, and music from patent coverage has been thought so obvious that it often remains only implicit. Still, while recent Supreme Court jurisprudence leaves a broad scope for patent subject matter, no one has suggested that the Court would extend section 101 to traditional art, music, or literature.

19. Thus, a machine programmed to play a new piece of music would be patent subject matter, but the music, as nonpatent subject matter, contributes nothing to the patentability analysis. Vincent Chiappetta, supra note 6, at 143, 172 (distinguishing software as computer-implementation technology, which is always patent subject matter, and software as language to communicate an algorithm or

process, which must be tested independently under the exclusions for abstract ideas or traditional art, music, and literature); Dennis S. Karjala, supra note 10, at 58–60 (arguing that computer implementation of an abstract theory would be patent subject matter but not patentable if the only advance is in the nontechnological theory). See also supra note 10.

20. Professor Moy has identified the problem of distinguishing patent-subject-matter information-based methodologies and nonpatent-subject-matter abstract ideas the fundamental problem facing the patent system. R. Carl Moy, supra note 10, at 1089. On the general problem of distinguishing unpatentable abstract ideas from patentable inventions in information processing, see Richard S. Gruner, *Intangible Inventions: Patentable Subject Matter for an Information Age*, 35 LOY. L.A. L. REV. 355 (2002).

21. E.g., Whelan Associates, Inc. v. Jaslow Dental Laboratory, Inc., 797 F.2d 1222 (3d Cir. 1986); Digital Communications Assocs. v. Softklone Distributing Corp., 659 F. Supp. 449 (N.D. Ga. 1987); Broderbund Software, Inc. v. Unison World, Inc., 648 F. Supp. 1127 (N.D. Cal. 1986).

22. E.g., Computer Associates Internat'l Inc. v. Altai, 982 F.2d 693 (2d Cir. 1992); Lotus Development Corp. v. Borland International, Inc., 49 F.3d 807 (1st Cir. 1995), *aff'd by an equally divided Court*, 516 U.S. 233 (1996); see generally Mark A. Lemley, *Convergence in the Law of Software Copyright*, 10 HIGH TECH L.J. 1 (1995).

23. *A Coherent Theory*, supra note 5, at 78–94; 99–116; Dennis S. Karjala, *Copyright Protection of Computer Program Structure*, 64 BROOKLYN L. REV. 519 (1998).

24. One federal district court long ago explicitly recognized the problem. At issue in Briggs v. New Hampshire Trotting & Breeding Association, Inc., 191 F. Supp. 234 (D.N.H. 1960), was whether copyright in a brochure describing a betting system for horse races covered the system itself. The court saw the relevance of *Baker v. Selden* but thought that *Baker* and its progeny involved processes that were clearly patentable, as opposed to sports, games, or similar systems:

But if games, sports, and similar systems and plans are neither copyrightable nor patentable, then there is a hiatus in the law which unjustly fails to offer protection to original inventors.

Id. at 236. The court denied copyright protection in the actual case, both because the copyright statute and the case law does not protect games, sports, and similar systems and also because the system involved in the case was "so elementary and ordinary that it is in the public domain." Id. at 236–37. The court strongly implied, however, that "original, complex, unique systems" would be protected. Id. at 236. Thus, categorical denial of patent protection could, under this reasoning, lead to copyright protection for complex, even though functional, systems.

25. Dennis S. Karjala, *Distinguishing Patent and Copyright Subject Matter*, 35 CONN. L. REV. ___ (2003).

26. E.g., *Relative Roles*, supra note 10, at 44–50; *A Coherent Theory*, supra note 5, at 56–66; Dennis S. Karjala, *Copyright Protection of Computer Software,*

Reverse Engineering, and Professor Miller, 19 DAYTON L. REV. 975, 976–83 (1994) [hereinafter referred to as *Copyright Protection*].

27. Cf. Ralph S. Brown, *Eligibility for Copyright Protection: A Search for Principled Standards*, 70 MINN. L. REV. 579, 604 (1985) (arguing that the patent/copyright boundary reflects the distinction in purpose between encouraging technological innovation and stimulating creative expression).

28. *A Coherent Theory*, supra note 5, at 56–58.

29. 17 U.S.C. '101 (definition of "useful article"). This definition makes clear that even highly "useful" things are not "useful articles," a point that is very often lost on commentators and courts who discuss functionality for purposes of copyright.

30. Professor Lunney has pointed out that the Federal Circuit has been lowering "nonobviousness" bar and seems now to be treating just about any novel advance as nonobvious. Glynn S. Lunney, Jr., *E-Obviousness*, 7 MICH. TELECOMM. & TECH. L. REV. 363, 373–74 and passim (2001). He cogently argues that this will encourage the investment of "creativity" in less socially desirable projects. Id. at 411–12. Judge Pauline Newman and former PTO director Q. Todd Dickinson both agreed recently that most patents cover only minor advances or improvements. Joint Hearings, Federal Trade Commission and Department of Justice, *Competition and Intellectual Property Law and Policy in the Knowledge-Based Economy*, Feb. 6, 2002, reported in 7 ELEC. COMMERCE & L. REP. (BNA) 138, 139 (Feb. 13, 2002). Professor Lunney, however, strongly disagrees with the notion of "functionality" as the touchstone for distinguishing patent and copyright subject matter. His objections are considered at length in the full version of this article.

31. Professor Gorman has reminded me that there is a sense in which patents have a broader scope of protection than copyrights, which is that patents can protect ideas and algorithms while copyright supposedly protects only the particular expression of these more abstract concepts. See also *Relative Roles*, supra note 10, at 45 n.8. As used herein with respect to the scope of copyright protection for traditional literary and artistic works, "broad protection" refers to the author's exclusive right to create derivative works that can be substantially different from the underlying work and the protection for "nonliteral" elements, like detailed plot sequence, that is afforded to these works. E.g., Sheldon v. Metro-Goldwyn Pictures Corp., 81 F.2d 49, 55 (2d Cir. 1936). The scope of a patent is strictly tied to the claim language and in that sense patents are more narrowly defined than copyrights. So, even though patents cover things that copyrights do not cover, they are not necessarily *broader* when measured from the respective starting points of the two rights: the claims in the case of patents and the fixation of the entire underlying work in the case of copyrights.

32. Pamela Samuelson, *CONTU Revisited: The Case Against Copyright Protection for Computer Programs in Machine-Readable Form*, 1984 DUKE L. J. 663, 749 (arguing that the essence of copyright, as opposed to other intellectual property systems, is that the content of a work has always had some nonfunctional aesthetic, informational, or entertaining qualities that are communicated to a human audience).

33. Whether and to what extent operating software, with the powerful network effects it engenders, should be protected by copyright is another question. Dennis S. Karjala, *Copyright Protection of Operating Software, Copyright Misuse, and Antitrust*, 9 CORN. J. L. & PUB. POL. 161 (1999).

34. National Commission on New Technological Uses of Copyrighted Works, Final Report (1978). That there is necessarily an exception to the functionality doctrine for program code by no means implies that other elements of programs, such as program structure or interfaces, also fall under the copyright protective umbrella. Indeed, these other elements are no more vulnerable to fast and inexpensive copying than more traditional works of technology.

35. E.g., A *Coherent Theory*, supra note 5, at 66–72.

36. 17 U.S.C. '102(b) ("In no case does copyright protection for an original work of authorship extend to any idea, procedure, process, system, method of operation, concept, principle, or discovery, regardless of the form in which it is described, explained, illustrated, or embodied in such work").

37. 101 U.S. 99, 102 (1880) ("To give to the author of the book an exclusive property in the art described therein, when no examination of its novelty has ever been officially made, would be a surprise and a fraud upon the public. That is the province of letters-patent, not of copyright").

38. J. H. Reichman, *Computer Programs as Applied Scientific Know-How: Implications of Copyright Protection for Commercialized University Research*, 42 VAND. L. REV. 639, 692–93 & n.288 (1989).

39. Nor does it mean that the functionality doctrine plays no role in deciding which noncode elements of programs, if any, are protected by the program copyright. Because noncode elements of programs like sequence, structure, and organization ("SSO") and user interfaces are not vulnerable to piracy in the same way that literal code is, there is no reason to interpret the program copyright broadly for these functional elements. Patent law works perfectly well for functional developments in these areas. *A Coherent Theory*, supra note 5, at 66–72.

40. See supra note 29 and accompanying text.

41. Blank forms and standardized test questions are functional works for which a thin copyright might be justified on antimisappropriation grounds. These and other examples are discussed in the full version of this article, supra note 25.

42. 17 U.S.C. ' 102(b); see supra note 36.

43. 35 U.S.C. ' 101 ("Whoever invents or discovers any new and useful process, machine, manufacture, or composition of matter . . . may obtain a patent therefor . . ."). For an excellent discussion of the history of process patents, see John R. Thomas, supra note 9.

44. *A Coherent Theory*, supra note 5, at 59 n.18.

45. *Relative Roles*, supra note 10, at 44–45; see supra text following note 5.

46. Some Irving Berlin copyrights date from the 1920s, and he did not die until 1989. If the life + 70 system had been in effect during his lifetime, some of his copyrights would have endured for 130–135 years.

47. Cf. Ralph S. Brown, supra note 27, at 609 (arguing that copyright is made easy to provide broad protection to works of enlightenment and diversion but can be easily bent to evade the limits of patent or to extend monopolies that have no support in any system).

48. *Relative Roles*, supra note 10, at 48; see also *A Coherent Theory*, supra note 5, at 61; *Copyright Protection*, supra note 26, at 979–80; cf. Julie E. Cohen & Mark A. Lemley, supra note 6, at 23 ("To an even greater extent than copyright law, patent law anticipates and even depends on one party improving another party's invention").

49. Diamond v. Diehr, 450 U.S. 175 (1981).

50. I am indebted to Professors Gorman and Weinreb for having made me aware of the need to be explicit about the objective verifiability of "improvements" in functional works, as opposed to artistic works in which "improvement" is largely a matter of taste. The text attempts to do that, but it must be admitted that there remain some types of information works that are also amenable to objectively verifiable improvement. Reference works like dictionaries can be improved with error correction, for example. See infra notes 51–52 & 55–57 and accompanying text. Consequently, the basic dichotomy for the patent/copyright borderline must remain the distinction between function on the one hand and information content on the other, where "function" in the context of systems and processes refers to the application of information to a task other than appreciation by human judgment.

51. Admittedly, the social policy underlying our harsher treatment of unauthorized "modifications" of traditional works of art, literature, and music inheres partially in noninstrumental, natural-rights notions that such works more reflect the personality of the author than functional works falling under the patent paradigm. Indeed, the former are constrained only by the imagination of the author, while the latter are constrained both by the laws of nature and by the functional goals the works are intended to achieve. Still, notions of instrumentalism enter the calculus as well. Society simply does not value a new version of a popular novel with a different final chapter as much as it does a steering wheel added to a motorized vehicle. There is therefore less downside risk in disallowing unauthorized "modifications" for a longer period than is given by patent. This is not to say, of course, that there is no social value in "improvements" to traditional copyright subject matter. After the (increasingly long) "limited time" for which copyright subsists, the market rather than the exclusive rights of copyright determines whether a modification is socially desirable, as some clearly are. One need only think of *West Side Story*, for example, or the many films based on works by Shakespeare, Jane Austen, Thomas Hardy, and other great authors. At least one commentator has argued on economic efficiency grounds that copyright goes too far in limiting the rights of "improvers." Mark A. Lemley, *The Economics of Improvement in Intellectual Property Law*, 75 TEX. L. REV. 989 (1997)

52. Mark A. Lemley, supra note 51; *but* see Suntrust Bank v. Houghton Mifflin Co., 268 F.3d 1257 (11th Cir.2001) (denying a preliminary injunction, on potential fair use "parody"grounds, of a new version of *Gone With the Wind* involving many of the same characters but told from a different perspective).

53. AT&T Corp. v. Excel Communications, Inc., 172 F.3d 1352, 1360 (Fed. Cir.1999).

54. Cf. Eldred v. Ashcroft, United States Supreme Court, No. 01–618, *Brief of George A. Akerlof, Kenneth Arrow et. al as Amici Curiae in Support of Petitioners*, at 14 ("In copyright [as opposed to patent], diverse, 'abundant' expression is the source of value, not successive refinements with respect to an agreed-upon metric of quality, and a large number of disparate innovators may be better at producing abundance"), *citing* Paul Goldstein, *Infringement of Copyright in Computer Programs*, 47 U. PITT. L. REV. 1119, 1123 (1986).

55. See supra note 29 and accompanying text.

56. Feist Pubs., Inc. v. Rural Tel. Serv. Co., 499 U.S. 340 (1991). Courts, of course, continue to protect facts as "opinions" or "estimates." CDN, Inc. v. Kapes, 197 F.3d 1256, 1259 (9th Cir. 1999); CCC Information Services, Inc. v. Maclean Hunter Market Reports, Inc., 44 F.3d 61 (2d Cir. 1994). The point here, though, is that copyright resolves the social policy balances involved in the protection of information, whether or not given information is actually protected by copyright. Patent law does not, and should not, have any role to play in the protection of information as such. That is reflected in the universally accepted, if rarely articulated, exclusion of traditional copyright subject matter from patent coverage. See supra note 18.

57. We are not considering here abstract ideas or theories of natural law, which are denied protection by *both* patent and copyright. E.g., State Street Bank & Trust Co. v. Signature Fin. Group, Inc., 149 F.3d 1368, 1376 (Fed. Cir. 1998); 17 U.S.C. ' 102(b); Baker v. Selden, 101 U.S. 99 (1879). Still, both abstract ideas and theories fit comfortably within the concept of "information" that should be and is denied status as patent subject matter. More problematic are functional processes in areas like politics or religion. See the full version of this article, supra note 25.

58. I believe it is correct to say that copyright, in the main, has eschewed the protection of "functionality" under my definition. Standardized tests are functional under this definition, however, and are routinely held copyright protected. Applied Innovations, Inc. v. Regents of the Univ. of Minnesota, 876 F.2d 626 (8th Cir. 1989); Educational Testing Services v. Katzman, 793 F.2d 533 (3d Cir. 1986); Educational Testing Service v. Simon, 95 F. Supp. 2d 1081 (C.D. Cal. 1999). Other functional works have occasionally fallen through the cracks into copyright protection as well. See *generally Relative Roles*, supra note 10, at 63–65 & n.36.

59. See supra note 18 and accompanying text.

60. The longer version of this article concludes that standardized taxonomies by medical and dental associations likely fall into this category. Patents in the competitive arts may be another. Vincent Chiappetta, supra note 10, at 320–24; Leo J. Raskind, supra note 11, at 92–93. Sports moves, too, would seem to be unlikely candidates for patents. Every major competitor has more than enough incentive to try to think of ways to improve performance. One need only witness the number of hours of training put in by Olympic hopefuls. Moreover, at least at the professional level, the relevant governing associations would likely ban any move that

was not fairly licensed to all competitors. It would make no more sense to allow only Dick Fosbury to use his famous Flop in the high jump than it would to allow the inventor of the glass pole to be the sole participant permitted to use that advance in pole vaulting. The event simply would no longer be interesting to spectators. That likelihood reduces the expected monetary return from a potential patentee's exclusive rights.

61. Fogerty v. Fantasy, Inc., 114 S.Ct. 1023, 1028 (1994) ("The primary objective of the Copyright Act is to encourage the production of original literary, artistic, and musical statement for the good of the public"); Feist Publications, Inc. v. Rural Tel. Serv. Co., Inc., 499 U.S. 340, 349 (1991) ("The primary objective of copyright is not to reward the labor of authors, but '[t]o promote the Progress of Science and useful Arts.'") (citations omitted); Bonito Boats, Inc. v. Thunder Craft Boats, Inc., 489 U.S. 141, 146 (1989) ("From their inception, the federal patent laws have embodied a careful balance between the need to promote innovation and the recognition that imitation and refinement through imitation are both necessary to invention itself and the very lifeblood of a competitive economy"); Sony Corp. of America v. Universal City Studios, Inc., 464 U.S. 417, 429 (1984) ("The monopoly privileges that Congress may authorize are neither unlimited nor primarily designed to provide a special private benefit. Rather, the limited grant is a means by which an important public purpose may be achieved"); Twentieth Century Music Corp. v. Aiken, 422 U.S. 151, 156 (1975) ("The immediate effect of our copyright law is to secure a fair return for an 'author's' creative labor. But the ultimate aim is, by this incentive, to stimulate artistic creativity for the general public good"); Graham v. John Deere Co., 383 U.S. 1, 9 (1966) (reviewing with approval Jefferson's view that "The patent monopoly was not designed to secure to the inventor his natural right in his discoveries. Rather, it was a reward, an inducement, to bring forth new knowledge"); United States v. Paramount Pictures, Inc., 334 U.S. 131, 158 (1948) ("The copyright law, like the patent statutes, makes reward to the owner a secondary consideration"); Fox Film Corp. v. Doyal, 286 U.S. 123, 127 (1932) ("The sole interest of the United States and the primary object in conferring the [copyright] monopoly lie in the general benefits derived by the public from the labors of authors").

62. See supra notes 29 and 45–57 and accompanying text.

63. Professor Gordon uses the term "restitutionary impulse" to refer to the tendency of courts to protect the fruits of intellectual labor whenever they sense that someone is unfairly benefitting from the efforts of another. Wendy J. Gordon, *On Owning Information: Intellectual Property and the Restitutionary Impulse*, 78 Va. L. Rev. 149, 277 (1992) ("[I]n the last twenty years the restitutionary impulse has acquired new force. . . . I speculate that judges may feel no need to examine the trend because the restitutionary notion that one deserves to keep the 'fruits of his labor' seems so evidently correct . . . that giving legal protection to intellectual products appears to require no special justification").

9

Advantage ISP: Terms of Service as Media Law

Sandra Braman and Stephanie Lynch

Although in the popular imagination the Internet remains a diffuse cloud, in practical terms those who use e-mail and surf the web do so through the Internet Service Providers (ISPs) through which they gain access (Blumenthal and Clark 1997). As the proportion of our communicative and informational lives conducted on-line steadily grows, the reality of ISPs as determinants of the conditions under which communicative activity takes place is transforming the de facto communication law environment. The abandonment of traditional First Amendment rights and forced transfer of the intellectual property rights of individuals to ISPs so far occur beneath the radar in contracts unread and lawsuits scattered throughout topic-specific analyses.

It has always been the case that any constitutional right—including those of the First Amendment—can be voluntarily yielded by contract. Historically, however, only occasionally were rights were so affected, few or only single rights were thus yielded up, and the contexts in which such a decision was made were those in which it was possible to make a choice among alternatives. In contrast, the contractually generated speech environment of ISPs restricts the rights of almost everyone and requires abandonment of a wide range of rights. As Acceptable Use Policies (AUPs) and Terms of Service (TOSs) increasingly harmonize with each other across ISPs, these restrictions on rights take place in a context in which the vast majority of users cannot choose an alternative.

The number of individuals who have found their speech constrained by ISPs is not known because there are no reporting requirements on ISPs and not all instances reach the courts, but anecdotal reports in extensive discussions in usenet groups suggests the problem is not infrequent. ISPs are beginning to act on their licenses to the intellectual property rights of content produced by their users in pay-for-access websites to, for example, salacious material garnered from user e-mail and further use of user-created content by ISPs must be expected. Meanwhile complementary elements of the legal environment that support ISPs in such moves are steadily being strengthened and a general climate of acceptance of some of the more fundamental features is being developed through systematic focus on single elements of the overall picture. The end result could well be a wholesale shift in the possibilities of free expression in the United States without public discussion and decision-making on the fundamental constitutional issues at stake. New laws and regulations put in place since 9/11 have significantly increased the responsibility of ISPs to act essentially as agents of the government, enhancing their quasi-regulatory role by adding a range of enforcement activities.

After a brief review of the development of the regulatory aspects of ISPs, this chapter presents the results of a study of ISP rules in place in spring of 2002 that have media law-like effect. Based on a close reading of texts in order to gain the greatest sense of differences in nuance and detail, the group of ISPs studied are largely commercial in nature but does include a range of types so that some comparison between U.S.-based vs. non-U.S.-based and commercial vs. noncommercial could be accomplished, as well as a comparison between ISP rules of today and those of the past. Analysis of the legal or quasi-legal effect of ISPs is important at this stage in their development because of the effects of path dependence on development of the industry and because the history of particular practices is an important element in constitutional analyses of the acceptability of those practices.

Doing so also provides a de facto list of policy issues in the contemporary internet environment that need attention by policy-makers. The chapter concludes with a look at possible responses to this situation.

The Development of ISP Regulatory-Like Functions

Discussion of ISPs as regulators began to appear in 1996 (Johnson and Post 1996; Reidenberg 1996). Since that time several factors encouraging the organizations to fill this function have been identified, but there has been almost no systematic analysis of the ways in which the function has been filled, effects of exercise of the function, or the constitutional acceptability of the practices.

Analyses of the ISP Regulatory Function

While media and telecommunications law casebooks and textbooks regularly organize their analyses of law and regulation as they pertain to specific media and industries, none of the cyberlaw books to date has yet explicitly treated communications law in the ISP environment (Baumer and Poindexter 2002; Ferrera et al. 2001; Girasa 2002; Lemley, et al. 2000; Lipschultz 2000; Maggs et al. 2001; Rosenoer 1997). The expansion of the ISP role in enforcement of state-made laws and regulations is receiving analysis (Frydman and Rorive 2002; Zittrain 2003), but there have been very few studies of ISP-made "law" in AUPs and TOSs.

One relatively shallow study of 11 common abuses of e-mail that combined acceptable use and general office policies that found attention to such abuses in every agreement and unsurprising differences only in degree of sensitivity to flames and newsgroup hosting problems (more in ISPs than in the other categories) and concern about leisure surfing (more in organization-specific access systems) (Siau et al. 2002). A study of the AUPs of local and state government agencies conducted in 1998 focused on hierarchical decisions regarding the identities of those permitted to access the internet (Menzel 1998). It did, however, yield a useful distinction among types of rhetorical approaches used in AUPs: some merely remind the user that the internet is no different from any other information technology and that the user is therefore subject to the same ethical and legal standards used elsewhere, some offered detailed statements of acceptable and unacceptable behaviors, and some offer general guidelines for internet-specific behavior. A study of the treatment of information security at ISPs concluded that policies alone are ineffective and need to be complemented with ongoing education and training not only of network administrators but also of users (Nosworthy 2000).

Development of the ISP Industry

As an industry, ISPs are still seeking their identity. Their early history was unusual, quickly achieving mass market status, becoming geographically pervasive, and offering a diverse spectrum of services. The nature of the ISP industry is still evolving: Firms in traditional media industries such as journalism experiment with providing ISP services to support other on-line activities, external entities such as courts have defined ISPs as a subset of telecommunications for purposes of legal analysis, and devoted ISPs continue to explore the economic possibilities of activities such as distributed computing that are utterly new (Greenstein 2000). ISP experimentation is currently underway with various business models and lines of activity, geographic reach and structure, and the role of ISPs vis a vis other approaches to integrating the "network of networks" (Downes and Greenstein 2002; Noam 2001). Meanwhile the ISP industry continues to expand, partially because the technical and financial barriers to entry have dropped and partially because the market has grown so explosively (Phillips 2001; Hallman and McClain 1999). The largest ISPs continue to grow through acquisition of start-ups and smaller entities.

Conditions are still open for experimentation for a variety of reasons: Operating conditions are no longer those of the internet's early "end to end" years in several dimensions of importance from a legal perspective for reasons that, as Blumenthal and Clark (1997) note, include loss of trust, the appearance of more demanding applications, a drop in the sophistication of users, and the desire of ISPs themselves to provide service differentiation. It is not yet clear whether ISPs should be identified as members of the "cyberspace community" more broadly defined, or as the powers that govern that community (Biegel 2001). Meanwhile very few users understand the functions of ISPs (Engel 1999), or the law dealing with them (Townsend et al. 2000).

Distinctions among ISPs

Of course ISPs are not all alike. Various approaches to distinguishing among them have been put forward. Doing so by geographic reach yields two distinctions: between local, national, and international ISPs; and between ISPs serving urban as opposed to rural areas (Greenstein 2000, 2001). Doing so by type of user results in a distinction between businesses and individuals as ISP customers (Engel 1999). Doing so by services offered and/or functions filled (Greenstein 2000) results in distinctions between those that provide basic access, high speed access, and complementary services; or in a finer articulation, between those that provide basic access (up to T-1), frontier access (faster than T-1), networking, hosting, and web page design. Doing so by architecture, placing the position of each ISP within the structure of the internet, results in distinguishing between ISPs with the functions of transit backbone, downstream ISPs, online service providers that package content, and firms that specialize in web site hosting (Gorman and Malecki 2000).

Each of these typologies highlights different dimensions of the ways in which ISPs operates as actors in the legal environment and as regulators or quasi-regulators in themselves. The geographic scope of an ISP could have implications for jurisdictional analyses (Bonnett 2001). User distinctions suggest economic implications of the ways in which ISPs intervene in use content and behaviors, but the one study of ISP user satisfaction (Wetzel 2000) did not include legal and quasi-legal elements in its survey. Service distinctions draw attention to the variable degree with which ISP Terms of Service (TOS) or Acceptable Use Policies (AUP) can have constitutional implications. And architectural distinctions such as the nature of the peering structure, the position of an ISP within the internet writ large, and innovation in technologies such as routers have implications both for interactions among regulatory approaches across ISPs and for the types of regulatory tools that are available to ISPs (Besen 2001).

Factors Stimulating the ISP Regulatory Function

To be generous, ISPs have come to fill the legal space as a consequence of their roles as what Greenstein (2001) describes as "technological mediators," providing necessary adaptations between a changing technical frontier and unique user needs. Many of the regulatory restrictions are the result of limits to bandwidth or, conversely, responses to the cost of expanding bandwidth. Even though ISP responses to these features of their situation are understandable, however, they do not make them constitutionally—or socially—acceptable. While these practices may be defended as responsible management, they often cross the line into manipulative control of content and applications of types long rejected in the larger communicative world. Some may argue that ISPs should be accepted as governance mechanisms, but they meet neither the regulatory criterion of being all-encompassing, (Biegel 2001) nor have they been developed in a democratic manner.

AUPs and TOSs have become more detailed and elaborate over time, stimulated by several factors. Doing so in many cases would be among the functions they serve their customers as technological mediators. In some cases they are responding to pressure from at least some of their users. intervention in network content and behaviors may be a natural outgrowth of the detailed, consistent, and regular monitoring of online activity carried out by network administrators. Ever-greater numbers of technical methods of intervention continue to appear. So do tools for enforcement, which now include slowing response time for a service an ISP wants to discourage, channeling surfers through advertisements it wants them to see, and identifying patterns of behavior that monitors can watch based on profiles (Lessig 1999). Tools that can be used by ISPs for governance include rules announced to members, stigmatizing behaviors as a way of triggering community norms to help regulate, manipulation of prices (increase, taxation of particular uses, or differential prices depending on user), changes in architecture, and monitoring of behaviors.

ISPs are also responding to developments within the law. Laws dealing with ISPs continue to grant ever-greater freedom to operate without fear or liability, (Patel 2002) while legislation has so far been unable to satisfactorily curb unwelcome on-line behaviors. Three types of liability regimes have been applied to ISPs:

(1) A negligence regime was first adopted in the United States and is presently employed in Europe that holds ISPs liable when they fail to exercise due care in monitoring of third party content.

(2) A strict liability regime was experimented with in both the U.S. and in Europe that holds ISPs liable for all injuries resulting from third party content.

(3) A no liability or conditional liability regime is presently used in the United States that holds ISPs almost completely immune from liability for third party content, conditioned only on minimal responsibilities as outlined in Sec. 230 of the Communications Decency Act (Schruers 2002).

User Responses

While theoretically any ISP can say to a user that s/he can "walk with one's fingers" to another ISP if there is dissatisfaction with the terms under which one may use the ISP, in reality several factors make this less and less practicable: First, terms of Service and Acceptable Use policies are increasingly becoming standardized so that there is less and less difference across ISPs; thus in most cases there is nowhere meaningfully different to go. Second, those ISPs that offer the greatest geographic scope tend to be those that are most restrictive—meaning that those who are most actively involved in public life beyond the local community level are those whose rights will be most restricted. The trend of mergers and acquisitions in the ISP industry exacerbates the first two factors.

Users and user groups are not without tools for response. A number of usenet groups provide venues for discussion of dissatisfactions with ISPs, though such conversations often tend toward discussion of the personalities of network administrators rather than the rule structure itself. Standards that reputable ISPs should meet are beginning to be discussed, though until now these have had to do with technical aspects of service rather than regulation of service. Those who bundle individual users into groups, such as managers of wired office or apartment buildings, are being provided with recommended terms of service, though again the only inclusion pertinent to quasi-legal interventions by ISPs is emphasis upon the nonexclusive license rights that make it possible for users to choose other ISPs if they are unhappy (Puentes and Rothenberg 2001). A consumer movement that joins together individuals concerned about threats to civil liberties presented by ISPs is beginning to appear (Akdeniz 2000). ISP management of communication content and behaviors receives a significant amount of discussion in usenet groups, though this discussion rarely moves into action. Some communication policy advocacy groups do things such as monitor websites that are blocked by filtering software. The most effective responses to date, however, appear to be individual- (Kevlin 2001) or ISP-level (Biegel 2001) use of technical methods for blocking content deemed inappropriate, though this has proven effective so far only for spam with the Realtime Blackhole List

Methodology

This study examined a relatively small number of cases in depth for maximum detail and nuance. While Greenstein grouped together a number of different approaches to the same problem in order to facilitate computerized content analysis of thousands of user agreements, this study employed close reading of texts in order to discern differences in stance presented by different ways of wording restrictions on restricted content and activities that are important from a legal perspective.

The study examined AUPs and TOSs of ISPs as they existed during the spring of 2002.[1] ISPs varied by geographic locus and reach, size, target market, and commercial nature. The study included the largest commercial ISPs as ranked by size, plus a small ISP from each region of the U.S. and an ISP from each of the other continents besides North America. "ISP" was defined for the purposes of this study as an institution that provides an ISP-like experience for the user. Thus examples of access to the web through K–12 schools, universities, and public libraries were included in the sample. A list of the ISPs studied can be found in table 9.1.

The result was identification of 59 rules that pertained to ISP-user relations (see table 9.2), divided into those dealing with general matters (about policies themselves, service limits, and limits to account use), identity, user liability, lack of ISP liability, and ISP treatment of privacy; 42 rules dealing with content (see table 9.3), divided into those dealing with illegal content, intellectual property, and other content restrictions; and 38 rules dealing with behavior (see table 9.4), divided into those dealing with illegal behavior, security, user treatment of privacy, and other behavior restrictions. The percentage of total ISPs with each of the specific rules identified can be found in tables 9.2, 9.3, and 9.4, along with a breakdown of commercial vs. noncommercial ISP rules.

While the original intent of the study was to compare the earliest versions of these agreements with their contemporary versions, only half a dozen of the earlier agreements were available through use of the Wayback Machine as others were not on-line but incorporated into software licenses off-line. These did provide an opportunity to make a first pass at examining the development of AUPs and TOSs across time.

Interpretations of user knowledge and responses were acquired via a review of the literature and through analysis of discussion of the regulatory practices of ISPs in usenet groups archived by Google located through use of search terms such as ISP, lawsuit, AUP, TOS, and privacy policy and from a sampling of messages in the news.admin.net-abuse groups that are dedicated to internet abuse issues.

Discussion

The relationship between users and ISPs as defined by AUPs and TOSs is best described currently as "advantage ISP" for a number of reasons described here in generalized form and discussed in more detail below. It is worth noting that while some of the rules—those dealing with infringement of intellectual property rights, invasions of privacy, and content and behavior so labeled—deal with matters treated as illegal under the laws of the United States, most forbid activities that are legal under U.S. law. In addition to the areas discussed in detail within the paper, the charts provide detail on rules put in place to protect personal privacy, encourage respect for intellectual property rights on the part of users, protect network security, and explicitly forbid content and behaviors already illegal under U.S. law.

- *Knowledge*: Users know little about rules or enforcement tools, practices, and history.
- *Liability*: Users are liable for the consequences of their uses of the ISP—whether or not intended—while ISP's have almost no liability even for failures of service for which telecommunications carriers were traditionally liable.
- *Intellectual property rights*: Users are forced to license all content to ISPs, and often publicity rights as well.
- *Abandonment of ponstitutional protections:* Agreements drafted by ISPs abandon constitutional standards for restrictions on speech of narrow tailoring, establishment of criteria to be met before restrictions can be deemed acceptable, and avoidance of vagueness and overbreadth, resulting in creation of a speech environment significantly more restrictive than that developed for society at large through judicial analysis of the aspects of constitutional law that deal with information, communication, and culture.
- *Comparative analyses:*

  (1) Commercial vs. noncommercial: Noncommercial ISPs provide greater protections for free speech and the intellectual property rights of users, but less in the way of privacy protections.

  (2) Current vs. past: The numbers of rules constraining communicative content and behaviors by ISPs is constantly growing. Some rules are dropped over time, but no pattern was discernible in this area

  (3) U.S.-based vs. non-U.S.-based: Non-U.S.-based ISPs provided less detail in the areas of intellectual property rights and privacy, but tended to restrict more areas of content and behavior that are legal in the United States and did not allow anonymity.

Knowledge

ISPs have the advantage in terms of knowledge about regulatory rules and practices in several ways.

Table 9.1
ISPs

AOL	By far the largest national ISP, with more than 26 million subscribers and 17.5% of market share. Offers TOS and privacy policies on its website, but maintains separate policies for subscribers to the service. Access to these appears to be limited to subscribers and requires the AOL software to access.
APCi	Provides internet access in the midwest United States, focusing on the St. Louis metro area.
AT&T	Dial-up service is ranked #10, with 1.4 million subscribers.
AT&T 1997	AT&T's policy as of June 6, 1997.
Bellsouth	Regional (Southeast) ISP offered through Bellsouth telephone company; ranked 15th with 730,000 subscribers.
Cablevision	Cable company providing national high-speed internet access; ranked 17th with 560,000 subscribers.
Charter	Communications company offering high-speed cable modem access. Ranked 14th, with 645,000 subscribers.
Chicago Public Schools	Chicago Public Schools
Columbia University	Columbia University
Compuserve	Owned by AOL, but operates separately. Ranked 6th, with 3,000,000 subscribers.
Earthlink	Offers various services (dial-up, broadband, etc.), each with its own policy. Only dial-up analyzed. Earthlink dial-up ranked 4th, with 4,800,000 subscribers.
Earthlink 2000	Earthlink's policy as of March 2 and 11, 2000
GOL-Japan	Provider based in Japan. GOL: Global OnLine.
Inter.net	Global provider, looked at Canadian branch.
Juno	Merged with NetZero to form United Online, but still maintains its own service. Merged company ranks 3rd and has 5,600,000 subscribers.

Juno 2000	Juno's policy as of March 1, 2000
Mindspring	Has been bought by Earthlink.
MSN	AOL's closest competitor, but still lags behind. Ranked 2nd, with 8,000,000 subscribers.
MSN 1996	MSN's policy as of October 26, 1996
M-Web	South Africa-based provider.
Naperville Public Library	Naperville Public Library
Pacific Internet	Singapore-based ISP.
Panix	Oldest commercial internet provider in New York.
Prodigy	Dial-up service ranked 5th, with 3,600,000 subscribers.
Prodigy 1999	Prodigy's policies as of April 22, 1999.
Rain	"Public Internet Broadcasting" service based in California.
San Francisco Public Library	San Francisco Public Library
Simplecom	West Alabama ISP.
Simplecom 1999	Simplecom's policy as of April 20, 1999.
Tuscaloosa City Schools	Tuscaloosa City Schools
University of Texas	University of Texas
Verizon	Communications company offering internet access among other services. DSL is ranked 11th, with 1,200,000 subscribers.
Worldcom	Commercial internet provider

Table 9.2
ISP-User Relations

	All (N=27)%	Com (N=21)%	NonCom (N=6)%
General			
Policies			
Rules in one place	55.56	42.86	100.00
Policy change alert	40.74	52.38	—
Where policy applies	37.04	42.86	16.67
Dispute res. process	29.63	33.33	16.67
May change ISP software	11.11	14.29	—
Member audit org.	11.11	14.29	—
Copyright infringement report process	22.22	28.57	—
Service limits			
Message retention	18.52	23.81	—
Disk space	25.93	28.57	16.67
On-line time	44.44	57.14	—
Website traffic	37.04	38.10	33.33
Number of sessions	07.41	09.52	—
No multiple logins	37.04	47.62	—
Limits to account use			
No unauthorized access	66.67	71.43	50.00
No use of other's acct	51.85	52.38	50.00
May not resell service	40.74	52.38	—
Pay for all transactions	11.11	14.29	—
Identity			
Anonymity allowed	11.11	14.29	—
No anonymity allowed*	—	—	—
No false ID	18.52	19.05	16.67
No false ID to mislead	14.81	19.05	—
No forging of headers	48.15	61.90	—
No impersonation	40.74	47.62	16.67
No use of vulgarity in screen names	07.41	09.52	—
User Liability			
ISP can remove material of concern	48.15	61.90	—
ISP indemnified against damage to user	48.15	57.14	16.67
User liable for account	40.74	47.62	16.67
User liable for damage to ISP	18.52	23.81	—

*Included in older AUPs

Table 9.2
(continued)

	All (N=27)%	Com (N=21)%	NonCom (N=6)%
No ISP Liability			
Accidental deletion/failure to store messages	25.93	33.33	—
Content/links	59.26	71.43	16.67
Copyright infringement by users	29.63	28.57	33.33
Transmission errors	37.04	47.62	—
Damage from material received	51.85	61.90	16.67
Damage from transactions	22.22	28.57	—
Failure/delay in removal of material	14.81	19.05	—
Interruption	44.44	57.14	—
Lack of timeliness	29.63	33.33	16.67
Loss due to unauthorized account use	14.81	19.05	—
Security lapse	33.33	42.86	—
Viruses, worms, etc	33.33	33.33	33.33
Privacy (ISP)			
Data collection techniques			
Cookies	25.93	33.33	—
Request info from user	29.63	38.10	—
User must update info	25.93	33.33	—
Other collection techniques	14.81	19.05	—
Stat techniques described	29.63	38.10	—
Types of info collected			
Personal information	37.04	47.62	—
Advertising presented	18.52	23.81	—
Computer	25.93	33.33	—
Computer use	22.22	28.57	—
Software	07.41	09.52	—
Use of information collected			
By function	37.04	47.62	—
No sale of personal info	22.22	28.57	—
Data sharing partners	40.74	52.38	—
Will cooperate with govt.	51.85	61.90	16.67
User options			
General opt-out	25.93	33.33	—
Directory listing opt-out	03.70	04.76	—
Detailed consent to uses	14.81	19.05	—
Other opt-out	03.70	04.76	—
Correction possible	29.63	38.10	—

Table 9.3
Content

	All (N=27)%	Com (N=21)%	NonCom (N=6)%
Illegal content			
No unlawful content	25.93	33.33	—
No defamation/libel/slander	44.44	57.14	—
No incitement to violence	11.11	09.52	16.67
No obscenity	62.96	66.67	50.00
Intellectual property (IP)			
IP rights claimed by ISP			
ISP has license to all postings	11.11	14.29	—
ISP has license to all postings to gen. public	25.93	33.33	—
May sub-license postings	07.41	09.52	—
May use postings for commercial purposes	11.11	14.29	—
May distribute postings	29.63	38.10	—
May produce derivative works	25.93	33.33	—
May publicly perform/ display postings	22.22	28.57	—
May reproduce postings	22.22	28.57	—
May use user's name in connection with postings	07.41	09.52	—
May delete submission	22.22	28.57	—
No compensation for use of material	07.41	09.52	—
IP rights infringement by user			
May not violate copyright	74.07	76.19	66.67
Download only one copy	03.70	04.76	—
May not create derivative works	22.22	28.57	—
May not delete/alter attribution	25.93	28.57	16.67
May not download w/o rights	40.74	38.10	50.00
May not post/upload w/o rights	62.96	71.43	33.33
May not reproduce other than for personal use	29.63	33.33	16.67

Table 9.3
(continued)

	All (N=27)%	Com (N=21)%	NonCom (N=6)%
Other content restrictions			
On non-personal objectionable content			
No inappropriate content	14.81	19.05	—
Use filters	03.70	04.76	—
No indecency/pornography	37.04	42.86	16.67
No material violating internet norms	14.81	19.05	—
No objectionable content	11.11	14.29	—
No posting off-topic (newsgroups)	44.44	57.14	—
No profanity	18.52	23.81	—
On personal abuse			
No harmful content	22.22	23.81	16.67
No abuse of others	33.33	42.86	—
No contesting crimes against humanity	03.70	04.76	—
No hate	37.04	42.86	16.67
No flaming (newsgroups)	11.11	14.29	—
No threat to person/property	66.67	85.71	—
On promotional efforts			
No chain letters	55.56	57.14	50.00
No charity requests	03.70	04.76	—
No contests	07.41	09.52	—
No petitions	03.70	04.76	—
No pyramid schemes	44.44	57.14	—
No spam	51.85	66.67	—
No surveys	07.41	09.52	—

Table 9.4
Behavior

	All (N=27)%	Com (N=21)%	NonCom (N=6)%
Illegal behavior			
No use for unlawful purposes	88.89	95.24	66.67
No fraud	18.52	23.81	—
No harassment	70.37	66.67	83.33
No stalking	14.81	19.05	—
Security			
General			
May not cause damage	29.63	23.81	50.00
May not intentionally cause damage	22.22	23.81	16.67
May not compromise security	70.37	76.19	50.00
Methods			
Use anti-virus software	03.70	—	16.67
No cancelbots	14.81	19.05	—
No trojan horses	22.22	28.57	—
No time bombs	03.70	04.76	—
No unauth use of third-party server	33.33	38.10	16.67
No viruses	44.44	52.38	16.67
No worms	25.93	33.33	—
No dist corrupted files	07.41	09.52	—
No dist tools for damaging security	25.93	28.57	16.67
Subject of damage			
Other user	29.63	38.10	—
Site	25.93	28.57	16.67
System	40.74	42.86	33.33

Table 9.4
(continued)

	All (N=27)%	Com (N=21)%	NonCom (N=6)%
Privacy (users)			
General			
No invasion of privacy	33.33	33.33	33.33
Maintain confidentiality	37.04	33.33	50.00
Care in dist personal info	18.52	14.29	33.33
Specific			
No coll of personal info	22.22	28.57	—
No receipt of passwords	11.11	14.29	—
No solicitation of passwords	25.93	33.33	—
No coll of email addresses	11.11	14.29	—
No coll of screen names	07.41	09.52	—
No coll of info re minors	07.41	09.52	—
Other behavior restrictions			
Limit cross-posting	51.85	61.90	16.67
No advertising	37.04	42.86	16.67
No automated queries	03.70	04.76	—
No commercial use	51.85	52.38	50.00
No gambling	03.70	—	16.67
No mail bombs	37.04	42.86	16.67
No mass mailing	40.74	47.62	16.67
No meta-searching site	03.70	04.76	—
No pinging	07.41	09.52	—
No restriction of use by others	62.96	71.43	33.33
No surveys	07.41	09.52	—

Knowledge of Rules While all of the noncommercial ISPs made it easy for users to become aware of the pertinent rules by publishing them all in one place, only approximately 43 percent of the commercial ISPs did so, otherwise requiring users to roam the site many clicks deep in order to gain all the knowledge needed. Just over half of the ISPs studied (all commercial) make clear that they will alert users to any changes in policy; otherwise users must continually check AUPs and TOSs to learn if there has been any change in policy. Some ISPs (11.10 percent) explicitly say they are free to change the ISP software at will and without alerting users to these changes; others may do the same but say nothing on this point. With many different kinds of activities taking place via ISPs, only just over a third indicate to users where specific types of policies apply.

Criteria for Decision-making While rules may be published, the criteria by which those rules are interpreted are not. All explication of administrative procedure, however—whether via the Administrative Procedures Act, regulation, or internal organizational rules and procedures—includes explicit and detailed discussion of the criteria of judgment used. Without making such decision-making rules clear, administrators can act arbitrarily and affected users have no grounds upon which to grieve or petition.

Regulatory Tools A wide range of regulatory-like tools is available to and used by ISPs while most users have neither knowledge of the ways in which those tools are used nor of ISP functions. Often modes of manipulation of content are not known or understood by nontechnical users of ISPs and so they may not be obvious—service may be slowed down, differential pricing may be established, or use habits monitored for development of profiles that can be used to justify further regulatory-like interventions. Means by which information is being gathered about users often are not understood by the users (e.g., clear gif., etc., let alone cookies). Noncommercial ISPs say almost nothing about their data collection practices.[2]

Enforcement Practices While rules are published, the range of possible ISP responses to infringements of rules are not. There is usually some threat of loss of service, but techniques of enforcement short of that and the steps through which decisions about loss of service are implemented and may be grieved are not detailed. Users do not know, for example, if they will be warned about behaviors they consider normal and acceptable but that are deemed unacceptable by the ISP before service is cut, or not. No means is provided for discussing the acceptability of various practices with the ISP. And in most cases there is no means by which ISP users can communicate with each other about petitioning the ISP for changes of rules felt to be unreasonable.

Enforcement History Unless there is conversation about it on usenet groups or other lists, there is also no public record of restrictions on speech. The importance of public knowledge about enforcement of law and regulation underlies the constitutional principle of public access to trials. Concern over loss of such knowledge is key among the issues raised by privatization of the law; by the 1980s, for example, newspaper companies began to examine their loss of access to records of decision-making of importance to the general public because of its impact as a result of the movement of corporate conflict resolution from the courts to modes of alternative dispute resolution. There is no systematic way of learning how ISP users are actually being treated in the broad areas included in AUPs and TOSs other than through anecdotal discussion on specialized bulletin boards of those cases that make it to the eye of others. Such knowledge is of course critical should user groups desire to seek a change in the rules under which they are permitted to communicate.

Liability

Today a "no," or "conditional," liability regime governs ISPs. For other media, *control = liability*, but ISPs effectively have *control without liability*. Of course neither control nor liability is a binary condition, but across media increases in control have meant an increase in liability. In the area of broadcasting, for example, not only has liability for matters such as libel long accompanied editorial control, but in recent years courts are increasingly insisting that broadcasters are liable for damage wrought by viewers inspired by or imitative of behaviors presented in television programming. While ISPs have pursued legal treatment as information distributors rather than content providers, the courts have not been consistent in this regard (Patel 2002).

Even more importantly, ISPs have insisted upon this identity while simultaneously claiming control over the intellectual property rights over material transmitted via their services through mandatory licensing. Thus there is a deep contradiction in ISP claims, on one hand, *not* to be content providers and, on the other, that they control all content. This contradiction has not yet received analysis in the courts because liability issues have been treated distinctly from intellectual property issues, but inclusion of the latter in analyses of the former should be expected in coming years. For the moment, however, ISPs have control without liability.

ISPs are not even held liable for service failures of the types for which more traditional types of telecommunications service providers are held accountable, such as interruption of service, security lapses, lack of timeliness of delivery of services, etc. They are able to escape such responsibilities because they are not classified as telecommunications companies. As organizations either new altogether or new to the telecommunications business, they do not have internal histories of concern about service provision. And because their relationships with the transmission network itself are varied but most often not that of ownership, it can be difficult to determine just who should be liable for failures of service. This issue is not likely to be resolved unless and until users demand greater commitments to reliability.

Users, however, are generally held accountable for any type of consequence of their uses of ISP services, whether to another user, a website, or the ISP itself. Importantly, many ISPs do not distinguish between causing damage intentionally and doing so unintentionally. (As we know, it may be possible to unwittingly cause technical damage either through ignorance regarding the technologies involved or through software applications so complex that they cannot be predicted.) The question of intentionality is of particular importance from a legal perspective, for intentionality is always key in constitutional analysis. Explorations of whether or not political speech constitutes clear and present danger, for example, includes the important criterion of intentionality, as does determination of fault in libel suits. Differences between ISP rules and constitutional law on this point mean that an ISP user accused of libel might be found innocent because of lack of intentionality by the courts but still lose service because under ISP rules that criterion is irrelevant.

Anonymity is one way of trying to avoid liability for communications that has been constitutionally protected in the United States because "liability" can translate into "punishment" for dissenting political speech or for corporate whistle-blowers even though both types of content have great social importance. ISPs have experimented with forbidding anonymity; those that explicitly permit it do so only in those environments in which it has been specifically described as acceptable such as in usenet groups or on lists that include permission for anonymity among their internal rules.

Property Rights

Users of some commercial ISPs are forced to grant the services licenses to all content uploaded or posted, with a slightly larger percentage insisting upon the same license only for content presented to the general public. Where they exist, ISP licenses include the entire bundle of intellectual property rights, with AUPs and TOSs specifying the various rights individually—reproduction, distribution, production of derivative works, performance, and display. While such licenses implicitly include the right of ISPs to make money from user content, two ISPs said they have the right to commercial use of what is posted, one emphasized that it would do so without compensation to the content producer user, and one further insisted on the right to offer further sub-licenses to others for commercial use. One ISP also included the right to use of the name of the person who originally posted or uploaded content, one element of the right to publicity.[3]

Interestingly, despite the protection from liability offered by the Digital Millennium Copyright Act (DMCA) if all of its terms are adhered to, (Wernick 2001) about three-quarters of ISPs in the study did *not* provide information regarding how to report alleged or suspected copyright violations to the organization.

Because ISPs strongly emphasize user adherence to copyright law, an asymmetry in potentials for use is created: ISPs, for example, are free to create derivative works based on content uploaded or posted by users, but users themselves are forbidden to do so.

Abandonment of Constitutional Protections

ISPs forbid many forms of constitutionally protected content both nonpersonal and personal in kind, though two of the ISPs in this study (APCi and BellSouth) insist that despite their rules they are not trying to censor or constrain the free flow of information.

In the area of nonpersonal content over 40 percent of commercial ISPs forbid posting off-topic, a third prohibit even those kinds of indecency and pornography that do not cross the legal line into constitutionally restrictable obscenity, and almost 20 percent forbid constitutionally acceptable profanity. Similarly, a number of communicative activities that are constitutionally protected, such as conducting surveys, requesting donations for charity, running contests, distributing chain letters, or circulating petitions, are forbidden by some ISPs. Though junk mail is not illegal when transmitted via the Post Office, over half of the ISPs surveyed treated spam as unacceptable. While impersonation and forgery are illegal under any conditions, many ISPs also forbid identity experimentation in screen IDs or message headers (as opposed to message content), even though such experimentation is one of the most noted features of internet use.

Several ISPs used very general terms to describe unacceptable nonpersonal content such as "inappropriate," "objectionable," or "material violating internet norms;" general prohibitions on more personal content included terms such as "harmful" and "flaming." Though hate speech unlinked to action is *not* illegal in the U.S. context, over a third of ISPs placed it on the unacceptable list, and one ISP specifically forbade "contesting crimes against humanity." Laws and regulations can be declared unconstitutional for vagueness (language so unclear that reasonable adults cannot agree on their meaning) and overbreadth (language that may be directed at specific types of unacceptable behavior or content but that is cast in terms so broad that many types of acceptable behavior or content are also included) of the kind exhibited by these types of general terms in AUPs and TOSs, particularly because no criteria are offered for determining when the bar had been crossed.

Behavioral limits on otherwise constitutional activities include restrictions on mass mailings (over 40 percent) and cross-posting of messages to more than one news group (almost 52 percent). A few ISPs forbade use of techniques such as automation of queries, meta-searching websites, or pinging. Because many ISPs distinguish between rates offered personal and business users, over a third forbid advertising and over half forbid use of the ISP for commercial use. Restrictions put in place by some of the ISPs presumably out of a good will effort to protect personal privacy are impracticable at best and offensive at worst, such as forbidding the collection of e-mail addresses, collection of screen names, and collection of e-mail addresses. Only one ISP, on the other hand, forbade collection of information about minors.

It is legal to restrict constitutional rights by contract, but by function ISPs serve as public fora. Contractual yielding of constitutional rights has historically been limited for two reasons, neither of which applies to the ISP context. Contracts have been undertaken only by individuals or small classes of people, while ISP agreements affect essentially all U.S. citizens. And contracts have previously been entered into only in situations in which doing so is a choice among alternatives—one can choose to take a particular job or not, for example. As ISP AUPs and TOSs become standardized, however, contractual abandonment of one's constitutional rights is taking place in a situation in which there are in fact no alternatives if one wants to communicate or receive information at all electronically. These differences between the use of contracts in other situations in which constitutional rights become limited and their use in the ISP environment suggests that contracts should be abandoned in favor of subjecting ISPs to public forum analysis.

Comparative Analyses

The inclusion of a range of types of ISPs in the study makes it possible to offer comparisons between different categories of ISPs, though the small numbers of each make the comments below only suggestive. Descriptions of the ISPs included in the study found in table 9.1 include detail about the comparative dimensions.

Commercial vs. Noncommercial Commercial ISPs tend to have many more rules than do noncommercial ISPs across categories. Most notably, none of the noncommercial ISPs in this study claimed a license to the intellectual property rights of content posted or uploaded. They also said far less about restrictions on content and behaviors. Noncommercial ISPs also, however, provided much less information about the kinds of data collected about users and ways in which that data is used and did not offer opt-out options for users concerned about protecting their privacy. The conclusion is that users appear to have greater protections for freedom of speech and much less fear about loss of their intellectual property rights when they use noncommercial ISPs, but in turn they may need to be more concerned about protections for their right to privacy.

Present vs. Past Though the original intent of the study was to compare original and contemporary AUPs and TOSs of every ISP in the study, it turned out that only six of sets of earlier agreements were accessible via the tool of The Wayback Machine, and these were not necessarily the first agreements used by each (MSN, 1996; AT&T, 1997; Prodigy, 1999; Simplecom, 1999; Juno, 2000; Earthlink, 2000). The most dramatic finding of the comparison was the explosion in the number of rules: while the average number of rules in the earlier agreements was 29, in spring of 2002 it was 60. Some earlier rules were dropped in current versions of AUPs and TOSs, but there was no discernible pattern in what disappeared. Possible explanations for changes in the terms of agreements over time include changes in ownership, legal impact, experience, and the desire to model examples set by other ISPs.

U.S.-based vs. Non-U.S.-based Four of the ISPs studied were based outside of the U.S. Of course non-U.S.-based ISPs did not include information on where to report infringements of U.S. copyright law. While one of the four did claim a license to all material posted, it did not detail the different elements of copyright as found in U.S. law; there was less concern about copyright infringement by users. There was not as much detail in non-U.S.-based ISPs in the area of privacy. Probably reflecting data privacy rules of the OECD, however, a larger percentage of non-U.S.-based ISPs did provide information about those with whom ISP user data collected would be shared. It is not surprising that non-U.S.-based ISPs included more restrictions on content and behaviors that are legal in the United States but often not elsewhere in the world. Anonymity was not allowed by non-U.S.-based ISPs.

Responses to ISP Regulation

Both economic and legal tools are available to those concerned about these trends in the development of a regulatory-like function for ISPs.

Economic Tools

Both general (consumer movement) and more specific (user group) techniques are available to users who are concerned about these trends, while entrepreneurs may choose to interpret such trends as a means of identifying a market niche.

Consumer Movements Consumer movements, now nascent, can serve several functions: they can educate consumers about issues, stimulate formation of groups large enough to have negotiation heft, and bring issues of concern into public discourse.

User Groups An economic tool of some potential strength is available to users who form into large groups of users. The example has been set by large office and apartment buildings that contract with specific ISPs to provide service to their tenants; while these agreements have tended to focus on reliability of service and marketing restrictions, they could be expanded to include negotiation over the conditions of communications. If all libraries in the American Library Association (ALA), all universities involved in EDUCAUSE, or all schools in a state or a school district, required certain features in their TOS or AUP as terms of a group contract the economics of the situation would force ISPs interested in the business to give up on unacceptable restrictions.

Market Niche On the entrepreneurial side, there is an as-yet-unfilled market niche for the ISP or ISPs that should choose to provide the widest possible protections for freedom of speech and other communicative, informational, and cultural rights as their distinguishing features. Two "public interest" ISPs were included in the study: Panix, which targets political activists in its marketing efforts, and Rain, which explicitly defines for itself a public service role. It is worth noting that while these two do forbid illegal manipulations of identity, they are more open to types of identity experimentation that would be considered legal under other circumstances; they do not insist upon a license to content posted or uploaded; they are more open than many ISPs to use of techniques such as meta-searching; and they have markedly fewer restrictions on content than other ISPs.

Legal Tools

The distinction between voluntary and involuntary communications is a legal tool that may be valuable in the court context, public forum analysis should be of value both in the courts and in Congress, and it is the responsibility of Congress to address violations of and potential changes to copyright law.

Voluntary vs. Involuntary Constitutional law has always distinguished between those situations in which one's speech conditions were voluntary and those in which they were involuntary, with much higher barriers to unacceptable speech in those conditions in which individuals cannot choose to avoid the communications or communicate through another means. The involuntary nature of the need to rely upon an ISP in order to communicate via the net provides an opening for legal analysis of restrictions on speech along the dimension of voluntariness.

Public Forum Analysis The distinction between public and private forums is used by the courts as a first question in determining when restrictions on speech are constitutional. Four types of forums have been distinguished: *Public forums* are publicly owned and controlled, with public functions and history (e.g., parks and sidewalks); restrictions on speech in public forums are subjected to the highest level of scrutiny. *Quasi-public forums* are publicly owned and controlled but are devoted to specific functions and have a history of restricted use (e.g., public universities and military bases); restrictions on speech in such contexts are justified if the speech in question would interfere with the functions to which the venues are devoted. *Quasi-private forums* are privately owned and controlled but serve public functions and have a history of public use (e.g., company towns, shopping malls, and airports); restrictions on speech in these venues are also acceptable if the speech in question would interfere with the functions to which the forums are devoted. *Private forums* are privately owned and controlled, serving private functions and with a history of private use (e.g., homes, personal offices); rules for speech in private forums are up to the discretion of those who own and control them.

Within this typology, ISPs should be considered quasi-private fora: Ownership and control may be public or private, but their functions are primarily public. Since expansion of the internet beyond the original research scientist users, the history of use is primarily public—but because that history is still being formed, bringing public forum analysis into evaluation of ISP acquisition of regulatory-like functions is particularly important right now. The concept of ISPs as quasi-public fora failed in the courts in the courts in cases in which users the concept was used as a defense for the practice of disseminating spam (*America OnLine v Cyber*

Promotions, 1996; *CompuServe v Cyber Promotions*, 1997). The harmonization of ISP Terms of Service and Acceptable Use Policies since that time, however, alters the facts sufficiently that this line of argument would be much more likely to succeed under contemporary conditions.

Copyright Law Mandatory licensing of the intellectual property rights of everyone who communicates via the internet is a de facto change in copyright law that should not have been permitted without explicit policy-making attention. The constitutionality of such a move under current conditions is a matter for the courts, and the question of whether or not such a move should be permitted should be directly addressed by Congress as a matter of statutory law. Neither of these legal processes need wait until ISPs begin taking advantage of the vast quantities of content over which they are asserting the right of commercial use.

Conclusions

This study of the kinds of rules being put in place to constrain communicative content and behaviors on the internet via user contracts with ISPs suggests that without public discussion and largely without public awareness a significant shift is taking place in the actual nature of the increasingly dominant electronic speech environment. There is a long history of carefully crafting constitutional law in such ways that when other social needs must be balanced against speech rights this is accomplished in ways that are as narrowly tailored as possible, specific to the end desired and effective in reaching that end, uses language that is clear and unambiguous, always tries to maximize the opening of alternative venues for speech, and respects the intellectual property rights of those who create content. ISP contracts appear to be ignoring that history and are putting place rules that restrict speech that are broad, vague, and ambiguous; specifically prohibit forms of speech that have been explicitly and repeatedly protected under U.S. law; exhibits no respect for the intellectual property rights of content creators; and does so in an environment in which increasingly there are no alternative venues for speech. Addressing these trends at this relatively early point in the history of use of the internet is critically important to ensure that all of the constitutional effort will not have been in vain. Tracking the development of a much larger number of ISP Acceptable Use Policies and Terms of Service agreements would be valuable as a means of determining whether or not such efforts are successful.

Notes

1. These agreements appear to change relatively quickly, and since the time of the study some of the smaller ISPs have already changed ownership.

2. The one exception is the mention by one noncommercial ISP of its compliance with Patriot Act requirements regarding surveillance of e-mail and web surfing practices.

3. While intellectual property rights provide a bundle of ownership rights to content produced, the right to publicity—which does not exist at the federal level in the United States but does in about half of the states—provides ownership rights in features of the individual such as the name, likeness, voice, and other identifying characteristics that may have commercial value.

References

Akdeniz, Y. 2000. "New Privacy Concerns: ISPs, Crime Prevention and Consumers' Rights," *International Review of Law, Computers & Technology* 14, no. 1:55–62.

America Online, Inc. v CyberPromotions, Inc., 948 F.Supp. 436 (E. D. Penn. 1996).

Baumer, D., and J. C. Poindexter. 2002. *Cyberlaw and E-commerce.* Boston: McGraw-Hill/Irwin.

Besen, S. 2001. "Advances in Routing Technologies and Internet Peering Agreements," *American Economic Review* 91, no. 2:292–297.

Biegel, S. 2001. *Beyond Our Control? Confronting the Limits of Our Legal System in the Age of Cyberspace.* Cambridge: The MIT Press.

Blumenthal, M. S., and D. D. Clark. 1997. "Rethinking the Design of the Internet: The End-to-End Arguments vs. the Brave New World," *ACM Transactions on Internet Technology* 1, no. 1:70–109.

Bonnett, T. W. 2001. "Is ISP-bound Traffic Local or Interstate?" *Federal Communications Law Journal* 53, no. 2:239–287.

CompuServe Incorporated v Cyber Promotions, Inc., 962 F.Supp. 1015 (S.D. Ohio 1997).

Downes, T., and S. Greenstein. 2002. "Universal Access and Local Internet Markets in the US," *Research Policy* 31, no.7: 1035–1052.

Engel, F. 1999. "The Role of Service Level Agreements in the Internet Service Provider Industry," *International Journal of Network Management* 9:299–301.

Ferrera, G. R., S. D. Lichtenstein, M. E. K. Reder, R. August, and W. T. Schiano. 2001. *Cyberlaw: Text and Cases*. Cincinnati: West/Thomson Learning.

Frydman, Benoit, and Rorive, Isabelle. 2002. "Regulating Internet Content through Intermediaries in Europe and the USA," *Zeitschrift für Rechtssoziologie* 23.

Girasa, R. J. 2002. *Cyberlaw: National and International Perspectives*. Upper Saddle River, N.J.: Prentice Hall.

Gorman, S. P., and E. J. Malecki. 2000. "The Networks of the Internet: An Analysis of Provider Networks in the USA," *Telecommunications Policy* 24:113–134.

Greenstein, S. 2000. "Building and Delivering the Virtual World: Commercializing Services for Internet Access," *Journal of Industrial Economics* 48, no. 4: 391–411.

Greenstein, S. 2001. "Technological Mediation and Commercial Development in the Early Internet Access Market," *California Management Review* 43, no. 2:75–95.

Hallman, G., and C. McClain. 1999. "Real Options Applications for Telecommunications Deregulation," in *The New Investment Theory of Real Options and Its Implications for Telecommunications Economics*, edited by J. Alleman and E. M. Noam, 139–158. Boston: Kluwer Academic Publishers.

Johnson, D. R., and D. G. Post 1996. "Law and Borders: The Rise of Law in Cyberspace," *Stanford Law Review* 48:1367–1402.

Lemley, M. A., P. S. Menell, R. P. Meges and P. Samuelson 2000. *Software and Internet Law*. Gaithersburg, N.Y.: Aspen Publishers.

Lipschultz, J. H. 2000. *Free Expression in the Age of the Internet: Social and Legal Boundaries*. Boulder: Westview Press.

Maggs, P. B., J. T. Soma, and J. A. Sprowl. 2001. *Internet and Computer Law: Cases, Comments, Questions*. St. Paul: West Group.

Menzel, D. C. (1998). www.cthics.gov: Issues and challenges facing public managers, *Public Administration Review*, 58(5), 445–52.

Noam, E. M. 2001. *Interconnecting the Network of Networks*. Cambridge: The MIT Press.

Nosworthy, J. D. 2000. "Implementing Information Security in the 21st Century," *Computers & Security* 19, no. 4:337–347.

Patel, S. K. 2002. "Immunizing Internet Service Providers from Third-Party Internet Defamation Claims: How Far Should Courts Go?" *Vanderbilt Law Review* 55, no. 2:647–91.

Phillips, J. T. 2000. "ISPs and ASPs Create New Records Issues," *Informa-tion Management Journal* 34, no. 4:61–63.

Puentes, J. P., and P. V. Rothenberg. 2001. "10 Essential Contract Terms for Broadband Service Agreements," *Journal of Property Management* 66, no. 6:43–46.

Reidenberg, J. 1996. "Governing Networks and Rule-Making in Cyberspace," *Emory Law Journal* 45:911–930.

Rosenoer, J. 1997. *Cyberlaw: The Law of the Internet.* New York: Springer.

Schruers, M. 2002. "The History and Economics of ISP Liability for Third Party Content," *Virginia Law Review* 88, no. 1:205–264.

Siau, Keng, F. F. Nahy, and L. Teng. 2002. "Acceptable Internet Use Policy: Surveying Use Policies of Three Organizations—Educational Institutions, ISPs, and Non-ISPs," *Communications of the ACM* 45, no. 1:75–80.

Townsend, A. M., R. J. Aalberts, and S. A. Gibson. 2000. "Libel and Slander on the Internet," *Communications of the ACM* 43, no. 6:16–21.

Wernick, A. S. 2001. "US Internet Service Provider Liability," *Computer Law & Security Report* 17, no. 4:247–249.

Wetzel, R. 2000) "ISP Customers Habits, Likes and Dislikes," *American Journal of Human Genetics* 7, no. 37:31–42.

Zittrain, J. 2003. "Internet Points of Control," in *The Emergent Global Information Policy Regime,* edited by S. Braman. Houndsmills, UK: Palgrave/ Macmillan.

10

Anticircumvention Misuse

Dan L. Burk

1 Introduction

The recent passage of the Digital Millennium Copyright Act in the United States has drastically altered the landscape of intellectual property. Copyright in digital media has in a very real sense been rendered obsolete, superseded by new technological anticircumvention rights that some have called "paracopyright."[1] Such paracopyright constitutes a separate set of rights, quite distinct from any copyright in the underlying content. These new rights are expansive and unprecedented. They allow control of uncopyrighted materials, and confer upon content owners a new exclusive right to control not only access to technologically protected works, but also to control ancillary technologies related to content protection.

Because such "paracopyright" is ripe for abuse, limits on anticompetitive overreaching are in order. Just as improper leveraging of patent and copyright may properly be curtailed by application of the misuse doctrine, so improper leveraging of paracopyright should properly be curtailed by application of misuse. This new application of misuse doctrine may be guided by the standards established in previous applications to patent and copyright law, and may serve a similar function in regulating the excesses invited by paracopyright protection.

2 Technical Measures

Copyright is to some extent the creature of technological change; the printing press, the camera, the phonograph, the photocopy machine, and other technological advances have all left their mark.[2] Copyright is typically justified as a legal measure designed to correct a "public goods" problem fostered by these technologies.[3] Art, music, software, and similar creative works are frequently expensive to create, but technological advances render them easily copied once they have been created. The provision of a legal right of exclusivity is intended to allow authors to recoup their investment in creative works by allowing them to deter unauthorized copying and related uses for a limited period while charging a fair return on authorized copies.

At the same time, the introduction of such a legal barrier artificially raises the cost of the work, placing it beyond the reach of some people who might have enjoyed or benefited from it at the lower price. Thus, the use of intellectual property law is always a balancing act between allowing the greatest number of people to enjoy works at low cost, without lowering the cost so much that the works will never be created in the first instance.[4] Indeed, this balance is constitutionally mandated in the United States, where the constitution provides Congress with the power to enact copyright laws only if such laws "promote the progress of science and the useful arts."[5]

Consequently, in the United States at least, the public is the intended beneficiary of intellectual property laws, although authors certainly may benefit in the process.[6] The scope of copyright is limited to original expression, excluding ideas, processes, and functional aspects of protected works.[7] Copyright is also rife with exemptions, exceptions, and user privileges of every type. Some of these are quite narrow, such as the exception for owners of commercial business establishments to publicly perform broadcast music on receivers of a particular type,[8] or the exception allowing public performance of music at agricultural fairs.[9] Other privileges are quite broad, such as the so-called "first sale" doctrine, which cuts off the copyright holder's right of distribution in individual copies after they have been purchased, effectively allowing the consumer to resell or dispose of the copy in almost any manner she wishes.[10]

In the United States, the most important exception to the copyright holder's rights may be the "fair use" exception, which facilitates a broad range of unauthorized uses.[11] This exception is highly flexible, allowing otherwise prohibited uses of all or part of a work, depending on the circumstances. Fair use plays a critical role in mediating between constitutionally mandated freedom of expression and copyright,[12] which might otherwise be used to suppress socially valuable criticism, commentary, or parody of copyrighted works.[13] As computer software has been added to the universe of copyrightable subject matter, courts have also looked to fair use theories to justify the temporary but unauthorized copying that inevitably occurs in reverse engineering software to create competitive or interoperable products.[14]

The presence of such exceptions and user privileges is often an annoyance to copyright holders, who might prefer either to suppress such uses or to profit by them. One strategy that has emerged for curtailing such uses is that of licensing; for example, the first sale doctrine does not apply if a copy of a work is leased rather than sold.[15] A licensing agreement might also provide that the user of a copy will surrender fair use or other privileges in return for use of the copy. In mass-market situations, negotiating such leases with every consumer who purchases a copy might be burdensome, but copyright holders have developed the strategy of the "shrink-wrap" license to deal with this problem.[16] Under this legal fiction, the consumer purportedly agrees to the terms of the license by opening the packaging or making use of the copy.[17]

However, the proliferation of digital technology makes monitoring and enforcement of the copyright holder's exclusive rights relatively more difficult.[18] Cheap and easy accessibility to computers and computer networks allows consumers to reproduce and distribute digitized materials, both in exercise of exceptions to copyright as well as in excess of those exceptions. The sheer volume of both permissible and infringing uses makes the task of detecting and censuring the latter, impermissible uses formidable. Licenses that purport to eliminate otherwise legal uses are made by the technology equally difficult to police and enforce.

Copyright holders might far prefer a world in which the rights granted them under statute or asserted via license became self-enforcing.[19] Something close to this can be achieved by the employment of technological devices that can accompany copies of a work as they are distributed, controlling uses of the work.[20] Such devices may take a variety of forms as hardware or software, or some combination of the two.[21] Technological control systems may be used in the first instance to prevent access to digital content without the permission of the content owner, for example, by provision of a password. Such access might be occasioned upon terms of payment or terms of usage for the protected content. The consumer might well be presented with an extended license, perhaps in the form of a "clickwrap," to which he must acquiesce before the control system permits access.[22]

Alternatively, the control system itself might be designed so that the terms of use or payment are embedded as constraints upon the degree of access. For example, rather than agreeing in a written license that as a condition of access, the user will make only one copy of the content, the technological controls may be built to allow only one copy to be made. Technological control systems may tie use of the work to a certain machine, or when attached to a network or other signaling device, monitor the degree and type of use of the work, perhaps to meter payment by the minute, by the bit, or by some other unit of usage. Indeed, where technological controls are used in combination with "clickwrap" licensing, the terms may be enforced by the control system itself.[23] They may allow different levels of use depending on the level of payment made. Contingent or alternative terms might be programmed into the system, allowing a single access for a certain fee, or unlimited access for a higher fee. Access might even be revoked automatically, or by remote command, if payments are not made in a timely fashion.[24]

As these examples demonstrate, technological controls can be scripted to incorporate restrictions that might otherwise be the subject matter of a written license. Lessig and Reidenberg have each observed that because of these characteristics, technological control and legal control may be substituted in a variety of instances.[25] But technological controls and legal controls also differ in certain aspects, notably in the degree of discretion afforded to the user. Where legal regulation constitutes the barrier to use of content, users may breach it at their discretion, avoiding penalty until they are apprehended and legal process is complete. Technological barriers may be less difficult for content owners to police and enforce: unless users are technologically sophisticated, unauthorized uses are simply impossible.

From the perspective of the content owner, the major drawback to reliance primarily upon technological controls is that the barrier erected by one programmer may be circumvented by another. Technically sophisticated users may find ways to circumvent or disable the control system. The majority of users are unlikely to have such skills, but might be supplied with user-friendly software "hacking tools" by others who are skilled.[26] The widespread availability of such skills, or of tools requiring little skill, could threaten the more complete control over content offered by technological management. Thus, while technological controls may increase the difficulty of unauthorized uses, technology alone cannot be expected to achieve complete control of protected content. Legal prohibitions against circumvention activity may be necessary to buttress the integrity and operation of the control system.

3 DMCA Anticircumvention

Indeed, such anticircumvention laws, acting as an adjunct to technological controls, confer upon content owners a degree of control never attainable under a regime of traditional copyright.[27] Content owners in the United States received just such an anticircumvention entitlement in the provisions of the Digital Millennium Copyright Act, or DMCA.[28] The statute was touted as legislation necessary to fulfill the United States' obligations under the World Intellectual Property Organization Copyright Treaty (WIPO Treaty).[29] However, the treaty requires only that signatory states provide "adequate legal protection and effective legal remedies" against circumvention of technological controls.[30] In the United States, such protection would already have been provided under the doctrine of contributory infringement, which attributes copyright liability to providers of technical devices that lack a substantial noninfringing use.[31]

Nonetheless, lobbying by content industries resulted in the enactment of so-called "implementing" legislation containing anticircumvention provisions that far exceed anything contemplated by the treaty.[32] Starkly put, the DMCA as enacted creates a new and unprecedented right to control access to copyrighted works. The statute outlaws the act of circumventing "a technological measure that effectively controls access to a work protected under this title."[33] It also prohibits "trafficking" or providing the means to circumvent either technological access controls or technological measures that control the exclusive rights of a copyright holder: that is to say, copy controls, display or performance controls, and so on. The act provides for a handful of exceptions for extremely narrowly defined purposes such as law enforcement, encryption research, and security testing. There is a narrow exception for the purposes of software interoperability, but not hardware. The statutory exceptions are confusing and somewhat contradictory, but are primarily directed to the prohibition on circumvention; exceptions to providing circumvention means are extremely limited.[34]

Thus, with very few and very limited exceptions, the statute penalizes both the circumvention of technical protection measures, and supplying the means for such circumvention. The exceptions to the act of anticircumvention by no means accommodate the range of uses permissible to consumers under copyright law. There is, for example, no explicit provision allowing the owner of a copy to make fair use of the work embodied in that copy, and at least one court has rejected the argument that such an exception should be read into the statute.[35] Fair use encompasses a wide range of legitimate uses, including quotation for criticism and commentary, many educational uses, and the reverse engineering of software for purposes of interoperability.

Presumably, then, if a user wishes to make fair use of a technologically protected work, she must first either locate an unsecured copy of the work, or in the absence of such a copy, ask the permission of the content owner. This has not been the rule where rights are secured by copyright rather than by technical measures, and for good reason. Many socially valuable fair uses might be deemed offensive or undesirable by the rights owner. We would expect permission to be denied by rational rights holders in many core instances of fair use, such where the fair user wishes to engage in criticism or parody of the work.[36] But in such cases, even where permission for the use has been explicitly declined by the rights owner, fair use has been permitted to proceed over the owner's objections.[37]

What should be clear from this description of the DMCA anticircumvention provisions is, first, that this statute enables a new form of exclusive right, a right of access.[38] Although they appear as part of the "Digital Millennium Copyright Act," are codified along with copyright in Title 17 of the United States Code, and are frequently mentioned in connection with copyright, these prohibitions on circumvention of technical protections are entirely separate from the exclusive rights under copyright. Violation of the technological protections on a copyrighted work is an infringement entirely separate from unauthorized reproduction, distribution, adaptation, public performance, public display, or digital transmission of the controlled material—the technological infringer need engage in none of these exclusive activities to violate the anticircumvention provision. The anticircumvention statute in fact makes this distinction explicit, stating that nothing in the anticircumvention statute is to "affect any rights, remedies, limitation, or defenses to copyright infringement."[39]

The separation between the anticircumvention right and copyright is most apparent when comparing the limitations on each. As described above, copyright contains numerous exceptions and user privileges, such as statutory provisions allowing unauthorized use of copyrighted works in classroom instruction,[40] in certain religious services,[41] creation of "back-up" copies of computer programs,[42] and so on. None of these uses is explicitly sanctioned by the anticircumvention provisions; if a work is protected by access controls, circumventing those controls to make a use privileged under the copyright act is still prohibited. Outside of circumvention for the few exceptions described above, the only statutorily sanctioned method for gaining access to technically protected works is with the permission of the content owner.

This second aspect of the anticircumvention right should also be apparent from the description above, that as a statutory matter, the anticircumvention provisions of the DMCA extend protection far beyond any exclusive right granted in the protected work. Indeed, they likely extend protection beyond any right that could lawfully be granted by Congress under the copyright clause of the United States constitution, causing some commentators to question the constitutionality of the statute.[43] For example, in accordance with the constitution, copyright expires after a "limited time" but the anticircumvention statutes contains no such provision for expiration. Similarly, the Supreme Court has held that copyright cannot constitutionally be extended to facts or unoriginal compilations[44], yet such information, if controlled by technical measures, would appear to protected by the anticircumvention right.

The DMCA does require that in order to qualify for the anticircumvention right, a technological system must control some copyrightable content.[45] But copyrightable content is typically mixed with uncopyrightable content, which will also be under the control of the technological protection system. Unauthorized extraction of unprotectable content from a copyrighted work has consistently been held not to violate copyright,[46] but extraction of such unprotectable content from a technically controlled copy would violate the anticircumvention right. The controlled content may include uncopyrightable facts, or public domain materials, or purely functional works, yet unauthorized access to such content will constitute a DMCA violation the same as for unauthorized access to copyrighted content Thus, the anticircumvention provisions permit exclusivity that copyright clearly does not permit, creating another set of rights altogether: a right to control access to technologically protected works.[47]

The corollary to these conclusions is a third unique aspect of the anticircumvention right: that this new right of access facilitates not merely the licensing of copyrighted materials, but also allows licensing of *access* to published but uncopyrightable materials. Control of access has always been the essence of trade secrecy, so long as information remained confidential, but the ability to deny access was lost when the trade secret became publicly available. Now, the owner of technologically controlled materials may authorize or deny access, which is to say that he may license access, while at that same time publicly circulating the material. Such licenses may be conditioned upon terms set by the rights holder. Certainly price and manner of payment should be expected among such terms, as will conditions of use and other restrictions. Such licenses may be presented in writing prior to access, or may be incorporated into the technological controls themselves.

Fourth, although Congress may not have fully appreciated this result, the antitrafficking provisions of the DMCA confer upon content owners an ancillary property right in circumvention technology, which is to say, a property right in the means of accessing content. The right to prevent importation, distribution, or provision of circumvention technology necessarily entails the right to authorize such activity by waiving suit. Because this dimension of "paracopyright" effectively grants copyright holders sweeping new ability to impose licensing terms upon the creators of access technologies, such terms may extend to markets and activities unrelated to the controlled content. This affords copyright holders a new method to control competition and innovation beyond the market for protected content. The first crop of cases enforcing the anticircumvention right suggests that the opportunity to exercise such control has not been lost on the beneficiaries of this new right.

4 Leveraging Paracopyright

In the relatively short time since their enactment, the DMCA anticircumvention provisions have been invoked in a handful of cases and reported incidents. Courts have typically been sympathetic to such claims when the incidents have reached the point of judicial action. Oddly, however, the facts of these cases have seldom involved misappropriation of technically protected content. Perhaps the most extreme example of this trend is found in *RealNetworks v. Streambox*,[48] where content ownership is altogether absent. Instead, the DMCA action was brought by the publisher of a popular software package used to receive music or video "streams" via the Internet. The RealPlayer receiver software, which would be typically installed on a user's desktop machine, achieves connection with a RealPlayer music or video server elsewhere on the network through a "secret handshake" protocol that allows the server and receiver to recognize one another.[49] Once a connection is achieved, the system contains a feature to determine whether the user of the receiver has obtained rights to copy the music files sent by the server, or only to listen to the music as it is sent.[50]

The defendant Streambox produced a competing receiver, as well as several other pieces of software designed to be interoperable with the RealPlayer system. In order to play RealPlayer signals, the Streambox receiving components connected with the RealPlayer server by emulating the "secret handshake" protocol.[51] However, once the connection was established, the Streambox product lacked the restriction feature that would prevent unauthorized copying of streamed music or video. RealNetworks brought suit against Streambox, alleging that their receiving components constituted a "circumvention device" under the DMCA. In an unpublished opinion, the court granted the preliminary injunction, holding that the emulation of the "secret handshake" protocol constituted a circumvention of the RealPlayer restriction features.[52]

The most striking feature of this opinion is that no content owner appears—although the DMCA was purportedly enacted to protect owners of copyrighted content, in the *RealNetworks* case, only producers of competing software technology were involved. No pirating or unauthorized reproduction of any copyrighted content was shown, only the production of an interoperable product that could have been used to produce unauthorized copies of content. At least one way to view the facts is as an attempt by a software publisher to impede or abolish the distribution of a rival product, and at a minimum the case demonstrates that the statute could be turned to such purposes.

Control over interoperable technology, rather than an explosion of unauthorized copying, lay similarly at the heart of the dispute in

Universal City Studios v. Reimerdes.[53] The *Reimerdes* suit arose over circumvention of a technical control system knows as the Content Scrambling System, or CSS, which was designed to secure access to DVD movie discs.[54] A key feature of the system allowed discs playback to restricted by geographic area.[55] Machines manufactured in different geographic areas were designed to allow access to the content of a given DVD only if the disc was coded to be played in that in corresponding geographic area, thus allowing significant control over the timing and distribution of movies released in different parts of the globe. A corollary effect of this control system is that DVDs may only be played on approved playback equipment, whose manufacturer has built the equipment for use with the CSS.[56]

A fifteen year old Norwegian developed a program called "DeCSS," designed to circumvent the access controls, purportedly in order to play DVDs on nonapproved playback systems.[57] The DeCSS program would thus allow DVDs purchased in one area of the world to be played on equipment that would otherwise be geographically incompatible. It would also allow DVDs to be played on a Linux-based playback system, for which no approved device existed.[58] The owners of DVD content—which is to say, movie studios—alleged that the DeCSS "hacking tool" violated the DMCA provisions prohibiting trafficking in circumvention devices, and successfully filed suit to prevent various web sites from either directly distributing the program or offering hypertext links to other sites where it might be found.[59]

A similar result was reached in *Sony v. Game Masters*,[60] where the alleged circumvention device was an add-on module "Game Enhancer" for the PlayStation videogame console. The Game Enhancer was sold with instructions on how to use the device not only to modify games, but also to use a U.S. marketed console to play games intended for sale only in Europe or Japan.[61] Much like the DVD CSS territory codes in *Reimerdes*, the PlayStation console was designed to operate when encrypted data from a game CD verified that the game was a Sony product authorized for distribution in the same geographical territory as the console. The Game Enhancer instructions allowed players to initialize a U.S. game, then temporarily turn control of the console over to the Game Enhancer while the U.S. game was removed and an import game inserted and loaded.[62] Control was then turned back over to the console's operating system, which would execute the game software based on the previous authorization. The court concluded that this constituted circumvention of a technological measure in violation of the DMCA, and that distribution of the Game Enhancer violated the DMCA trafficking provisions.[63]

These incidents suggest that the anticircumvention right lends itself to use in ways that may be entirely unrelated to preventing unauthorized copying or distribution of copyrighted works. In the *Streambox* case, for example, the anticircumvention right has arguably been employed to suppress competing technology by preventing interoperability with products that include technical protections. The *Reimerdes* case suggests the use of paracopyright to force users to purchase or employ related products; DVD access controls require that the disc be played on approved hardware, effectively dictating the consumer's purchase of particular playback equipment.

The geographic "terms of use" for DVD or PlayStation access are embedded in the devices themselves, but paracopyright seems additionally positioned to facilitate anticompetitive terms in explicit licenses. Where a particular use would be permissible under copyright law, content owners may be able to forbid that use as a condition of access. For example, a content owner might as a condition of access contractually require a user to agree not to engage in fair use or reverse engineering as a condition of access—circumventing technical controls to engage in such uses without agreeing to the contract would constitute an anticircumvention violation. Judging by current trends in copyright "shrinkwrap" licensing, other access licensing terms might include noncompetition provisions or restraints on resale, lease, or lending of the controlled copy.

At some point, such leveraging of access control seems certain to overstep the bounds militated by sound policy or intended by Congress. In the past, abuse of intellectual property rights has been to some extent restrained by judicial application of the doctrine of misuse. The history of this doctrine suggests that it may be adaptable to use in new situations, and admirably suited to curtailing overreaching uses of "paracopyright."

5 Patent Misuse

The legal doctrine of misuse first arose in patent law, where patent rights might be leveraged into licensing terms that exceed the proper scope of the patent grant.[64] Misuse is an equitable doctrine, closely related to the doctrine of unclean hands.[65] Under these doctrines, a plaintiff who seeks the aid of a court to enforce his rights must not himself have been guilty of violating others' rights.[66] A defendant may therefore raise misuse as an equitable defense to infringement even though he himself is not necessarily the victim or target of the misuse. The court finding misuse exercises its discretion by refusing to aid the wrongdoer. Consequently, the court will decline to enforce the misuser's right against any party, whether or not harmed by the misuse, until the misuse has been "purged"; that is, until the misuser has reversed the effects of the conduct.[67]

Patent misuse has typically been found where a defendant can show some attempt by the patent holder to obtain greater economic advantage than Congress intended, or to restrain trade in ways not contemplated by the patent grant.[68] Such behavior frequently, although not exclusively, involves licensing, and classic cases of patent misuse typically concerned cases of "tying," that is, requiring purchase of an unpatented item in conjunction with purchase or license of a patented item.[69] For example, in the *Morton Salt* case where the United States Supreme Court firmly established the misuse defense, the patent holder was found to have exceeded the patent granted on its machine for depositing salt tablets, as it required licensees of the machines to use the machines only with salt tablets purchased from the patent holder.[70] According to the Court, such use of patent rights to leverage sales in an unpatented item tends to thwart the public policy underlying the patent grant, even if the patentee does not violate the antitrust statutes.[71]

In the decades subsequent to the *Morton Salt* case, patent misuse expanded to encompass a wide range of anticompetitive activities. Many of these activities coincided with violations of the antitrust laws; others were uniquely patent policy violations. For example, in *Brulotte v. Thys*, a patent holder's attempt to collect royalties from licensees beyond the term of the patent grant was declared contrary to public policy, and so constituted misuse.[72] If such licensing were permitted, the Court reasoned, the movement of the claimed invention into the public domain after expiration of the patent would be frustrated.[73] The Court therefore held such attempt to leverage the patent beyond the set term per se unlawful, establishing a new category of per se patent misuse.

The holding in *Brulotte* established federal patent policy as a basis for finding misuse, although the Court remained a bit vague about the exact parameters of the policy involved. Clearly, private attempts to relegislate the scope of a patent grant would constitute misuse, although it was the temporal scope of the grant—the statutory period of the patent—that could be precisely determined. This federal policy argument was extended in a later ruling, *Lear v. Adkins*,[74] where a patent holder raised the contract doctrine of licensee estoppel to prevent a licensee from challenging the validity of the licensed patent. In *Lear*, unlike *Brulotte*, the license required payment of royalties on a potentially invalid patent, rather than upon an expired patent. The Supreme Court extended its holding in *Brulotte* to rule that the federal policy favoring elimination of invalid patents preempted the state contract law doctrine of licensee estoppel. These decisions formed the basis for a separate line of cases delineating the proper role of state and federal law in the protection of proprietary rights, ultimately explicating rules that constrain state intellectual property law by both federal public policy and constitutional dimensions.[75]

The doctrine of misuse proper developed independently of the *Brulotte/Lear* line of cases, but the legal milieu of its development, together with the lack of clear guidance as to the limits of federal patent policy, fueled an unwarranted proliferation of the doctrine.[76] In a period when courts tended toward an expansive interpretation of antitrust law, patents were frequently regarded as "monopolies" to be voided at any opportunity. Because it frequently overlapped with real or perceived antitrust violations, patent misuse became a favorite tool to implement many courts' general hostility to patents. Misuse eventually became viewed, with some justification, as a bargain-basement all-purpose claim against patent enforcement.[77]

In response, Congress statutorily limited the scope of patent misuse, especially where it may overlap with antitrust violations.[78] The patent statute now catalogs a variety of patent related activities, such as refusals to license, that may at one time have been considered misuse, but which are now statutorily approved.[79] Tying arrangements between patented inventions and other items specifically adapted for use with the patented invention are similarly approved.[80] Even tying between patented inventions and unpatented staple articles of commerce are to be prohibited only where the tie meets the antitrust test of market power in the tying item.[81]

Historical antipathy toward patenting has also receded, to be replaced over the last two decades by an effusive new attitude, not merely of tolerance, but of nearly unbounded enthusiasm toward patents.[82] This patent fervor has in part been fueled by the creation of the United States Court of Appeals for the Federal Circuit, a body invested by Congress with exclusive appellate jurisdiction over patent cases,[83] and with a perceived mandate to produce a uniform body of U.S. patent law.[84] The court has generally taken this as a charge to support and expand patent law, and in the course of doing so has drastically limited the scope of misuse. Outside of those practices explicitly declared by Congress not to be misuse,[85] and a few categorical instances of "per se" misuse, the Federal Circuit has elaborated a "reasonableness" standard for evaluating misuse,[86] and seems inclined to find almost any activity engaged in by a patentee to be reasonable.

Despite this general contraction of the patent misuse doctrine, the Federal Circuit does continue to recognize misuse as applicable where the patentee unreasonably attempts to extend his rights beyond the statutory limits of the patent.[87] Even under the permissive standard articulated by the court, truly outrageous practices on the part of patentees would be found unreasonable.[88] And although in many cases such unreasonable or overreaching behavior in patent licensing will constitute an antitrust violation, even where it does not, the behavior may still constitute misuse. This is certainly the case in the "per se" categories recognized misuse, such as extension of royalties beyond the patent term.

6 Copyright Misuse

Although the importance of the misuse defense has waned in patent law, it has experienced a somewhat surprisingly renaissance elsewhere, within the law of copyright. This new flourishing may be due to the recent employment of copyright to cover technological inventions. Early commentary considering the inclusion of software within copyright predicted that copyright might need to borrow doctrines such as misuse from patent law in order to accommodate the characteristics of functional works.[89] True to prediction, the typical setting for copyright misuse has been that involving computer software, and courts developing this relatively new claim have drawn heavily on older patent cases for their rationale. As in classic patent misuse, to establish copyright misuse, a defendant must establish either (1) that the plaintiff violated the antitrust laws, or (2) that the plaintiff illegally extended its monopoly beyond the scope of the copyright or violated the public policies underlying the copyright laws.[90]

Although parallels between patent law and copyright law have long lent themselves to claims of copyright misuse, acceptance of the claim came slowly. In 1948 a district court in Minnesota recognized the defense of copyright misuse, but the decision was reversed on appeal.[91] For the next forty years, defendants unsuccessfully asserted the defense.[92] More recently, courts have begun to actively apply misuse principles to overreaching in copyright licensing. The germinal case in this line of doctrinal development was *Lasercomb America, Inc. v. Reynolds*,[93] where the court found that a license attempting to prevent the licensees from independently innovating a competing product amounted to copyright misuse. The defendant in *Lasercomb* had licensed four copies of a die-cutting computer program from the plaintiff, then circumvented the software's protective devices and made three unauthorized copies of the program.[94] When the copyright holder sued for infringement, the defendant asserted copyright misuse on the basis of Lasercomb's standard licensing agreement, which provided that licensees were barred from independently innovating a competing product for ninety-nine years.[95]

In accepting the defendant's claim of misuse, the Fourth Circuit Court of Appeals relied heavily on the patent misuse reasoning from *Morton Salt Co. v. G.S. Suppinger*. Of particular concern to the court was the copyright holder's attempt to withdraw its competitor's creative abilities from the public; the agreement not only attempted to suppress any independent implementation of the idea, but forbade the licensee from developing any kind of computer-assisted die making software.[96] The licensee was required to prevent all its directors, officers and employees from assisting in any manner to develop computer-assisted die-making software.[97] In a nod to the *Brulotte* line of patent cases, the court noted that the license's ninety-nine year prohibition could outlast the copyright itself.[98] Significantly, the court held that the defendant need not itself be subject to the egregious licensing term—as indeed the defendant was not—in order to assert the defense.[99]

Following *Lasercomb*, several other courts recognized claims of copyright misuse. Notably, the Ninth Circuit Court of Appeals in *Practice Management Information Corporation v. American Medical Association* held that a licensing agreement not to use competing products comprised copyright misuse.[100] At issue was a medical procedure indexing code in which the AMA claimed copyright, and which it licensed to a governmental agency, the Health Care Financing Agency (HCFA) on condition that HCFA would promote the use of the AMA code and agree not to use any competing system.[101] The Ninth Circuit agreed with a misuse claim brought by a publisher of medical texts, holding that public policy was offended by the AMA's attempt to license its code by imposing an anticompetitive exclusivity restriction.[102]

In a third leading case, *Alcatel USA, Inc. v. DGI Technologies, Inc.*, the Fifth Circuit held that a license limiting the use of operating system software to hardware produced by the copyright owner constituted copyright misuse.[103] The *Alcatel* case provides a particularly important example of copyright misuse doctrine in the context of computer interoperability. The plaintiff Alcatel produced equipment for telephone switching systems, which were controlled by a copyrighted operating system software.[104] The operating system was licensed to customers under terms that allowed use of the operating system only in conjunction with Alcatel's hardware. Customers frequently wished to add microprocessor cards to expand capacity of their switches. When an Alcatel competitor copied Alcatel's software in order to design a competing microprocessor card, Alcatel sued for copyright infringement. However, the court upheld the defendant's claim of copyright misuse, reasoning that Alcatel was leveraging its software copyright to obtain patent-like control over its unpatented microprocessor cards.[105]

Following these three leading cases, other circuit courts indicated acceptance of copyright misuse, and the doctrine appears to have become firmly ensconced in the law of copyright. The discussion about copyright misuse now focuses primarily on its proper application and extent, rather than on its existence as a defense to infringement. However, the precise contours of the doctrine are still not clear, and the exact border between copyright misuse and antitrust remains particularly vague and controversial.[106] Much as in the previous development of patent misuse, antitrust violations may constitute copyright misuse, but misuse is not limited to the antitrust context.[107] While the lack of congruity between copyright misuse and antitrust has frustrated certain commentators,[108] these features of the doctrine in fact position it to play the gap-filling, coordinating, and policy preserving roles discussed in the previous section.

7 Anticrcumvention Misuse

The history of misuse reviewed here underscores the past importance of misuse in curtailing the forays of intellectual property rights holders beyond the uses intended for their rights. The consistent theme of misuse cases is limitation on private extensions of IP rights contrary to public policy—not simply to ward off antitrust violations, or even prevent economically anticompetitive activity. Some types of intellectual property overreaching will surely create anticompetitive effects, and curtailing overreaching may cure some anticompetitive behaviors. To the extent that misuse doctrine does so, it may complement or overlap antitrust law, and might sometimes be taken for antitrust law. But the most recent copyright misuse cases make clear, as do the early patent cases, that the unique role of misuse is to police the constitutional and statutory limitations on exclusive rights. The extension of misuse to "paracopyright" is appropriate to deter such inappropriate leveraging of these new rights.

The question to be addressed regarding anticircumvention misuse will not simply be whether the particular application of the right is anticompetitive, although some—and perhaps all—anticompetitive uses will surely be misuses. Rather, a finding of misuse would be proper where the ends to which the anticircumvention right is employed exceed the reasonable grant of the right. For this standard to have any definite structure, it will be necessary to determine what the bounds of the anticircumvention grant might be. That in turn entails some determination of Congressional intention in creating the right in the first place.

Fortunately, the legislative history behind the DMCA anticircumvention provisions is fairly clear regarding Congressional intent, and indeed somewhat repetitious on this point. The legislative record of the DMCA is replete with references to the need for anticircumvention measures to prevent "piracy."[109] While this pejorative is used rather loosely by the content industries who backed the DMCA—to include even legitimate though unauthorized copying of a work[110]—the legislative record reflects a use of the term by both legislators and industry proponents of anticircumvention provisions most often to refer to large-scale, unauthorized commercial reproduction and distribution of copyrighted works in competition with the legitimate copyright owner. Some uses of the term also refer to widespread but private unauthorized reproduction and distribution that might not be explicitly commercial, but which would adversely affect the commercial market for authorized copies of the work. Content industries lobbying Congress for circumvention protection repeatedly emphasized the potential of digital piracy as the threat motivating their appeal to the legislature. Legislators who sponsored or favored the DMCA repeatedly cited the threat of piracy as the motivation for their support.

Thus, the legislative record suggests that the anticircumvention right was intended by Congress as a shield rather than a sword, intended as a means to prevent wholesale misappropriation of copyrighted content, rather than as a means to extend content owners' exclusivity to adjacent technologies. Yet the cases brought by rights holders thus far, as described above, have been characterized by a decided lack of anything resembling "piracy" or unauthorized copying. Rather, the common employment of the anticircumvention right in the cases described above has been, certainly in effect if not in intent, directed to suppressing competitive products. Such use of the anticircumvention right is strikingly similar to use of copyright in the copyright misuse cases such as *Alcatel*.

Established doctrines of patent or copyright misuse will be inadequate to limit overreaching digital content licenses, in part because such licenses need not be based on a release from copyright or patent liability, but on release from circumvention liability. One can perhaps already see the precursor to such a case in the current dispute over access controls in a case such as *Reimerdes*. Note that in that case, the CSS was used to limit the playback of technologically controlled works to certain approved DVD players. This is in essence a form of tying, using the technological control system, and the legal sanctions backing it, to force customers who purchase DVDs to use only particular DVD players. It is only a very short step from these facts to an anticompetitive situation in which customers are being required to purchase particular players to play discs that, absent the access control, could be played using the DVD machines manufactured by other, unapproved manufacturers. Even if such tying did not rise to the level of an antitrust violation, there would arguably be a misuse of the anticircumvention right: leveraging the right granted in the technological control system to control unpatented players, much as the copyright in *Alcatel* was leveraged to control telephone switching hardware.

The holdings of copyright misuse cases such as *Lasercomb* similarly suggest that employment of the anticircumvention right to facilitate contractual overreaching should be equally suspect. Where a particular use would be permissible under copyright law, content owners may attempt to exclude the use as a condition of access. For example, reverse engineering or fair use of content, while permissible under the copyright act would not constitute an acceptable reason to circumvent technical controls, so the content owner's permission must be sought to obtain access to the work. But the owner may as a condition of access contractually require a user to agree not to engage in fair use or reverse engineering. Judging by current trends in copyright "shrinkwrap" licensing, other access licensing terms might include noncompetition provisions or restraints on resale, lease, or lending of the controlled copy. Such access conditioned upon noncompetition agreements, or upon agreement not to use competing products, should give rise to a misuse defense as it has in copyright misuse.

The *Reimerdes* case foreshadows an additional consideration, not previously encountered in the history of misuse, that anticircumvention rights may be leveraged via terms built into the technological control system itself. The geographic limitation of the DVD playback system was not written out in a license, but was instead built into a technical standard—indeed, were such limitations incorporated into a written license, they might well create antitrust issues.[111] Neither should it matter whether such contractual prohibitions are instantiated as text accompanying the authorized access to content, or whether they are built into the technology that controls the use of the content once it has been accessed. Such overreaching is equivalent, whether as text backed by law or as software backed by law, and either is a candidate for limitation via misuse.

As anticircumvention misuse doctrine develops, there will be questions regarding the contours of anticircumvention misuse and antitrust law. The trafficking provisions of the DMCA, in particular, confer not simply an extra measure of content control, but broad power to dictate technological format and interoperability. The very concept of a secure or managed digital environment contemplates that only approved or certified interoperation will occur: unapproved devices or applications potentially compromise the security of the system. This type of interoperability control is a version of the technical standards problem that has been identified in other commentaries on computer technology, and a full analysis of the issue lies beyond the scope of this chapter.[112] But examples such as the

DVD CSS or RealAudio "secret handshake" serve to illustrate the general point that control of a dominant technical protection standard can allow a firm or group of firms to dictate who will be allowed to offer competing or complementary products in a given market.

The anticircumvention statute serves to extend and cement such technical control. Exceptions to the DMCA anticircumvention provisions may permit competitors to reverse engineer secure platforms and products to produce interoperable software, but the reverse engineering exception does not extend to reverse engineering hardware or data, nor does it allow reverse engineering for any purpose other than software interoperability.[113] Neither does it allow consumers to use the interoperable product produced, particularly if the product does not itself comply with the security standard. This places firms under additional pressure to adopt the dominant standard, as they may face DMCA liability if they promulgate competing or complementary products under a different standard.

The ability to police and control technical standards for content management may concentrate enormous market power in the hands of a small number of companies. In the hands of already dominant software and hardware manufacturers—here both Microsoft and Intel come quickly to mind[114]—such control over technical compatibility could be used curtail innovation and deter the development of alternative technologies. To the extent that manufacturers with a large installed user base can use anticircumvention rights to prevent reverse engineering, and maintain licensing exclusivity to their products, the DMCA represents an enormous windfall advantage in maintaining their current position.[115] Some such activity will fall within the reasonable extent of the anticircumvention right given by Congress—the grant of the right may have been unwise, but that is the legislature's failure, not the recipient firms'.

But protection of copyrighted content, not maintenance of market dominance, was the stated legislative intent behind granting the anticircumvention right. This suggests that sooner or later, "paracopyright" leveraging will cross the line into antitrust violation. Stated differently, there will inevitably arise some friction between exercise of anticircumvention rights legitimately granted by Congress and the restrictions on anticompetitive behavior imposed by antitrust law. As in it has in patent and copyright law, misuse doctrine may serve to coordinate anticircumvention with antitrust, helping to reconcile the requirements of the two bodies of law. Misuse may also serve a common-law "gap-filling" function to cover anticompetitive behaviors that may not rise to level of a formal antitrust violation. These functions may be particularly necessary for anticircumvention, given that the right is new, the statute creating it is more than a little vague, and Congress seems to have given little thought to the competitive effects of the statute.

Additionally, misuse may assist in coordinating between the anticircumvention right and other more established forms of intellectual property, particularly patent and copyright, when these rights are all extant in the same technical system. This type of coordination may be critical with regard to the technical protections themselves, rather than with regard to the content protected. Rights management systems are themselves likely to be covered by various forms of intellectual property: copyright for rights management software; patent for rights management software and hardware, as well as for processes related to the system.[116] Portions of the technology may be covered by combinations of patent or copyright or the anticircumvention right, or by no proprietary right at all. Use of these rights, or of combinations of these rights, to improperly deter analysis or duplication of the unprotected elements of the technology should be subject to an appropriate combination of patent, copyright, and anticircumvention misuse claims.

9 Conclusion

The anticircumvention provisions of the DMCA have been justly criticized for the chilling effect they will have on use of technologically protected content. But the more serious effects of these provisions may be their impact on technologies adjacent to digitized content. Control of such technologies seems to be a consequence of the statute that was unanticipated by Congress, but which is quickly emerging as the statute's primary employment. Given the introduction of this sweeping new right, whose parameters are unclear and whose relationship to adjacent law is ill defined, misuse is needed to serve a limiting and coordinating role. Recent anticircumvention cases suggest that the need for a doctrine of anticircumvention misuse is real, and the time for its application will arrive shortly, if indeed it has not already arrived.

Notes

1. *See* H.R. Rep. No 105–551, pt. 2 at 24 (1998).

2. *See* PAUL GOLDSTEIN, COPYRIGHT'S HIGHWAY: THE LAW AND LORE OF COPYRIGHT FROM GUTENBERG TO THE CELESTIAL JUKEBOX (1994).

3. *See* William M. Landes & Richard E. Posner, *An Economic Analysis of Copyright Law*, 18 J. LEGAL STUD. 325 (1985).

4. *See id.*

5. U.S. Const. Art. I, sec. 8. cl. 8

6. Sony Corp. of America v. Universal City Studios, 464 U.S. 417 (1984); United States v. Paramount Pictures, 334 U.S. 131, 158 (1941).

7. 17 U.S.C. § 102 (b).

8. 17 U.S.C. § 110 (5).

9. 17 U.S.C. § 110 (6).

10. 17 U.S.C. § 109.

11. 17 U.S.C. § 107.

12. *See* L. Ray Patterson, *Free Speech, Copyright, and Fair Use*, 40 VAND. L. REV. 1, 3 (1987); Harry N. Rosenfield, *The Constitutional Dimensions of Fair Use in Copyright Law*, 50 NOTRE DAME L. REV. 790, 796–98 (1975); Lionel S. Sobel, *Copyright and the First Amendment: A Gathering Storm?*, 19 COPYRIGHT L. SYMP. (ASCAP) 43, 66 (1971); *see also* Robert Denicola, *Copyright and Free Speech: Constitutional Limitations on the Protection of Expression*, 67 CAL. L. REV. 283, 289–99 (1979); Melville B. Nimmer, *Does Copyright Abridge the First Amendment Guarantees of Free Speech and Press?*, 17 UCLA L. REV. 1180, 1190 (1970).

13. *See* Campbell v. Acuff-Rose, 510 U.S. 569 (1994); *see also* Richard A. Posner, *When is Parody Fair Use?*, 21 J. LEGAL STUD. 67 (1992); Robert P. Merges, *Are You Making Fun of Me?: Notes on Market Failure and the Parody Defense in Copyright*, 21 AIPLA Q.J. 305 (1993).

14. *See, e.g.*, Sony Computer Entm't, Inc. v. Connectix Corp., 203 F.3d 596, 602–08 (9th Cir. 2000); DSC Communictions Corp. v. DGI Techs., 81 F.3d 597, 601 (5th Cir. 1996); Bateman v. Mnemonics, 79 F.3d 1532, 1539 fn. 18 (11th Cir. 1996); Sega Enters., Ltd. v. Accolade Inc., 977 F.2d 1510, 1520 (9th Cir. 1992); Atari Games Corp. v. Nintendo of Am. Inc., 977 F.2d 832, 843–44 (Fed. Cir. 1992).

15. David Nimmer *et al.*, *The Metamorphosis of Contract Into Expand*, 53 PITT. L. REV. 543 (1992).

16. *See* Charles R. McManis, *The Privatization (or "Shrink-Wrapping") of American Copyright Law*, 87 CAL. L. REV. 173 (1999); Michael J. Madison, *Legal-Ware: Contract and Copyright in the Digital Age*, 67 FORDHAM L. REV. 1025 (1998); Mark A. Lemley, *Intellectual Property and Shrinkwrap Licenses*, 68 SO. CAL. L. REV. 1249 (1995).

17. *See* David A. Rice, *Public Goods, Private Contract, and Public Policy: Federal Preemption of Software License Prohibitions Against Reverse Engineering*, 53 Pitt. L. Rev. 543 (1992); Mark Lemley, *Beyond Preemption: The Law and Policy of Intellectual Property Licensing*, 87 Cal. L. Rev. 111 (1999); J. H. Reichman & Jonathan A. Franklin, *Privately Legislated Intellectual Property Rights: Reconciling Freedom of Contract with Public Good Uses of Information*, 147 U. Penn. L. Rev. 876 (1999).

18. *See* Computer Sci & Telecomm. Bd., Nat'l Research Council, the Digital Dilemma: Intellectual Property in the Information Age (2000).

19. *See* Charles Clark, *The Answer to the Machine is the Machine in* The Future of Copyright in a Digital Environment 149 (P. Bernt Hugenholz, ed., 1996); Kenneth W. Dam, *Self-Help in the Digital Jungle*, 28 J. Legal Stud. 393 (1999).

20. *See generally* The Digital Dilemma *supra* note 18 at 153–76; Mark Stefik, *Shifting the Possible: How Digital Property Rights Challenge Us to Rethink Digital Publishing*, 12 Berkeley Tech. L. J. 138 (1997).

21. *See* Eric Schlachter, *The Intellectual Property Reniassance in Cyberspace: Why Copyright Law Could be Unimportant on the Internet*, 12 Berkeley Tech. L. J. 15, 38–45.

22. *See* Tom W. Bell, *Fair Use v. Fared Use: The Impact of Automated Rights Management in Copyright's Fair Use Doctrine*, 76 N.C. L. Rev. 557 (1998).

23. *See id.*; Dean S. Marks & Bruce H. Turnbull, *Technical Protection Measures: The Intersection of Technology, Law, and Commercial Licenses*, 22 Eur. Intell. Prop. Rep. 198 (2000)

24. *See* Julie E. Cohen, *Copyright and the Jurisprudence of Self-Help*, 13 Berkeley Tech. L.J. 1089 (1998).

25. See Lawrence Lessig, Code and Other Laws of Cyberspace (1999); Joel Reidenberg, *Lex Informatica*, 76 Tex. L. Rev. 553 (1998).

26. *See* James Raymond Davis, *On Self-Enforcing Contracts, the Right to Hack, and Willfully Ignorant Agents*, 13 Berkeley Tech. L.J. 1145, 1147 (1998).

27. *See* Schlachter, *supra* note 21, *see also* Glynn S. Lunney, Jr., *Death of Copyright: Digital Technology, Private Copying, and the Digital Millennium Copyright Act*, 87 Va. L. Rev. 813 (2001); Pamela Samuelson, *Will the Copyright Office Be Obsolete in the 21st Century?*, 13 Cardoza Arts & Ent. L.J. 55 (1994).

28. Digital Millennium Copyright Act, Pub. L. No. 105–304, Title I, 112 Stat. 2860 (1998) codified at 17 U.S.C. 1201(a)-(b) (1999).

29. *See* Pamela Samuelson, *The U.S. Digital Agenda at WIPO*, 37 Va. J. Int'l L. 369, (1997).

30. World Intellectual Property Organization: Copyright Treaty, December 20, 1996, art. 11, 36 I.L.M. 65

31. *See* Sony v. Universal Studios, 464 U.S. 417 (1984).

32. *See* Samuelson, *supra* note 29 at 413–15.

33. 17 U.S.C. § 1201 (a)(1)(A).

34. *See* Pamela Samuelson, *Intellectual Property and the Digital Economy: Why the Anti-Circumvention Regulations Need to Be Revised*, 14 BERKELEY TECH. L.J. 539 (1999).

35. Universal City Studio v. Corley, 273 F.3d 429 (2d Cir. 2001).

36. *See* Alfred P. Yen, *When Authors Won't Sell: Parody, Fair Use, and Efficiency in Copyright Law*, 62 U. COLO. L. REV. 79 (1991).

37. *See, e.g.*, Campbell v. Acuff-Rose Music, Inc., 510 U.S. 569 (1994).

38. *See* Jane Ginsburg, *Copyright Legislation for the "Digital Millennium,"* 23 COLUM-VLA J.L & ARTS 137, 140–43 (1999).

39. 17 U.S.C. § 1201 (c)(1).

40. 17 U.S.C. § 110 (1).

41. 17 U.S.C. § 110 (3).

42. 17 U.S.C. § 117.

43. *See* Cohen, *supra* note 24; *see also* Yochai Benkler, *Free as the Air to Common Use: First Amendment Constraints on the Enclosure of the Public Domain*, 74 N.Y.U. L. REV. 354 (1999).

44. Feist Publications Inc. v. Rural Telephone Service Co., Inc., 499 U.S. 340 (1991).

45. *See, e.g.*, § 1201 (a)(1)(A).

46. *See, e.g.*, Sony Computer Entm't, Inc. v. Connectix Corp., 203 F.3d 596, 602–08 (9th Cir. 2000); DSC Communictions Corp. v. DGI Techs., 81 F.3d 597, 601 (5th Cir. 1996); Bateman v. Mnemonics, 79 F.3d 1532, 1539 fn. 18 (11th Cir. 1996); Sega Enters., Ltd. v. Accolade Inc., 977 F.2d 1510, 1520 (9th Cir. 1992); Atari Games Corp. v. Nintendo of Am. Inc., 977 F.2d 832, 843–44 (Fed. Cir. 1992).

47. *See* Jane Ginsburg, *Copyright and Control*, 101 COLUMB. L. REV. (2001); Jane Ginsburg, *From Having Copies to Experiencing Works in* U.S. INTELLECTUAL PROPERTY: LAW & POLICY (Hugh Hansen, ed., 2000).

48. No. C99–2070P (W.D. Wash. Jan. 18, 2000), 2000 U.S. Dist. LEXIS 1889.

49. 2000 U.S. Dist LEXIS 1889 at *6.

50. *Id.*

51. *Id.* at *11.

52. *Id.* at *19–20.

53. 82 F. Supp. 211 (S.D.N.Y. 2000) *aff'd sub. nom.* Universal City Studio v. Corley, 273 F.3d 429 (2d Cir. 2001).

54. *Id.* at 214.

55. 111 F. Supp.2d 294, 308 (S.D.N.Y. 2000).

56. *Id.*

57. *Id.* at 311.

58. *Id.*

59. 82 F.Supp.2d 211.

60. 87 F.Supp.2d 976 (1999).

61. *Id.* at 981.

62. *Id.* at 981–82.

63. *Id.* at 987.

64. Richard Calkins, *Patent Law: The Impact of the 1988 Patent Misuse Reform Act and the Noerr-Pennington Doctrine on Misuse Defenses and Antitrust Counterclaims*, 38 DRAKE L. REV 175, 187 (1988–89).

65. *See* Robert P. Merges, *Reflections on Current Legislation Affecting Patent Misuse*, 70 J. PAT. & TRADEMARK OFF. SOC'Y 793, 796 (1988).

66. B. Braun Med. Inc. v. Abbott Labs., 124 F.3d 1419, 1427 (Fed. Cir. 19970).

67. *See* United States Gypsum Co. v. National Gypsum Co., 352 U.S. 457, 465 (1957); White Cap Co. v. Owens-Illinois Glass Co., 352 F.2d 694, 698 (6th Cir.) *cert denied* 346 U.S. 876 (1953); *In re* Yarn Processing Patent Validity Litig., 472 F.Supp. 180, 190–91 (S.D. Fla. 1979); *see generally* 6 D. CHISUM, PATENTS § 19.04[4].

68. *See generally* 6 CHISUM *supra* note 67 § 19.04.

69. *See generally* Kenneth J. Burchfiel, *Patent Misuse and Antitrust Reform: "Blessed Be The Tie,"* 4 HARV. J.L. & TECH. 1 (1991).

70. Morton Salt Co. v. G.S. Suppinger Co., 314 U.S. 488 (1942); *see also* Carbice Corporation v. American Patents Development Corporation, 283 U.S. 27 (1931).

71. *Id.* at 493.

72. 379 U.S. 29 (1964).

73. *Id.*

74. 395 U.S. 653 (1969).

75. *See* Kewanee Oil Co. v. Bicron Corp., 416 U.S. 470 (1974); Aronson v. Quick Point Pencil Co., 440 U.S. 257 (1979); Bonito Boats, Inc. v. Thunder Craft Boats, Inc., 489 U.S. 141 (1989).

76. *See* Tom Arnold & Louis Riley, *Contributory Infringement and Patent Misuse: The Enactment of § 271 and Its Subsequent Amendments*, 76 J. PAT. & TRADEMARK OFF. SOC'Y 357, 365 (1994).

77. *See, e.g.*, C.R. Bard Inc. v. M3 Sys., 157 F.3d 1340, 1373 (Fed. Cir. 1998).

78. Pub. L. No. 100–703, § 201, 102 Stat 4676 (1988).

79. 35 U.S.C. § 271(d).

80. *Id.*

81. *Id.*; *see also* Virginia Panel Corp. v. MAC Panel Co., 133 F.3d 860 (Fed. Cir. 1997).

82. *See* John Barton, *Patents and Antitrust: A Rethinking in Light of Patent Breadth and Sequential Innovation*, 65 ANTITRUST L.J. 449, 449 (1997).

83. 28 U.S.C § 1295.

84. *See* Rochelle Cooper Dreyfuss, *The Federal Circuit: A Case Study in Specialized Courts*, 64 N.Y.U. L. Rev. 1 (1989); Donald R. Dunner, *The United States Court of Appeals for the Federal Circuit: Its First Three Years*, 13 AIPLA Q. J. 185 (1985).

85. 35 U.S.C. § 271 (d).

86. *See* Mallinckrodt, Inc. v. Medipart, Inc., 976 F.2d 700 (Fed. Cir. 1992); Windsurfing Int'l Inc. v. AMF Inc., 782 F.2d 995 (Fed. Cir. 1986).

87. *See* 976 F.2d 700.

88. *See, e.g.*, Janice M. Mueller, *Patent Misuse Through the Capture of Industry Standards*, 623 (2002).

89. *See* Paul Goldstein, *Infringement of Copyright in Computer Programs*, 47 U. Pitt. L. Rev. 1119 (1986).

90. *See* Brett Frischmann and Dan Moylan, *The Evolving Common Law Doctrine of Copyright Misuse: A Unified Theory and its Application to Software*, 15 Berkeley Tech. L.J. 865, 881 (2000).

91. M. Witmark & Sons v. Jensen, 80 F.Supp. 843 (D. Minn. 1948), *appeal dismissed*, M. Witmark & Sons v. Berger Amusement Co., 177 F.2d 515 (8th Cir. 1949).

92. *See* Harms, Inc. v. Sansom House Enterprises, Inc., 162 F.Supp. 129, 135 (E.D. Ps. 1958), *aff'd*, Leo Feist, Inc. v. Lew Tendler Tavern, Inc., 267 F.2d 494 (3d Cir. 1959; *see also* Orth-O-Vision, Inc. v. Home Box Office, 474 F.Supp. 672, 686 (S.D.N.Y. 1979); Foreign Car Parts, Inc. v. Auto World, Inc., 366 F.Supp. 977, 979 (M.D. Pa. 1973).

93. 91 F.2d 970 (9th Cir. 1990).

94. *Id.* at 971

95. *Id.* at 973.

96. *Id.* at 978–79.

97. *Id.* at 973, 978.

98. *Id.* at 978.

99. *Id.* at 979.

100. 121 F.3d 516 (9th Cir. 1997).

101. *Id.* at 517.

102. *Id.* at 520–21.

103. 166 F3d 772 (5th Cir. 1999).

104. *Id.* at 775.

105. *Id.* at 778.

106. *See* Xavier Fellmeth, *Copyright Misuseand the Limits of the Intellectual Property Monopoly*, 6 J. Intell. Prop. L. 1 (1998); Troy Paredes, *Copyright Misuse and Tying: Will Courts Stop Misusing Misuse?*, 9 High Tech. L.J. 271

(1994); Ramsey Hannah, Note, *Misusing Antitrust: The Search for Functional Copyright Misuse Standards*, 46 STAN. L. REV. 401 (1994).

107. *See Lasercomb*, 911 F.2d at 978.

108. *See, e.g.*, Saturday Evening Post Co. v. Rumbleseat Press, Inc. 816 F.2d 1191, 1200 (7th Cir. 1987); *see also* Scheiber v. Dolby Laboratories, 293 F.3d 1014, 1017 (7th Cir. 2002); USM Corp. v. SPS Techs., Inc., 694 F.2d 505, 510 (7th Cir. 1982).

109. *See, e.g., NII Copyright Protection Act of 1995: Hearings on H.R. 2441 Before the Subcommittee on Courts and Intellectual Property of the House Committee on the Judiciary*, 104th Cong. 2d Sess. 22 (1996) (Statement of Jack Valenti, President and CEO, Motion Picture Association of America, Inc.); *WIPO Treaties Implementation Act; and Online Copyright Liability Limitation Act; Hearing on H.R. 2281 and H.R. 2280 Before the Subcommittee on Courts and Intellectual Property of the House Committee on the Judiciary*, 105th Cong., 1st Sess. 200 (1998) (Statement of Hilary Rosen, President and CEO, Recording Industry Association of America).

110. *See* JESSICA LITMAN, DIGITAL COPYRIGHT 85–86 (2001)

111. *See* IIA P. AREEDA & H. HOVENKAMP, ANTITRUST ¶¶ 556.

112. *See* Pamela Samuelson & Suzanne Scotchmer, *The Law and Economics of Reverse Engineering*, 111 YALE L.J. 1575, 1623–25 (2002); *see generally*, Mark Lemley, *Antitrust and the Internet Standardization Problem*, 28 CONN. L. REV. 1041 (1996), Mark Lemley & David McGowan, *Legal Implications of Network Economic Effects*, 86 CAL L. REV. 479 (1998).

113. 17 U.S.C. § 1201(f)(1),(2).

114. *See* United States v. Microsoft Corp., 253 F.3d 34, 54–58 (D.C. Cir. 2001); *see also* Franklin M. Fisher & Daniel L. Rubinfeld, U.S. v. Microsoft—*An Economic Analysis*, 46 ANTITRUST BULL. 1 (2001).

115. *See* William E. Cohen, *Competition and Foreclosure in the Context of Installed Base and Compatibility Effects*, 64 ANTITRUST L.J. 535 (1996); Samuelson & Scotchmer, *supra* note 111 at 1617.

116. *See* Julie Cohen, *Reverse Engineering and The Rise of Electronic Vigilantism: Intellectual Property Implications of "Lock-Out" Programs*,68 S. CAL. L. REV. 1091 (1995).

III

Regulatory Innovation and Responses to Technological Change

11

Improving Network Reliability—Liability Rules Must Recognize Investor Risk/Reward Strategies*

Barbara A. Cherry

I. Introduction

Network reliability of telecommunications systems is affected not only by technological capabilities and limitations but also by the carriers' economic incentives to invest in the performance of their systems. These economic incentives depend, in part, on the liability rules applied to telecommunications carriers for damages arising from service interruptions and outages. The liability regime for telecommunications carriers is shifting from one based on an absolute limit on liability in tariffs to a form of strict liability under the common law. By mandatorily detariffing interstate, interexchange telecommunications services, the FCC has contributed to the acceleration of this process but without adequate consideration of the likely impacts on the achievability of other public policy goals for the telecommunications industry, such as universal service, broadband deployment or homeland security.

The federal government's inattention to the changing—potentially catastrophic levels of—liabilities of telecommunications carriers under deregulatory policies stands in start contrast to its treatment of other common carriers and public utilities. Yet, recent events also reveal the need for federal government attention to the liability rules facing the telecommunications industry. For example, the terrorist attacks on the World Trade Center on September 11, 2001, highlight some of the vulnerabilities of the U.S. transportation and communications infrastructures that can be exploited to adversely affect national security and the economy. Future Congressional intervention regarding telecommunications carriers' liability may

*The opinions expressed in this paper are those of the author, and do not necessarily represent the views of the FCC or the federal government.

be necessary not only because telecommunications networks constitute a critical infrastructure, but also because network security is dependent upon the financial viability of the carriers. The recent economic downturn in the telecommunication sector also underscores the need for prompt and comprehensive evaluation of the factors that may threaten the financial viability of telecommunications carriers.

This chapter is organized as follows. Section II describes the changing liability regime for telecommunications carriers that is being accelerated by the federal detariffing process. Section III discusses the impact on telecommunications carriers' economic incentives and why federal government action is necessary to prevent the shift in liability rules from undermining other public policy objectives. Section IV contrasts the continuity in and attention to liability regimes for transportation carriers under deregulation with the shifting liability regime for telecommunications carriers and the paucity of government attention to its effects on the public interest. Section V suggests the types of liability rules that may be required for telecommunications carriers given the benefit of experience with other common carrier and public utility regimes. The chapter ends with a statement of conclusions.

II. The Shifting Liability Regime for Telecommunications Carriers

The manner in which the liability regime for telecommunications carriers has developed differently from that of other common carriers in the United States is discussed in detail by the author in Cherry (1999).[1] Some critical historical developments are described in this section, so that the ramifications of detariffing and the need for corrective policy action can be better understood.

For over one hundred years, telecommunications carriers (telephone companies) have limited their liability for damages from service interruption or outages to a pro rata credit of the customer's service charge. Under a *tariffing* regime, the traditional justifications for upholding such limited liability provisions have been both legally and factually flawed, creating a liability regime different from all other common carriers, including telegraph companies. As a result, proper legal analyses based on fundamental tort and contract law principles applicable to common carriers have never been conducted for telecommunications carriers. For this reason, as this section explains, the effects of *detariffing* on the telecommunications carrier-customer relationship have not been properly understood by policy makers. Whereas detariffing did not fundamentally alter the liability regimes of the airline, railroad and surface transportation carriers, it *does* alter the regime for telecommunications carriers. Yet, ironically, only the liability rules for transportation carriers received careful consideration during passage of federal deregulatory laws, as discussed in section IV.

Common Law Liability Rules for Common Carriers

The duties of common carriers are based on tort, not contract, law. They originated from the duties placed on "public callings" in a feudal economy during the Middle Ages in England.[2] The duties of public callings were to serve without discrimination at reasonable rates and with adequate care. Public callings were also strictly liable for damages for which their conduct was the proximate cause, with exceptions only for acts of God and of enemies of the king. Common carriers were held strictly liable for reasons of public policy: "[E]lse these carriers might have an opportunity of undoing all persons who had any dealings with them, by commingling with thieves &c. and yet doing it in such a clandestine manner, as would not be possible to be discovered." *Coggs v. Bernard*, 2 Ld. Raym. 909, 918 (1703).

The common law of common carriers in the United States further developed based on these English common law principles. In the nineteenth century, with the rise of contract law, common carriers attempted to limit their tort liability in contracts with customers. Important U.S. Supreme Court cases decided the validity of such contracts in the context of railroads. In *Railroad Co. v. Lockwood*, 84 U.S. 357 (1873), the Court held that a railroad company could not *exempt* itself from liability for negligence. However, in *Hart v. Pennsylvania Railroad Co.*, 112 U.S. 331 (1884), the Court held that a common carrier could *limit* its common liability upon an agreed valuation by the parties. Shortly thereafter, Congress passed the Interstate Commerce Act (ICA) of 1887, creating the Interstate Commerce Commission (ICC) with regulatory powers over railroads. Under the ICA, railroads were to file tariffs with the ICC that contained the rates and regulations governing the conditions of service.

After the ICA was amended in 1910 to extend its jurisdiction over telegraph and telephone companies, the U.S. Supreme Court further elaborated upon the validity of valuation agreements to limit common carrier liability. In *Union Pacific R.R. Co. v. Burke*, 255 U.S. 317 (1921), the

Court held that under the common law—not altered by the filed rate doctrine applicable to tariffs under the ICA—a valuation agreement was valid only where the customer is given a *choice of rates under which full liability is an option* and the rate is tied to the level of liability accepted by the carrier. This common law rule is often referred to as the *released value doctrine*. The holding in *Union Pacific R.R. Co. v. Burke* has been followed and endorsed by the Court in subsequent transportation common carrier cases.[3]

In the same year, the U.S. Supreme Court considered the validity of a limitation on liability clause for a telegraph company in *Western Union v. Esteve Bros.*, 256 U.S. 566 (1921). In this case the Court upheld the limited liability clause in a tariff for an unrepeated message because the plaintiff could have chosen a repeated message at a higher rate and higher level of *limited* liability. However, the Court specifically declined to determine whether the rule in *Union Pacific R.R. Co. v. Burke* should be applied to telegraph companies to invalidate the higher limit on liability for a repeated message—as there was no rate available for the sender to insure full value—because the facts of the case did not require it to rule on the matter. The Court has never addressed this question in a subsequent case, whether for telegraph or telephone companies.

Yet, a month before *Union v. Esteve Bros.* was decided but after it was argued before the Court, the ICC found the limitations on liability provisions contained in the tariffs of telegraph companies to be unreasonable in *Limitations on Liability in Transmitting Telegrams*, 61 I.C.C. 541 (1921), which is also known as the *Second Unrepeated Message Case*.[4] More specifically, the ICC found the maximum limit of $50 for both repeated and unrepeated messages to be unreasonable, and raised the limit to $500 for unrepeated messages and $5000 for repeated messages. More importantly, the ICC required all telegraph companies to offer the sender the option to value a message in excess of $5000—with no cap on the value that could be declared—at a cost of one-tenth of 1 per cent of the stated value. Even though such valued messages would likely be infrequently used, the ICC found it important that telegraph carriers be potentially liable for the full amount consistent with common law liability.

Thus, the ICC found that, even under tariffs, telegraph companies' limitations on liability provisions were valid only if the customer had the option to declare the full value for which the carrier would be liable. This

ICC determination is consistent with the common law rule set forth in *Union Pacific R.R. Co. v. Burke*, and has never been overruled. Therefore, even though the U.S. Supreme Court has never addressed the applicability of the released value doctrine to telegraph or telephone companies, the ICC had done so for telegraph companies.

Misapplication of *Western Union v. Esteve Bros.* and the Filed Rate Doctrine

As discussed more fully in Cherry (1999), limitations of liability provisions for telephone companies originated in the subscriber contracts of the Bell companies in the nineteenth century prior to both the expiration of the telephone patents and regulation by any federal or state agencies.[5] These provisions were not created by statute or by any state or federal commission, but were permitted to persist after telephone companies were brought under the jurisdiction of the ICC (later the FCC) and the state regulatory commissions.

No federal commission orders addressing the validity of limitations on liability provisions in telephone company tariffs were issued until the late 1970s and early 1980s, after enactment of the Communications Act of 1934 and the creation of the FCC. The first order was in *American Satellite Corp. v. Southwestern Bell Telephone Co.*, 64 F.C.C. 2d 503 (1977), in which the FCC simply stated that Southwestern Bell Telephone Co. may limit its liability in the absence of willful misconduct. It cited *Western Union v. Esteve Bros.* in support, but provided no discussion. Then *In the Matter of AT&T*, 76 F.C.C. 2d 195 (1980), the FCC upheld AT&T's limited liability provision, again citing *Western Union v. Esteve Bros.* in a footnote but providing no discussion. Fifteen years later, in *Richman Bros. Records, Inc. v. U.S. Sprint Communications Corp.*, 10 F.C.C Rcd 13, 639 (1995) the Common Carrier Bureau of the FCC upheld the limited liability provision of Sprint. It based its decision on the filed rate doctrine—not mentioned in the previous two orders—citing *Western Union v. Esteve Bros.* as authority.

This reliance on *Western Union v. Esteve Bros.*, and later the filed rate doctrine, for upholding telephone companies' limited liability provisions is fundamentally flawed. First, the holding in *Western Union v. Esteve Bros.* was based on a set of facts that did not exist in any of the telephone cases. In *Western Union v. Esteve Bros.* the limited liability provision for the unrepeated message was held valid because there was another rate with a higher limit on liability (the repeated message) that the customer could have chosen. In the telephone company cases, no alternative rate and level of liability was offered to the customers.

Second, in *Western Union v. Esteve Bros.* the U.S. Supreme Court declined to address whether the released value doctrine applied to telegraph companies because the facts of the case did not require it to. However, in the telephone cases, the provision of only one rate and one level of liability meant that the case could not be properly decided without considering the applicability of the released value doctrine established in *Union Pacific R.R. Co. v. Burke*. In the FCC telephone cases, the *Union Pacific R.R. Co. v. Burke* case is not even mentioned.

Third, the ICC had already decided in the *Second Repeated Message Case* that telegraph companies must offer the customer the option to declare full value for which the company would be liable—in essence applying the released value doctrine. In each of the above FCC telephone cases, the FCC has never even acknowledged the existence of the *Second Unrepeated Message Case* of its predecessor, the ICC.

Fourth, the filed rate doctrine means that a given rate, and associated terms and conditions, must be applied nondiscriminatorily *among customers*. However, the released value doctrine determines whether the limitation on liability provision associated with a given rate is valid *for any customer*. Thus, the filed rate doctrine only means that a valid tariff rate or provision must be applied nondiscriminatorily *across customers;* however, it does not address the underlying validity of a limitation on liability provision under the common law.

For all of these reasons, the FCC's historical reliance on *Western Union v. Esteve Bros.* and the filed rate doctrine to uphold the validity of telephone company's limitations on liability provisions in tariffs has been fundamentally flawed. Unfortunately, proper evaluation as to the applicability of the released value doctrine to telephone companies has never been addressed by any agency or court, whether state or federal.[6]

Continuing Flawed Reasoning Under Detariffing

As just described, improper reliance on *Western Union v. Esteve Bros.* and the filed rate doctrine has been the basis for the FCC's *upholding* the validity of limitations on liability tariff provisions for telephone companies—now called telecommunications carriers. The FCC has recently used the same flawed reasoning to achieve the opposite result—for *rejecting* telecommunications carriers' limitations on liability provisions—and without fully understanding the consequences.

In its *Domestic Detariffing Order,*[7] the FCC concluded that tariffs were not necessary to ensure that the rates, practices and classifications of nondominant interexchange carriers for interstate domestic interexchange services are just, reasonable and nondiscriminatory. In particular, the FCC eliminated tariffs in order to prevent carriers from invoking the filed rate doctrine to unilaterally change terms and conditions in their contractual relationships with customers in a manner not available in most commercial relationships. Believing the filed rate doctrine to be the basis for upholding the validity of telecommunications carrier's limitations on liability provisions, the FCC did make a cursory reference to the effect on the liability of telecommunications carriers. Dedicating only one sentence to the issue, the FCC stated that "[i]n addition, complete detariffing would further the public interest by preventing carriers from unilaterally limiting their liability for damages."[8] In support of this statement, the FCC merely cited cases in a footnote, including *Western Union v. Esteve Bros.* and *Richman Bros. Records, Inc. v. U.S. Sprint Communications.*

In this one statement and footnote, the FCC committed several errors. It repeated its flawed reliance on *Western Union v. Esteve Bros.* and the filed rate doctrine, but now to justify the opposite result—to reject limitations on liability provisions. Furthermore, the FCC made no effort to determine whether elimination of limitations of liability provisions was in fact in the public interest. Having traditionally relied on limitations on liability provisions, at least in part, to keep rates low,[9] the FCC gave no explanation as to why the elimination of such provisions and the likely impact on rates were now in the public interest. Finally, the FCC failed to consider the effect that elimination of limitations on liability provisions might have on the carriers' financial abilities to fulfill other regulatory obligations, such as universal service.

As a result, the FCC has *never* conducted a proper analysis of the limitations on liability provisions for telecommunications carriers. It did not conduct the analysis required under the released value doctrine while these provisions were filed as part of the telecommunications carriers' tariffs, and effectively created an absolute limit on liability of a pro rata credit of the customer service charge. Nor did it consider the impact of eliminating this absolute limit on liability—by eliminating the filed rate doctrine, which, although improperly so, had been used to uphold telecommunications carriers' limitations on liability provisions—and replacing it with litigation under the common law and state consumer protection laws.

Rediscovering Unconscionability Since Detariffing

Under the common law of contracts, the general principle of unconscionability can be used to invalidate a provision of a contract.[10] Unconscionability is usually found when "gross inequality of bargaining power, together with terms unreasonably favorable to the stronger party, . . . show that the weaker party had no meaningful choice, no real alternative, or did not in fact assent or appear to assent to the unfair terms."[11] The application of the principle of unconscionability in the specific context of common carriers is the underlying basis of the U.S. Supreme Court cases invalidating exculpatory clauses and establishing the released value doctrine for transportation common carriers discussed earlier in this section.[12]

With the elimination of the filed rate doctrine under detariffing, the provisions of telecommunications carriers' contracts for interstate services are now being challenged under common law rules of contract law.[13] Significantly, in *Ting v. AT&T*, 182 F. Supp. 902 (N.D. Cal. 2002), *affirmed,* 319 F.3d 1126 (9th Cir. 2003), a class action suit was brought, claiming that AT&T's limited liability provisions in its Consumer Services Agreement violated the state Consumer Legal Remedies Act. Finding the filed rate doctrine to be inapplicable in a detariffed regime, the court held that the provisions were unconscionable as a contract of adhesion with harsh and one-sided provisions.

Challenges of unconscionability are also being made with regard to contract provisions for intrastate services. For example, in *Association of Communication Enterprises v. Ameritech Illinois*, 2002 WL 226889 (I.C.C. 2002), the Illinois Commerce Commission found the termination charges in Ameritech's service agreement with resellers under its ValueLink Services Tariff to be unlawful and unconscionable under Illinois law.

These challenges of unconscionability are in addition to state court and commission cases—predating federal detariffing—that foreshadowed the need to revisit the limitations on liability provisions for the provision of intrastate telecommunications.[14] However, these cases deferred reevaluation of the propriety of limitations on liability provisions to some unspecified future time.[15] More recently, the Indiana Utility Regulatory Commission has opened an industry-wide public utility investigation to consider the propriety of Ameritech Indiana's limited liability provisions for service interruption tariffs.[16]

Thus, federal detariffing is triggering litigation as to the validity of interstate contract provisions—including limitations on liability—that had previously been blocked by invocation of the filed rate doctrine. State commissions are also starting to consider the validity of contract or tariff provisions—actions that had previously been hinted at but not pursued. The continuing legitimacy of the traditional liability regime of an absolute limit on liability (based on a pro rata credit of the customer's service charge) is now under serious threat. According to the common law principles of common carrier liability discussed throughout section II, telecommunications carriers could now be facing a regime based on strict liability with contracts to limit liability arguably subject to the released value doctrine.

III. Economic Ramifications of a Shifting Liability Regime

Change in liability rules governing telecommunications carriers will change carriers' economic incentives through both direct and indirect effects.[17] The direct effects arise from the increase in carriers' costs, such as the need to invest in greater precautions and to pay higher damage claims. They also include the transaction costs of parties negotiating differing levels of liability, as required by applicability of the released value doctrine. Indirect effects result from the interaction of the new liability rules with other regulatory rules and public policy goals.

Rate Levels and Universal Service

The increased costs will need, at least in part, to be passed on to customers in terms of higher rates. This is because there is a limit to how much of the increased costs can be absorbed by shareholders in order to continue to attract capital and investment, particularly given the recent downturn in the telecommunications sector.

Higher rate levels may pose some public policy concerns. First is the longstanding obligation of telecommunications carriers to provide services at just and reasonable rates. Depending upon how the increased rates are allocated among services and customer classes, there may be some customers for whom basic telecommunications services are no longer available at reasonable rates. Second, additional funding may be required for the universal service support mechanisms established under section 254 of the Telecommunications Act of 1996, to meet the rising rate levels for existing beneficiaries. Merely delegating a greater financial burden to the universal service programs may not be a sustainable proposition, however, given concerns as to the viability of maintaining the existing funding levels based on contributions from carriers.[18]

There may be additional ramifications for universal service goals with the maintenance of asymmetric regulation among carriers when the liability for damages is no longer capped at the traditionally low levels. For example, the costs of fulfilling carrier of last resort obligations imposed on incumbent local exchange companies under state law—which are usually also the eligible carriers with an obligation to serve the entire serving area[19]—may significantly rise in light of the increased risk of providing service. To address those costs, universal service funding again may need to rise, or carriers will look for opportunities to avoid the obligation to serve unprofitable areas. In either event, maintenance of continuous, high quality service at reasonable rates may be difficult to maintain on a ubiquitous basis.

Unique Risks of High Risk, High Reliability Organizations

Telecommunications carriers also face certain risks of catastrophic potential that may dramatically alter the provision of telecommunications services with the shifting liability regime discussed in section II. This is because carriers bear the characteristics of high risk, high reliability organizations. Special attention to liability rules is essential for such organizations, because "no matter how effective conventional safety devices are, there is a form of accident that is inevitable."[20] For these organizations, reliability is a more pressing issue than efficiency.

High risk, high reliability organizations (HRHR's) are those bearing certain characteristics for which the occurrence of accidents of catastrophic potential is inevitable, or "normal."[21] The inevitability of accidents is due to system characteristics of the organization, interactive complexity and tight coupling. Interactive complexity refers to the inability to anticipate and address in advance the circumstances giving rise to reliability problems. Tight coupling refers to the inability to intervene to constrain the operation of the system when problems do arise. Finally, what makes HRHR's unique is the catastrophic consequences of their failures. This is why such organizations are referred to as high reliability organizations.

Examples of HRHR's are aircraft carriers, nuclear power plants, nuclear weapons systems, space missions, chemical plants, and military early warning systems. Telecommunications systems can also be considered HRHR's, particularly in light of recent technological advances in communications and information systems. The telecommunications network is based on sophisticated computer systems, complex software programs, and tight coupling between steps in the transmission of information both within and among providers. The consequences of outages can be extensive and potentially catastrophic, as demonstrated by AT&T's nationwide outage on Martin Luther King's birthday in 1990[22] and Ameritech's loss of an entire 5ESS switch in the Hinsdale fire in 1988.[23]

The importance of considering telecommunications carriers as HRHR's is understanding the need for an aggregate cap on liability in order to ensure the provision of service. For example, in 1957, Congress limited the tort liability of nuclear power plant operators to $560 million for a given event in the Price-Anderson Act.[24] Legislation was deemed necessary because full liability coverage was unavailable from the private insurance industry and members of the power industry were unwilling to self-insure.[25] Recently, Congress passed bills to further update the Price-Anderson Act consistent with its current structure.[26]

By detariffing, the FCC may have unwittingly eliminated a cap on liability that has played an important role in the widespread deployment and continuing operation of the telecommunications infrastructure. Therefore, it is certainly relevant to consider whether a cap on the aggregate liability for a given event is necessary to ensure continued operation and investment in the telecommunications industry—particularly at the level desired to meet existing universal service goals, much less the deployment of a broadband infrastructure.

Terrorism and Economic Downturn of the Telecommunications Sector

The severity of the economic effects of a shifting liability regime for telecommunications carriers described above will differ as circumstances change. In this regard, recent events heighten the need for prompt public policy attention for the liability rules affecting the telecommunications industry.

The terrorist acts of September 11, 2001, not only inflicted enormous damage and loss of life but also demonstrated widespread economic effects of catastrophic events. In order to prevent potentially devastating financial impacts on air carriers, property owners of the World Trade Center and New York City, Congress passed the Air Transportation Safety and System Stabilization Act,[27] and the Aviation and Transportation and Security Act.[28] The terrorist acts also seriously damaged some telecommunications facilities, but no legislation was passed by Congress to address the potential liability or financial losses suffered by telecommunications carriers. Most likely this is because telecommunications carriers sought no legislative relief, but for reasons related to prevailing circumstances that may not persist over time. First, the most severely impacted carrier, Verizon, still files tariffs for its intrastate services and would expect continued protection of the tariff limited liability provisions. Second, any affected carrier would likely claim, if necessary, the common law defense of an act of the public enemy to exempt it from liability. Third, the events of September 11, 2001, predated the decision in *Ting v. AT&T,* discussed in section II, which was the first case after federal detariffing to invalidate as unconscionable AT&T's limitation on liability provisions in its Consumer Services Agreement.[29] The circumstances underlying each of these reasons may not prevail in future instances of third party actions that seriously damage telecommunications facilities.

Although Congressional intervention may not have been necessary to address financial consequences for telecommunications carriers with respect to the specific terrorist acts of September 11, 2001, it may be needed to better ensure national security in the future. This is because telecommunications is a critical infrastructure for national security.[30] Furthermore, the telecommunications network security is dependent upon the financial viability of the carriers. The importance of this interdependence is evident given the recent downturn in the telecommunications sector, the rash of competitive local exchange company bankruptcies, questionable accounting practices, and the bankruptcy of Worldcom.[31]

For all these reasons, the federal government should be comprehensive in its evaluation of factors that may threaten the financial viability of telecommunications carriers. The shifting liability regime for telecommunications carriers—particularly when coupled with HRHR characteristics of telecommunications systems, the heightened risk of terrorism, and the compromised financial state of the telecommunications sector—is a significant change in circumstances warranting specific policy attention.

IV. Continuity in the Liability Regimes for Transportation Carriers

This section discusses how the shifting liability regime for telecommunications carriers described in section II is contrary to the experience of transportation carriers.[32] First, common law principles of common carrier liability were retained under the federal statutory frameworks for railroads, motor carriers and air carriers, for which primary jurisdiction was given to regulatory agencies to determine the reasonableness of the rates and terms of service. Second, upon deregulation of these carriers, the same common law principles remained in effect with (in most respects) enforcement of common carrier liability simply returned in the first instance to the courts. Importantly, the released value doctrine clearly applied both before and after deregulation. Furthermore, even after deregulation, the federal government has continued to monitor and modify, where necessary, the liability regimes of transportation carriers.

Liability Regime Under Federal Agency Regulation

The liability of railroads has been governed by the Carmack Amendment (modifying section 20(11) of the ICA) in its varying forms since 1906.[33] The Carmack Amendment was initially enacted in 1906 to provide uniformity that had been lacking in litigation, as railroads sought ways to avoid their common law liability, and did not increase a carrier's common law liability nor prohibit carriers from limiting liability to an agreed value. Under the Carmack Amendment framework, released rates were permitted only if authorized by the ICC.[34]

Similar regulatory frameworks were later established for other surface transportation and air carriers. ICC regulation, the Carmack Amendment (as amended over time), and the released rate doctrine were made applicable to motor carriers in the Motor Carrier Act of 1935 and subsequently extended to surface freight forwarders.[35] In addition, the Civil Aeronautics Act of 1938 and the Federal Aviation Act of 1958 established a regulatory framework for air carriers similar to that for railroads. Although these acts did not contain language equivalent to the Carmack Amendment, they did preserve the federal common law relating to agreed valuation tariffs under the released value doctrine established in *Union Pacific R.R. Co. v. Burke*.[36]

Liability Regime After "Deregulation"

Several deregulatory acts affecting surface transportation carriers occurred during the 1980s and 1990s. They include the Motor Carrier Act of 1980, the Staggers Rail Act of 1980, the Surface Freight Forwarder Deregulation Act of 1986, the Trucking Industry Regulatory Reform Act of 1994, and the ICC Termination Act of 1995.

As a result of these acts, railroads are no longer required to file tariffs, and there is no government agency oversight of the reasonableness of rates. However, the liability of railroads is still governed by a modified form of the Carmack Amendment, codified at 49 U.S.C.A. sec. 11706.[37] Railroads are still liable for actual loss or injury to property, are prohibited from exempting themselves from liability, and are permitted to limit their liability upon written declaration of the shipper. Furthermore, section 10502(e) prohibits the Surface Transportation Board[38] from exempting rail carriers from its obligations under section 11706. The primary change under deregulation is that enforcement of section 11706 is solely by the courts.

Pursuant to the ICC Termination Act of 1995, the liability of certain motor carriers and freight forwarders is still governed by the Carmack Amendment, now codified as 49 U.S.C.A. 14706.[39] Section 14706 still imposes liability for actual loss or injury to property. Released value contracts are permitted, and their reasonableness is determined by the courts under federal common law. Under section 14101(b), shippers are also allowed to contract out of Carmack Amendment liability; but this provision is primarily used by large volume shippers.

Some motor carriers' activities are still subject to stricter regulation. For movement of household goods, carriers must file tariffs subject to invalidation by the Surface Transportation Board (STB),[40] and must petition the STB to modify or eliminate released value rates. Congress continued federal oversight for the carriage of household goods because it believed that consumers continued to need protective regulation.[41]

Airline deregulation occurred with the passage of two federal acts, the Air Cargo Deregulation Act of 1977 and the Airline Deregulation Act of 1978.[42] The air carriers were exempt from the tariff filing provisions and the Civil Aeronautics Board was abolished. However, importantly, "deregulation . . . had no impact upon the applicability of the federal common law's released value doctrine. . . . It merely did away with the applicability of the doctrine of primary jurisdiction. After deregulation, the validity of the agreed value provision . . . became a purely judicial question for determination by application of the federal common law patterned upon the policy of the Carmack amendment."[43]

Thus, both before and after the deregulatory acts, surface transportation and air carriers' ability to limit their liability has been constrained by the released value doctrine. Only the venue changed in which consumers could first challenge the reasonableness of limitations of liability—from the relevant agency under a tariff structure to the courts with the elimination of tariffs. This is dramatically different from the effect of detariffing on the liability regime for telecommunications carriers discussed in section II.

Cargo Liability Study of 1998

Ironically, notwithstanding the greater continuity in liability regimes for transportation carriers than telecommunications carriers, it is the U.S. Department of Transportation (DOT)—not the FCC—that has studied the impact of deregulatory legislation on liability rules. The DOT has conducted two studies of cargo liability, one before and one during the period of deregulation. The first was conducted in 1975 to assist the U.S. Government in formulating an intermodal liability regime. The second is the *Cargo Liability Study of 1998*,[44] which was mandated by Congress in the ICC Termination Act of 1995 to determine whether any modifications or reforms should be made to the loss and damage provisions of section 14706 (the Carmack Amendment applicable to motor carriers).[45] The *Cargo Liability Study of 1998* also explored liability issues on an intermodal basis, both domestically and internationally.

In the *Cargo Liability Study of 1998*, most of the recommendations for change relate to intermodal liability issues. With regard to (intramodal) liability under section 14706, DOT generally concludes "that the current liability system functions reasonably well and that it requires only modest adjustment to assure fairness to all parties."[46] As for allowing liability to vary by released rates or contracts, DOT does examine differences in application among motor carriers that has led to lack of uniformity. However, DOT recommends that the current system of released rates continue until shippers and carriers come closer to agreement on an alternative liability regime, such as a set liability limit with the option that the shipper could declare a higher value and purchase excess valuation coverage.[47]

In contrast with DOT, the FCC has never conducted a study of the carrier liability regime for telecommunications carriers, whether before or after detariffing. As discussed in section II, the FCC has issued orders related to limitations on liability using cursory and flawed legal analysis. A proper evaluation of what liability rules are in the public interest is long overdue—after all, the ICC concluded such an investigation for telegraph companies in 1921.[48]

V. Toward a Revised Liability Regime for Telecommunications Carriers

To evaluate what liability rules are in the public interest requires an understanding of how liability rules affect economic incentives so that an appropriate balance can be reached between the interests of carriers and customers. Historically, in evaluating the appropriateness of liability regimes for transportation carriers, the federal government has valued uniformity of federal guidelines "to create a measure of predictability for interstate carriers in the exposure to damages they face."[49] Furthermore, at times it may be necessary to protect carriers "against catastrophic, crippling liability by establishing monetary caps on awards and restricting the types of claims that may be brought against carriers, while accommodating the interests of injured [customers] by creating a presumption of liability against the carrier."[50]

In balancing the interests of carriers and shippers, the liability regimes among the transportation carriers are remarkably similar in terms of the higher level principles. Table 11.1 provides a summary of these principles after deregulation.[51]

As shown in table 11.1, all carriers are required to provide reasonable, nondiscriminatory service to the public.[52] In addition, all regimes impose strict liability—a presumption of fault—on the carrier, subject to some

Table 11.1
Cargo Liability Regimes of Transportation Carriers

Regime	Motor carrier	Railroad	Domestic air	International air
Relevant law	ICC Termination Act (1995)	Stagger's Rail Act (1980)	Airline Deregulation Act of 1978	Warsaw Convention, amended by Montreal Protocol (1999)
Carrier obligations	Reasonable, non-discriminatory service to the public	Reasonable, non-discriminatory service to the public	Reasonable, non-discriminatory service to the public	Reasonable, non-discriminatory service to the public
Basis of liability	Strict liability	Strict liability	"Strict accountability" (presumed fault of carrier)	Presumed fault of carrier, but court may exonerate wholly or partly on finding claimant negligent
Burden of proof	On carrier as to defenses	On carrier as to defenses	On carrier as to defenses	On carrier to prove that it took all necessary measures or that it was impossible to take such measures
Limitations on liability	Actual loss, except released rates and contract rates	Actual loss, except released rates and contract rates	None by law; released value doctrine	$23/kg, but shipper may declare up to full value
Carrier defenses	Common law defenses; bill of lading exceptions	Common law defenses; bill of lading exceptions	Common law defenses; defenses in air waybill	Negligent piloting or navigation

defenses for which the carrier bears the burden of proof. For domestic transportation, limitations on liability are permitted under a released value system for motor carriers, railroads and air carriers; and contracting out of the Carmack Amendment requirements is permitted for railroads and motor carriers under certain circumstances. For international cargo, a presumptive level of liability is set at $23/kg, but the shipper has the option to declare an amount up to the full value. Intermodal differences arise, not from differences in the underlying legal principles, but in the detail of application. For example, the terms in bills of lading and the dollar levels in released rates or contracts vary greatly among the types of carriers, and sometimes among types of cargo within a given mode of transportation.[53]

Cherry (1999) develops preliminary conclusions for designing liability rules for telecommunications carriers. They are based on an economic analysis of liability rules, for which there is a well-developed literature,[54] in the context of telecommunications industry characteristics.[55] The analysis also incorporates the effects that liability rules will likely have on the achievement of other regulatory rules and public policy objectives, such as those discussed in section III. These preliminary conclusions are outlined here to illustrate their similarities to rules developed for transportation carriers.

First, to reduce transaction costs in achieving an optimal level of care by the carrier and to provide uniformity of results across jurisdictions and similarly situated customers, telecommunications carriers should be subject to a standard of strict liability. In recognition of events beyond the carrier's control, the common law defenses of acts of God, the public enemy, and public authority should remain available. Liability based on strict liability with certain common law defenses is consistent with the legal principles applied to transportation carriers.

Second, to address problems of moral hazard, the strict liability standard should be modified to induce appropriate precautions by customers. To induce due care by the customer ex ante to an event of service interruption or outage, the carrier should be able to invoke the defense of contributory negligence. This is consistent with the common law defense of acts considered the fault of the shipper (sender).[56] In addition, to induce due care ex post to the event, customers should bear a duty to mitigate damages. The availability of competitive alternatives would be a relevant factor in both situations, and may differ among services and/or classes of customers.

Third, the level of liability to which a telecommunications carrier should be strictly liable should be presumptively limited, not unlimited. In this regard, a presumption of limited liability would exclude carrier liability for extraordinary damages unless specifically negotiated among the parties.[57] The economic efficiency properties of imposing a presumptively limited or unlimited liability rule depend upon the ability of the parties to contract around the initial liability rule and the transaction costs of doing so. Bebchuk and Shavell (1991) show that a presumption of limited liability is generally superior, and unambiguously so when the number of customers likely to suffer extraordinary damages is a minority (so that transaction costs are economized by requiring such customers to communicate their special circumstances to the carrier).[58] A limited liability rule would also be more consistent with achieving other existing regulatory obligations and objectives, such as those discussed in section III. A presumption of limited liability is consistent with the statutory requirement that liability of railroads and motor carriers is limited to actual loss, which has been interpreted to exclude recovery for punitive damages, emotional distress, or state-imposed obligations.[59]

Fourth, circumstances will vary among services and classes of customers as to whether bargaining should be permitted around a limited liability rule. If bargaining is permitted, then the liability rule would be presumptively limited; if no bargaining is permitted, then liability is absolutely limited to some prespecified amount. For larger volume customers, there is more equality in bargaining power with carriers, and carriers can differentiate levels of reliability among such customers. Therefore, bargaining should be permitted on some form of released value or contract basis. This is consistent with the use of released value and contract rates by transportation carriers. However, for the mass market—generally residential and small business customers—such circumstances do not prevail. The vast inequality of bargaining power among the customers and carriers, and the difficulties in providing differing levels of reliability to customers (particularly those in the same geographic area), indicate the need to consider an absolute limit on liability for the mass market.[60] This is consistent with the absolute limit on liability that has historically prevailed under tariffs for telecommunications carriers.

However, at what prespecified amount liability should be limited requires empirical evaluation of social costs and benefits in light of other policy objectives discussed in section III. Furthermore, such evaluation may indicate the need for liability limits to vary among serving areas to reflect significant differences in costs of providing service.

Fifth, the characteristics of telecommunications carriers as high risk, high reliability organizations indicate the need to consider an absolute limit on liability for the aggregate level of damages for which a carrier would be liable to all customers arising from a single event of service interruption or outage. A limit on aggregate liability would protect carriers from the uncertainty of catastrophic levels of liability and thereby be more consistent with achievement of ubiquitous availability of service at reasonable rates. Such a rule would be consistent with Congressional treatment of nuclear power plants, air carriers, and the terrorist acts on September 11, 2001. However, setting an appropriate threshold for such an aggregate limit on liability would again require empirical evaluation of factors specific to the telecommunications sector.

Finally, because of asymmetric regulatory burdens borne among carriers, liability rules may need to differ among types of carriers. For example, carriers of last resort obligations and eligible carriers—usually incumbent local exchange carriers—may require different liability rules to enable such carriers to financially meet their obligations. The potential for differing liability rules among carriers has already been foreseen by the Michigan Public Service Commission.[61] It is also consistent with the varying application of the Carmack Amendment among motor carriers.[62]

These preliminary conclusions provide a useful framework for the changes in liability rules that should be contemplated for telecommunications carriers. The anomalous inattention to liability rules of telecommunications carriers in a deregulatory environment needs to be rectified through an appropriate balancing of interests between carriers and customers. This will require a systematic evaluation of the economic effects of liability rules in the context of specific telecommunications industry characteristics and public policy objectives. In this regard, the treatment of transportation carriers, as well as appreciation of the historical role that limited liability has played in the telecommunications industry, are important sources of insights.

VI. Conclusion

For over a hundred years, telecommunications carriers have limited their liability for damages to customers arising from service interruptions and outages to a pro rata credit of the customer's service charge. This absolute limit on liability is contrary to the common law of common carrier liability and the statutory liability regimes imposed on other common carriers, such as transportation carriers and even telegraph companies. Furthermore, its persistence for telecommunications carriers is due to the misapplication of prior case law, particularly *Western Union v. Esteve Bros.*, and the filed rate doctrine. With federal detariffing of interstate interexchange telecommunications services, the validity of traditional limitations on liability provisions is being successfully challenged on the grounds of unconscionability. As a result, the liability regime of telecommunications is now shifting from an absolute limit on liability toward one of strict liability under the common law of common carriers.

Given the economic effects of this shift in liability rules, the continuing achievement of other public policy goals may be jeopardized in the long run. These goals include reasonable rate levels, the ubiquitous availability of service, and the sustainability of federal universal service support mechanisms. The characteristics of telecommunications carriers as high risk, high reliability organizations, coupled with vulnerabilities to acts of terrorism and the recent economic downturn in the telecommunications sector, further exacerbate the risks of failing to fulfill these public policy objectives.

This shift in the liability regime for telecommunications carriers is occurring, however, without any explicit consideration of whether or not it is in the public interest. This inattention is contrary to that which has been given to other carriers. Since 1906, Congress has been actively involved in codifying liability rules for transportation carriers. Furthermore, both during periods of regulation and deregulation, the U.S. Department of Transportation has conducted cargo liability studies to determine whether liability rules properly balance the interests of carrier and shippers. An investigation regarding appropriate liability rules for telegraph companies was also done by the Interstate Commerce Commission in 1921, and the liability rules established then still exist today.

Yet, no such analysis has ever been conducted by the FCC nor considered by Congress for telecommunications carriers.

Cherry (1999) provides preliminary conclusions for designing liability rules for telecommunications carriers. These rules are similar to those already in place for transportation carriers. For example, carrier liability should be based on a standard of strict liability, presumptively limited to nonextraordinary damages, with contributory negligence and common law defenses. Released value rates or contract rates should be available for certain services and customer classes, such as large volume users with telecommunications intensive businesses. However, an absolute limit on liability may need to be retained for other services and customer classes, such as residential and small business customers, where carriers retain much greater bargaining power but limited ability to differentiate standards of service reliability. Furthermore, given the characteristics of telecommunications carriers as high risk, high reliability organizations, an absolute limit on the aggregate level of damages for a given event may be in the public interest. Finally, owing to asymmetric regulatory burdens, liability rules may also need to vary among differently situated carriers. In any event, empirical assessment of the factors specific to the telecommunications sector are required, and long overdue, for determining appropriate liability rules for telecommunications carriers.

Notes

1. B. Cherry, *The Crisis in Telecommunications Carrier Liability* (Norwell, MA: Kluwer Academic Publishers) (1999).

2. For a discussion of the development of public callings, see M. Glaeser, *Public Utilities in American Capitalism* (New York: The Macmillan Co.) (1957); and E. Adler, "Business Jurisprudence," 28 Harv. L. Rev. 135.

3. See, e.g., *New York, New Haven & Hartford Railroad Co. v. Nothnagle*, 346 U.S. 128 (1953); *Bisso v. Inland Waterways Corp.*, 349 U.S. 85 (1955).

4. The *Second Unrepeated Message* Case was a general investigation of the limitation on liability provisions for telegraph companies, triggered by an earlier complaint case, *Unrepeated Message Case*, 44 I.C.C. 670 (1917).

5. See note 1, *supra,* at pp. 20–23.

6. See Cherry (1999), note 1, *supra,* for an in-depth discussion of why the common law liability of common carriers has remained unaddressed at both the federal and state levels.

7. *Second Report and Order*, Policy and Rules Concerning the Interstate, Interexchange Marketplace, 11 F.C.C. Rcd 20730 (1996) (hereinafter the "*Domestic*

Detariffing Order"), upheld in *MCI Worldcom, Inc. v. FCC*, 209 F.2d 760 (D.C. Cir. 2000). Detariffing finally took effect on August 1, 2001. *Common Carrier Bureau Extends Transition Period for Detariffing Consumer Domestic Long Distance Services*, 16 F.C.C. Rcd. 2906 (2001).

8. 11 FC.C. Rcd at par. 55.

9. See *In the Matter of AT&T*, 76 F.C.C. 2d 195 (1980).

10. *Restatement (Second) of Contracts*, section 208.

11. Id. at Comment d.

12. See, e.g., *Restatement (Second) of Contracts*, section 195 (1979 Main Vol.) Comment a; *Restatement (Third) of Torts: Apportionment of Liability*, Topic 1, Basis Rules of Comparative Responsibility (1999 Main Vol.); *Henningsen v. Bloomfield Motors, Inc.*, 161 A. 2d 69 (N.J. Sup. Ct. 1960).

13. *Frontline Communications International, Inc. v. Sprint Communications Co.*, 178 F. Supp. 2d 432 (S.D.N.Y. 2001).

14. See Cherry (1999), note 1, *supra,* at 30–39, 43–44.

15. *In the Matter of the Investigation into Limitation of Liability Clauses Contained in Utility Tariffs*, Case no. 85–1406-AU-COI, Public Utilities Commission of Ohio (Oct. 6, 1987), entry on rehearing (Nov. 24, 1987); *In the Matter of the Proceedings, on the Commission's Own Motion, to Consider Revisions to the Commission's Rule Limiting the Liability of Telephone Companies*, Case no. U–8035, Michigan Public Service Commission (1986). See also *In Re Illinois Bell Switching Station Litigation*, 641 N.E.2d 440 (Ill. Sup. Ct. 1994) (in concurring opinion, Chief Justice questions enforceability of limited liability provisions as regulation changes with technological advances and increased competition).

16. *In Re Propriety of Tariff Provisions*, Cause no. 42002, 2201 WL 797976 (Ind. U.R.C. 2001).

17. An analysis of how to design economically efficient liability rules for telecommunications carriers is provided in Cherry (1999), note 1, *supra,* at pp. 67–117.

18. See *Further Notice of Proposed Rulemaking and Report and Order*, Federal-State Joint Board on Universal Service, FCC 02–43 (released Feb. 26, 2002).

19. Only eligible carriers, defined under section 214(e) of the Telecommunications Act of 1996, may receive funding from federal universal service support mechanisms under section 254.

20. C. Perrow, *Normal Accidents* (New York: Basic Books, 1984) p. 3.

21. See C. Perrow, *Normal Accidents* (New York: Basic Books, 1984).

22. See L. Lee, *The Day the Phones Stopped* (New York: Donald I. Fine) (1991).

23. See *In Re Illinois Bell Switching Station Litigation*, 641 N.E.2d 440 (Ill. Sup. Ct. 1994).

24. The current liability provisions of the Price-Anderson Act are found at 42 U.S.C.A. sec. 2210 (1994 & Supp. 2002).

25. See D. R. Anderson, "The Price-Anderson Act: Its Importance in the Development of Nuclear Power," 30 *Chartered Property Casualty Underwriters Annals*, 253–264 (1977).

26. The House has passed H.R. 2983, 107th Congress, 1st Session; the Senate has passed H.R. 4.

27. 49 U.S.C.A 40101 *et seq* (1997 & Supp. 2002), P.L. 107–42, enacted September 22, 2001.

28. 49 U.S.C.A. sec. 114 *et seq* (1997 & Supp. 2002), P.L. 107–71, enacted November 19, 2001.

29. The events of September 11, 2001 also predated the decision in *Frontline Communications International, Inc. v. Sprint Communications Co.*, see note 13, *supra.*

30. *National Strategy for Homeland Security*, Office of Homeland Security, The White House, Washington, D.C. (July 2002), pp. 29–35.

31. See "Industry and Govt. Eye Role of Finances in Network Security," *Communications Daily* 22, no. 148 (August 1, 2002): 1

32. This section discusses the liability regime of transportation carriers with regard to loss or injury to property, not for injury or death of passengers. For a discussion of cargo liability among all transportation carriers, see *Cargo Liability Study*, U.S. Department of Commerce (August 1998) (hereinafter "*Cargo Liability Study of 1998*").

33. See R. Sigmon, *Miller's Law of Freight Loss and Damage Claims* (4th edition) (Dubuque, IA: Wm. C. Brown Co., 1976) pp. 9, 15–16.

34. *Union Pacific R.R. Co. v. Burke* was decided in 1921, after the passage of the initial Carmack Amendment and subsequent amendments in 1915 and 1916.

35. Sigmon, note 33, *supra,* at p. 10.

36. *First Pennsylvania Bank v. Eastern Airlines*, 731 F.2d 1113, 1117 & fn 4 (3rd Cir. 1984).

37. (1997 & Supp. 2002). See *Cargo Liability Study of 1998*, note 32, *supra,* at p. 10.

38. The ICC Termination Act of 1995 created the Surface Transportation Board within the U.S. Department of Transportation.

39. (1997 & Supp. 2002). Some activities of motor carriers are exempt from section 14706 liability under section 13506.

40. 49 U.S.C.A. section 13702 (1997).

41. *Cargo Liability Study of 1998*, note 32, *supra,* at p. 6.

42. For the effect of deregulation on airline liability, see B. Leto, "Administrative Law—Airline Deregulation—Deregulatory Scheme had no Effect on the Applicable Substantive Law to Determine Liability of Shipper for Lost Shipment of Goods," 30 Vill. L. Rev. 890 (1985).

43. *First Pennsylvania Bank v. Eastern Airlines*, 731 F.2d 1113, 1122 (1984). Accord, *Mauseth v. American Airlines*, 24 Fed. Appx. 809 (9th Cir. 2001).

44. See note 32, *supra.*

45. 49 U.S.C.A. sec. 14706(g) (1997).

46. *Cargo Liability Study of 1998*, note 32, *supra,* at p. 57.

47. Ibid. at pp. 58–59.

48. See note 4 and accompanying text, *supra.*

49. *Gordon v. United Van Lines*, 130 F.3d 282,287 (7th Cir. 1997) (punitive and emotional distress damages are nonrecoverable against motor carriers). See also *American Airlines v. Wolens*, 513 U.S. 219 (1995) (airline deregulation bars actions for violation of state-imposed obligations).

50. *King v. American Airlines, Inc.*, 284 F.3d 352, 357 (2d Cir. 2002) (refers to the liability system for international air transportation under the Warsaw Convention). *Accord El Al Israel Airlines, Ltd. v. Tsui Yuan Tseng*, 525 U.S. 155 (1999).

51. Table 11.1 is based on a table provided in the *Cargo Liability Study of 1998,* note 32, *supra,* at p. 15. For common law defenses, see Sigmon, note 33, *supra,* at pp. 76–133.

52. These are the tort obligations of common carriers. See note 2, *supra.*

53. See *Cargo Liability Study of 1998*, note 32, *supra,* at pp. 39–54.

54. Modern literature of the economic analysis of liability rules began with an article written by R. H. Coase, "The Problem of Social Cost," 3 Journal of Law & Economics, no. 3, 1–44.

55. See Cherry (1999), note 1, *supra,* at pp. 83–95.

56. See Sigmon, note 33, *supra,* at pp. 76–133.

57. A presumption of limited liability, excluding liability for extraordinary damages, is the common law rule developed under contract law in *Hadley v. Baxendale*, 9 Ex. 341 (1854).

58. L. Bebchuk & S. Shavell, "Information and the Scope of Liability for Breach of Contract: The Rule of *Hadley v. Baxendale*," 9 J. of Law, Economics & Organization 98 (1991).

59. See note 49, *supra.*

60. This is consistent with placing stricter regulation on motor carriers' transport of household goods due to unequal bargaining power. See notes 40 and 41, *supra.*

61. See note 15, *supra.*

62. See notes 39–41 and accompanying text, *supra,* and *Cargo Liability Study of 1998*, note 32, *supra.*

12

Emergent Locations: Implementing Wireless 9-1-1 in Texas, Virginia, and Ontario

David J. Phillips, Priscilla M. Regan, and Colin J. Bennett

Introduction

In both the United States and Canada, regulatory efforts are underway to incorporate wireless systems into the existing emergency response framework. From a technological perspective this requires that wireless providers find a way to communicate to public safety personnel the geographic location of a cell phone with some degree of accuracy. Wireless companies, equipment manufacturers, emergency response providers, individual citizens and civil liberties groups have all contributed to the policy discussions about how this can best be implemented. In both countries, implementation of what is termed "wireless enhanced 9-1-1" is very much in progress.[1]

The particular method of implementation has specific privacy implications, including whether or not the disclosure of locational information is limited to 9-1-1 calls, whether anonymous 9-1-1 calls are possible, and whether locational data, once disclosed to 9-1-1, are subject to secondary use. A host of inter-related concerns inform implementation decisions, and so structure the privacy environment. For example, differences in the source of funding for 9-1-1 infrastructure changes may affect the degree to which wireless service providers depend on secondary uses of locational information. Infrastructure decisions, such as how a caller's location is determined, where databases are stored, what entities maintain them, and who has access to them, are influenced by many factors, including regional geography, population density, and technical compatibility with legacy wireless protocols and legacy 9-1-1 systems. The specific regional competitive environment may promote conflicts or alliances among wireless carriers, incumbent local exchange carriers (ILECs), and Public Safety Answering Points (PSAPs).

The U.S. and Canada have, so far, taken divergent regulatory approaches to the enhanced 9-1-1 problem. A mandate by the Federal Communications Commissioner (FCC) contrasts with a less interventionist approach by the Canadian Radio-Television and Communications Commission (CRTC). In the United States, the FCC took the lead in putting the issue of wireless E9-1-1 on the agenda, initiated policy alternatives, and provided an incentive and a forum for industry and public safety representatives to cooperate. In Canada, federal governmental agencies have so far stepped back to allow the affected parties to cooperate, under the auspices of the Canadian Wireless Telecommunications Association (CWTA), to perform trials and, they hope, to develop a workable solution without regulatory intervention.[2] The question for this chapter is whether these policy differences make any difference to the actual implementation of cellular 9-1-1 systems 'on the ground,' especially as it is related to the implications for personal privacy.

In this chapter, we analyze the implementation of E9-1-1 in three regions—rural Texas, Virginia and Ontario—and outline the roles played by the various public and private sector actors. These cases permit comparison across several axes. Privacy laws, "sunshine" laws, and telecommunication policy mechanisms vary from state to state and between Canada and the U.S. The corporate interests and industry structures of the ILECs and wireless carriers can be compared, as well as corporate linkages among those entities and PSAPs. Data for this analysis includes interviews with representatives from PSAPs, ILECs, and wireless carriers, as well as trade and popular press reports and public records.

Because implementation of wireless E9-1-1 is still in its early stages and because implementation has spawned new players and new relationships among players, we are examining privacy issues at a somewhat awkward but nevertheless important time. Although privacy is not always explicitly on the agenda we will see that the implementation processes and frameworks may affect privacy in some important and enduring ways.

Several factors are common across all three cases. First, they share a legacy technical configuration of emergency call processing. Also, implementation both in the United States and Canada are influenced, to a greater or lesser degree, by the FCC's E9-1-1 mandates. These contexts are discussed below before the examination of individual cases.

The Technical Configuration of the E9-1-1 System

Since the 1980s, a relatively stable configuration has been in place for determining the location of the source of an emergency call, and routing that call to the appropriate response center. However, this configuration was designed only for wired line telephones. At the heart of this system are Automatic Number Identification (ANI) and Automatic Location Identification (ALI). It is therefore sometimes known as the ANI/ALI system.

ANI has been a feature of wired phone systems since the 1980s. In addition to the voice signal, phone calls carry a data signal that usually includes the number of the phone originating the call. This data signal makes features such as caller-ID possible. ALI is essentially a database that links phone numbers to physical addresses. Unlike ANI, it is used only in the E9-1-1 system. Historically, it has been populated from the local phone company's service records.

A simplified version of the routing of a typical 9-1-1 call from a wired line goes something like this: A customer dials "911" from a wired line. The Local Exchange Carrier (LEC) recognizes the call as an emergency call and forwards it to a tandem switch. The switch splits the call into a voice portion and a data portion. The call's ANI is analyzed to determine the Public Service Answering Point (PSAP) serving the region associated with that phone number, and the voice portion of the call is forwarded, via a special purpose selective router, to that PSAP. The ANI is also sent, along a separate path, to the PSAP. The ANI is displayed when the PSAP operator answers the call. Simultaneously, the PSAP sends the ANI along a dedicated circuit to the ALI database, which returns to the PSAP the address associated with the caller's phone number. The PSAP operator is then in possession of a voice connection with the emergency caller, the caller's location, and a call back number in case the voice connection fails. The operator then determines the appropriate response to the call, and dispatches emergency services as necessary.

This degree of functionality, known as "Phase 0 E9-1-1," was developed when wired phone systems operated by near-monopoly LECs were overwhelmingly the norm. The emergence of wireless telephony and telecommunications competition have necessitated several changes to this technical paradigm. With the advent of telecommunications competition, issues of inter-industry competitive strategy intervene in and complicate efforts to establish efficient, equitable and cooperative methods of access to and control of the tandem switches, selective routers, and databases that comprise the 9-1-1 system. Mobile telephony radically complicates the maintenance of the ALI database. Since phone numbers can no longer be statically linked with their location, the ALI database must be reconfigured to permit dynamic updates. Moreover, LEC service records are no longer sufficient to provide the data to populate the ALI database. Other means of determining the location of wireless calls must be found.

"Phase 1 E9-1-1" constitutes the first attempt to update the ANI/ALI system to accommodate wireless telephony. Functionally, to be Phase 1 compliant, the ANI/ALI system must deliver to the PSAP a call back number as well as the location of the cell tower from which a wireless 9-1-1 call originates. In practice, Phase 1 compliance is implemented by dynamically updating the ANI/ALI database with the location of the calling cell tower before the PSAP queries the database. Each cell tower is assigned a unique Emergency Service Routing Key (ESRK), and the ANI/ALI database has records for each ESRK. When a wireless caller dials "911," the call goes first to the nearest cell tower, then to the mobile company's Mobile Telephone Switching Office (MTSO). The MTSO handles all switching for the wireless system. The MTSO recognizes the call as a 9-1-1 call, and queries a database both for the ESRK of the cell site handling the call, and for the PSAP that handles calls for that ESRK. The MTSO splits the call into a voice path and a data path, and sends the voice path and the ESRK, via the selective router, to the PSAP. The MTSO also updates the ESRK's record in the ANI/ALI database with the caller's call back number and the address of the cell site handling the 9-1-1 call. When the voice call and the ESRK arrive at the PSAP, the PSAP queries the ANI/ALI database with the ESRK. The database then returns the call back number and the cell site location to the PSAP.

"Phase 2 E9-1-1" requires that mobile telephony operators deliver the coordinates of the caller to the PSAP. This entails much more radical infrastructure changes than Phase 1 does, including, at the very least, techniques for determining those coordinates. Except in a very few areas, Phase 2 deployment is still in the experimental stage, and no standard configuration has been agreed upon.

Phase 0 ANI/ALI is common across North America. It is the technical legacy upon which all of the cases studied below depend. Phase 1 implementation is fairly standard in the United States, though our cases will show how the contexts of its implementation in Texas and Virginia have had different implications for privacy and surveillance. These same distinctions between Phase I and Phase 2 implementation have been adopted in Canada, though at a different pace and, as we shall see, under different regulatory conditions.

The Regulatory Context of E9-1-1

In the United States, FCC mandates have been the driving force behind E9-1-1 deployment. In 1996 the FCC adopted rules and a framework for including wireless calls in the E9-1-1 system.[3] The FCC has moved into this area because of its responsibility under the Communications Act of 1934 to promote "safety of life and property through the use of wire and radio communication."[4] In its documents and press releases, the FCC defines its role not as one of compelling or directing, but as one of "promoting cooperative efforts," "prompting the accelerated delivery," and "making sure that ongoing processes are in place."[5]

The goal of the FCC was to provide wireless callers with the same level of access to 9-1-1 services that is available to wireline callers. The FCC's 1996 final rules required wireless carriers to provide Phase I service within a year. They also established a four-year schedule for Phase II implementation beginning October 1, 2001, and ending December 31, 2005.[6]

Much of the FCC's time has gone to considering waivers by wireless carriers regarding the Phase I and Phase II deadlines for compliance. The first Phase II deadline, October 1, 2001, witnessed a number of waiver requests by five nationwide wireless carriers—Nextel, Sprint PCS, Verizon, Cingular, and AT&T. On October 19th the FCC also provided guidance to small and mid-size carriers seeking relief from Phase I ALI rules.[7] Without question implementation of the FCC rules has not proceeded smoothly.

The U.S. Congress has supported FCC 9-1-1 mandates. The Wireless Communications and Public Safety Act of 1999 (WCPSA), designates "911" as the universal emergency phone number for wireless and wireline calls and required the FCC, states, and local governments to coordinate wireless E9-1-1. Section 4 of the law gives wireless carriers, users, providers and PSAPs the same liability protection regardless of whether a call originated from a wireless or wireline phone. Section 5 of WCPSA requires that a customer give prior authorization for any disclosure of call location information for a purpose other than the delivery of emergency services.[8]

These FCC and Congressional actions have had a direct bearing on Canadian telecommunication policy, because in many respects Canadian stakeholders have tried to grapple with these similar issues without having to resort to direct intervention from the Canadian Radio and Telecommunications Commission (CRTC). Indeed, direct observation of U.S. events prompted the Canadian PSAPs, wireless industry and the ILECs to seek a more cooperative and consensual approach. Moreover, the development of E9-1-1 policy in Canada cannot, of course, be separated from a relatively dependent economic relationship, shaped in part by U.S.-dominated corporate actors in the telecommunications industry. With this brief exposition of the common technical and political contexts of Canadian and U.S. E9-1-1 systems, we turn to specific cases.

The Implementation of Wireless E9-1-1 in Texas

Three environmental factors have affected the privacy implications of E9-1-1 implementation in Texas. These are the consolidation of rural emergency operations into a state-wide system, a legislative commitment to telecommunications industry competition, and a legislative commitment to data sharing across state operations.

Administration of the E9-1-1 system in Texas is shared among three bodies—home rule cities, Emergency Communication Districts (ECDs), and regional Councils of Government (COGs). Home rule cities and ECDs are those municipalities and regions whose emergency response systems (ERSs) predate state initiatives, and so are grandfathered out of state regulation. The grandfathered systems tend to coincide with the most densely populated regions of the state. The state Legislature established the Commission on State Emergency Communications (CSEC) in 1985 to coordinate ERS in the remaining, predominantly rural, areas of the state. CSEC's purview extends over 2/3 of the Texas land mass, but only 1/3 of its population.[9] This research concerns only those regions under CSEC's jurisdiction—that is, rural Texas.

CSEC controls the state budget for emergency communication systems, and so has enormous influence on the implementation of those systems. Formally, all decisions regarding the implementation of emergency communications fall to the local government at the level of either municipality or county. However, coordination of many state initiatives, including 9-1-1 initiatives, occurs at the level of the regional Council of Government, or COG. COGs are formally voluntary associations of local governments with no regulatory power or other authority. Decisions of COGs are not binding on member governments. However, in order to receive state funding for 9-1-1 programs, local governments must cooperate in a regional plan through their COG. The COGs, in turn, report to CSEC. If the regional plans adhere to standards promulgated by CSEC, they will be eligible for funding.[10]

CSEC is funded by three types of surcharges on wired and wireless phone lines, including 9-1-1 service fees, 9-1-1 equalization surcharges, and wireless communication fees. These fees are set by the state legislature and are applied immediately to the state's general fund. The legislature then appropriates funds to CSEC. Historically, these fees have generated more revenue than is appropriated to CSEC.[11] Also, within each COG, funds collected from urban areas subsidize ERS in rural areas. At least one urban region under CSEC purview, Corpus Christi, has bridled at subsidizing other regions, and has withdrawn from the CSEC program to become a Home Rule City.[12] Home Rule Cities may set their own surcharge rates and spend the revenues as they wish.

By statute, CSEC revenue can only be used for activities associated with the delivery of a 9-1-1 call. These include network, database, mapping, addressing, and PSAP equipment costs. They do not include costs associated with the provision of emergency services.

This structure of 9-1-1 funding and coordination has two corollaries. First, the state E9-1-1 mandate funds general state operations. This is a significant source of revenue in a state that devoutly eschews income taxes. Second, by harnessing the purchasing power of rural PSAPs though CSEC, the Texas legislature has created a significantly powerful market actor in the E9-1-1 communications industry. As will be explored in the following section, this has had important implications for the shape of the locational surveillance industry in general.

In 1997, the Texas legislature, disturbed by inadequacies in the local ILEC's handling of the E9-1-1 system, solicited competitive bids for managing the state wireline and wireless 9-1-1 ALI database. The contract was awarded to SCC, Inc., which has since changed its name to Intrado.

Intrado's core business had been in inter-industry billing management. After the breakup of the AT&T monopoly, number portability, local service unbundling and local service resale had required intricate billing arrangements among competitive local exchange carriers (CLECs), long distance carriers, and wireless operators. Intrado has used this expertise in inter-industry information management in developing and managing a new ANI/ALI database. The Texas contract is for database services only; Intrado does not route or deliver 9-1-1 calls in Texas. Instead, they populate the ANI/ALI database and send routing directions to the selective routers still run, for the most part, by ILECs.

Though the contract is for database services only, it is a significant step toward parlaying Intrado's core database and inter-industry network expertise into more general commercial applications. Specifically, the contract has allowed Intrado to develop in two ways. Intrado has incorporated geographic elements into its database and routing systems, and has established itself as a player in the public safety community. These have been leveraged into emergency notification systems, which are in turn leveraged into "location-based telecommuniations services, telematics and permission-based mobile messaging."[13] This shift in Intrado's strategic industry position, from emergency response systems to more general mobile communication management, can be seen in the rhetoric of its corporate reports. In 2000, these reports referred to Intrado's core offering as "SafetyNet." By 2001, this had become "Informed Response."[14]

Intrado's entry as an industrial player in the 9-1-1 infrastructure could not have been possible without regulatory intervention. Intrado's contract was conditional on a successful test of their system in the Greater Harris County 9-1-1 Emergency Network (GHCEN), the ECD serving the Houston area. Southwestern Bell (SWB) was (and is) the Houston area ILEC. They had also been the incumbent provider of 9-1-1 services, including both the ANI/ALI database and the selective routers used for emergency call delivery. In 1999, Intrado attempted to connect their new ALI database to SWB's tandem switching network. SWB refused to interconnect, claiming, among other things, that Intrado's system was incompatible with SWB's switches, and that interconnection posed a threat to the reliability of the existing 9-1-1 network. CSEC, GHCEN, and Intrado filed a complaint with the Texas Public Utilities Commission, which ordered SWB to unbundle tariffs for 9-1-1 databases and 9-1-1 call delivery. In the face of this order, SWB also agreed to a timetable and technical specifications for interconnection with their selective routers.[15] The TX PUC 9-1-1 tariff unbundling rules were some of the first in the nation.

Successful emergency services depend on actually getting personnel to a particular place rapidly. Standardized addressing and standardized maps are an essential part of the system. They also have privacy implications.

While Texas law gives counties and municipalities the authority to standardize rural addressing within their jurisdictions,[16] CSEC administers state funds for local addressing programs. To receive these funds, counties must meet standard vetting criteria. Thus, as with other 9-1-1 services, addressing is officially controlled by local authorities, but actually standardized through a state agency.[17] Even after standardized statewide procedures were established, addressing has been a notoriously difficult process. Establishing what constitutes a road, a parcel, or a dwelling has been an interpretive nightmare. Moreover, there has been local opposition to addressing projects. This resistance is sometimes because residents oppose standardized addresses in principle, but more often residents simply resist changing their established address.[18]

Addressing projects are usually performed in conjunction with mapping projects. Most, but not all, maps generated in conjunction with addressing projects are digital. However, some localities still use paper plat maps.[19] Digital base maps are obtained from private mapping vendors, public utilities, or the Texas Department of Information Resources. To this base map are added "attribute data," including customer location information obtained from "USPS, local government[s], telephone companies, . . . electric, gas or water utilit[ies], . . . county appraisal district[s], public school district[s], or other public entit[ies]."[20]

Mapping is expensive. For example, Hays County was mapped in 1987 at a cost of $78,000. However, information held by state and county agencies is subject to public disclosure. Once the mapping occurs with public funds, private citizens or companies can and do obtain the maps for only the cost of copying them. Map publishers regularly buy subsidized maps from counties, and use them as the raw material for consumer market products.[21]

State law makes certain information exempt from disclosure requirements.[22] Personal information is redacted from maps before they are released to the public.[23] However, state confidentiality law applies only to release to the public. COGs may release E9-1-1 addressing information to county agencies "for the purpose of sending tax notices and voter registration notices."[24] In fact, such a sharing of information among governmental bodies is encouraged "in the interest of the efficient and economical administration of statutory duties."[25]

In summary, Texas 9-1-1 implementation has had several far reaching implications. Firstly, a significant national market actor—CSEC—was created by consolidating the purchasing power of many small operators throughout a vast region. Secondly, CSEC's actions in the 9-1-1 arena have supported not only new market actors, but entirely new markets, notably the markets for third party location databases and third party location delivery networks. This was encouraged by an explicit legislative agenda promoting competition in the Texas telecommunications industry. Thirdly, Texas state funding of 9-1-1 addressing and mapping programs, and the structure of Texas privacy and freedom of information laws have had significant market and governmental repercussions. State agencies, including police and tax authorities, have ready access to personal information gathered in these addressing and mapping projects. The projects also make base maps, with personal information redacted, available to any private entity at relatively low cost.

The Implementation of Wireless 9-1-1 in Virginia

Administration of 9-1-1 in Virginia has traditionally been the province of the county-level public service answering points (PSAPs), which dispatch calls to police, fire and ambulance for response, and the local phone companies. The heart of the E9-1-1 operations has been, and continues to be, the dispatch centers that receive all 9-1-1 calls, identify the location and person making the call, and send the appropriate response team. With the introduction of Enhanced 9-1-1 (E9-1-1), the technological sophistication of the ERS operation, the jurisdictional boundaries involved in responding to a 9-1-1-call, and the number of players have all become more complex. The dispatch center is heavily dependent on computer and communications technologies not just for receiving and relaying calls but also for identifying where a call is coming from and where rescue personnel are located.[26] Real time mapping displayed on computer aided dispatch terminals (CADs) are the most visible component in an E9-1-1 dispatch center. Indeed the sophistication and timeliness of the maps is a primary distinction between traditional 9-1-1, enhanced 9-1-1, and now wireless E9-1-1. And with the move to wireless E9-1-1, the number of stakeholders and organizational complexity required for implementation increases.

In Virginia the implementation of wireless E9-1-1 is taking place in the three phases outlined earlier:

- Phase 0—a PSAP directly receives all wireless calls made within their jurisdiction at their communications center rather than at the respective State police district headquarters. All PSAPs that were E9-1-1 capable by July 1, 2000 are required by the General Assembly to start receiving wireless calls by July 1, 2002. All other PSAPs must start receiving wireless calls by July 1, 2003.
- Phase I—PSAP receives a caller's Call Back Number or ANI (automatic number identification), the identity of the wireless provider, and the location of the cell antenna receiving the call.
- Phase II—PSAP receives the actual location of the caller by longitude and latitude (ALI), as well as the ANI.

The organizational complexity involved in these implementation phases is well illustrated by the "Scope of Work" for one of the wireless workgroups in Virginia.[27] For Phase I, implementation requires that the operations of three entities—the local exchange carriers (LEC), the wireless service providers (WSPs), and the PSAPs—are managed and coordinated. The 9-1-1 service provider, which in this case is the local exchange carrier (Verizon) , must have ready the 9-1-1 selective router and the ALI database. The WSPs, which are quite numerous and vary throughout the state, need to provide maps showing all cell sector addresses and routing information for all the PSAPs that they serve. The PSAPs must install wireless 9-1-1 trunks, upgrades to their communications equipment, and upgrades in the dispatch center.

The memorandum of understanding (MOU) between AT&T Wireless (AWS) and Fairfax County specifies a similar but more detailed division of responsibilities for Phase I E9-1-1 implementation. AWS is responsible for: the design, installation, operation, maintenance and provision of all network elements; for interconnecting the Mobile Switching Center to the Selective Router/E9-1-1 Tandems and for interconnecting the SCP (Intrado-type services) to the ALI database; for working with vendors on performance measures; and for negotiating with the LEC, in cooperation with the PSAP, for appropriate interconnection agreements. The MOU notes that ALI Database functions lie outside the responsibility of AWS. The PSAPs are responsible for: validating and updating PSAP Jurisdiction Area Boundary Maps; providing and verifying needed data about each PSAP's existing infrastructure; identifying appropriate Routing Codes; and, informing third party vendors, such as Computer Aided Dispatch (CAD) providers of data to be delivered with 9-1-1 calls for coordination with PSAP premise-based systems. The MOU also recognizes, but does not specify the responsibilities of, several third parties including the 9-1-1 provider/LEC, the host ALI provider, and the SCP software developers and hardware providers.

To cover the costs of implementing Phase I and Phase II, PSAPs can apply to the Wireless E9-1-1 Services Board for funds to install necessary equipment upgrades and the wireless service providers can apply for funds to support the PSAP requests. In 1998, the General Assembly passed legislation that placed a $0.75 surcharge on every wireless telephone billed by a wireless service provider in Virginia. It also created a seven-member (three from local government, three from the telecommunications industry, and the state Comptroller as chair) Wireless E9-1-1 Services Board to administer the fund. In 2000, the size of the Board was increased to fourteen to allow representation from the police chiefs, fire chiefs, EMS chiefs, sheriffs, State Police, and emergency management.

By the end of FY2001, the Wireless E9-1-1 Fund generated approximately $1.8 million each month and had a balance of about $38 million at the end of FY 2001. In FY 2001, $6.7 million was paid to the PSAPs and $2.4 million was paid to the wireless service providers. The annual PSAP costs for statewide implementation of Phase I of wireless E9-1-1 is estimated at $11.5 million. Accurate estimates of Phase II costs are not available; however, it appears that Phase II costs are primarily nonrecurring while Phase I costs will continue into Phase II.[28]

Implementation of both Phase I and Phase II in Virginia is complicated by the number of wireless service providers and the number of PSAPs. Wireless service providers include: Alltel; AT&T Wireless; Cingular; Devon; Cellular One; Nextel; Ntelos; Shenandoah Cellular; Sprint PCS; Triton PCS; U.S. Cellular; Verizon; and Virginia Cellular. As of the end of FY 2001, At&T Wireless and Cingular are the only two national providers yet to complete a Phase I deployment. All of the other national providers had completed several deployments.

According to the *FY2001 Annual Report*[29] of the Wireless E9-1-1 Services Board, the implementation of wireless E9-1-1 Phase I, which requires that the PSAP be provided with the caller's phone number and the address of the cell site receiving the call, has "progressed, though not quickly." In a memo to all emergency communications coordinators in Virginia the Coordinator of Public Safety Communications points out that one factor that has proven important to the success of Phase I implementation is if PSAPs initiate implementation as a regional group.[30] Most of the PSAPs in Virginia have formed such regional groups.

The *Report* notes that some providers have been aggressive in their implementation but that technical problems and corporate issues have slowed other providers. The localities were not blamed with the lack of progress, instead it was the lack of preparedness by the wireless service providers and the local exchange carriers.[31] The *Memorandum of Understanding*[32] between AT&T Wireless (AWS) and Fairfax County discussed above illustrates the complexity and detail of the process.

An additional complication in implementing wireless E9-1-1 is the development of accurate and useable maps. For E9-1-1 the ALI database that Verizon maintains is not map-based but instead is a tabular, text-based system.[33] The PSAP has the responsibility for geo-coding the address. Many counties have invested local resources in the creation of their own GIS systems, which can be used to geo-code addresses. In some cases the PSAP geo-coding is well synchronized with the ALI information and in other cases it is not. Many PSAPs have taken the basic county GIS map, included it as part of their CAD system, and then maintain and update it separately from the county GIS map. In wireless E9-1-1, the PSAP believes that it is querying Verizon (LEC) for the ALI information. The ALI database recognizes that it is a cell phone call and passes the number on to a third party, often Intrado or TCS, to get the mapping coordinates that it needs.

Phase II implementation, requiring that PSAPs be provided with the actual location of the caller by longitude and latitude, has been delayed by waivers in Virginia as it has throughout the country. According to the *Annual Report*, "the primary difference between Phase I and Phase II is the addition of mapping within the PSAP. To provide this, the Board has included funding for the creation of mapping data for the locality and for the mapping display system to be used by the E9-1-1 call takers."[34] Instead of an incremental approach involving funding for each locality, the Commonwealth and the Board have decided on a statewide approach, involving a partnership with the Virginia Geographic Information Network (VGIN). VGIN had previously sought funding for a project of this nature but until the wireless E9-1-1 mandate there had not been a compelling need for such mapping. There are three elements to the state-based mapping project: 1) a digital orthophotography, which was conducted by aerial photography of the state; 2) centerline street location, for which the Virginia Department of Transportation provided the most reliable and updated information; and, 3) addressing, which was derived from local public and private sector entities.

According to Guidelines of the Wireless E9-1-1 Services Board, the PSAP can implement the mapping display system as part of the CPE equipment, computer-aided dispatch (CAD) system, or as a separate stand-alone system.[35] The Wireless E9-1-1 Services Board estimates that Phase II costs will be as follows:

- Statewide Base Mapping (Imagery) $10 million
- Statewide Base Mapping (street centerline) $10 million
- PSAP Mapping Display systems $ 5 million
- Wireless Provider Phase II Cost (nonrecurring) $10 million

The General Assembly recognized the revenue generating potential of the VGIN project and exempted it from FOIA requirements. Private sector companies may license VGIN maps but may not resell and must abide by the use terms of the license agreement. Any public entity (federal, state or local) may get access to VGIN maps but may not sell the information.

The implementation of Virginia's Wireless E–9-1-1 has begun in a largely cooperative manner, particularly on the part of the public sector actors. The Wireless E–9-1-1 Services Board has provided a focus not only for funding decisions but also for generating cooperative regional arrangements and sharing experiences and learning. As has been true in Texas, implementation has spawned new market actors and new market opportunities for third party location databases and location delivery networks. Implementation has also provided an opportunity to develop a statewide mapping project for which there had previously been interest but not sufficient need and support. With implementation of Phase II questions about privacy and access to information generated for and in the Wireless E–9-1-1 environment may become more apparent.

The Implementation of Wireless 9-1-1 in Ontario

On the face of it, the implementation of wireless E9-1-1 in Ontario, as well as in other parts of Canada, stands in stark contrast to the Texan and Virginian experiences outlined above. The differences are rooted in the historical emergence of a standardized platform for the routing of 9-1-1 calls, which has made the implementation of wireless enhanced emergency service in Canada somewhat less complicated than in the United States.

Canada has a similar system of primary and secondary Public Service Answering Points (PSAPs), sometimes called Central Emergency Reporting Bureaus, funded and operated by municipal governments. Primary PSAPs serve to screen calls and to route them to the appropriate secondary PSAPs operated by the respective fire, police and ambulance services.[36] Provincial E9-1-1 service provides for the transport of all customer dialed 9-1-1 calls to the appropriate local 9-1-1 PSAP. So when a customer dials 9-1-1, the wired carrier being used, transports the call to the dedicated switch operated by the Incumbent Local Exchange Carrier (ILEC).[37] The call is then transferred to the appropriate police, fire or ambulance dispatch centre. As in the United States, the original 9-1-1 system was designed as a simple voice connection to an operator who would verify location verbally and dispatch the appropriate service; from the outset Basic 9-1-1 has included an automatic ringback facility. This kind of basic service has been in existence in Canada since the mid-1970s, although the pace of development has varied from region to region.

As more sophisticated telecommunications services developed, a second generation of 9-1-1 service emerged in the mid-1980s. The so-called "Enhanced 9-1-1" service permits the transfer of Automatic Number Identification (ANI) *and* Automatic Location Identification (ALI) to the PSAPs. With call-back number and location information displayed on a video display terminal, emergency operators can still assist the caller where verbal communication was impossible, or connection was terminated. Over time, ANI data was matched against street addresses in the ALI database, and emergency personnel directed to the exact location from which the call originated. Until the 1980s, therefore, it appeared that there was a good deal of convergence between the Canadian and American systems. The ANI/ALI system had become "the benchmark feature of E9-1-1 across North America."[38]

In the Ontario context, as elsewhere in Canada, however, the implementation of the ANI/ALI system was not without difficulty. In every region, there were "underlaps" and "overlaps" caused by the wide variety of addressing systems used by local communities, and the fact that these systems rarely conformed with the addressing and billing information held by the respective ILEC.[39] Thus, some addresses fell through the gaps and had no access to 9-1-1 service. Other addresses fell into more than one 9-1-1 region, causing confusion as to the closest emergency dispatch center. Moreover, the cost for 9-1-1 had always come from the property tax base of local communities. The extra expense of administering an enhanced 9-1-1 system meant that the cost had to be passed along to subscribers, raising the question for the ILEC of who to bill when the telephone exchanges and municipal boundaries were not consonant.

A third generation of E9-1-1 emerged in the early 1990s, therefore, as a result of the need to standardize the addressing system for emergency response services and to deal with billing anomalies. The Public Emergency Reporting Service (PERS) is a cooperative effort by the ILECs and local municipalities. In Ontario, local governments essentially apply to the ILEC (Bell Canada) to be included in the 9-1-1-PERS system. In return, Bell requires that all streets be numbered in a consistent and accurate format. The system requires that every household shall have an address, and every street a name. Therefore, the address, rather than the phone number, drives the PERS system and the boundary conflict problem is, one hopes, eliminated. Of course, as a result of historical accident, municipal reorganization, or amalgamation, there may be duplicate street addresses in some communities. So when a municipality decides to introduce 9-1-1 PERS, it must ensure that each household or business is assigned a unique civil address—a number, a street name, a street suffix, and in some cases a directional indicator. So, as we saw in Texas, some municipalities have been undertaking quite extensive efforts at renumbering, and are having to convince residents that the changes are worthwhile.[40]

Once the municipalities have drawn new maps based on a standardized street address, they then add the location and coverage area of emergency response dispatch centers. These maps may arrive at Bell's 9-1-1 Service office in a variety of digitized, or nondigitized, forms. Bell then digitizes in standard format, and overlays the telephone number, and an Emergency Service Number (ESN) that is associated with a clear Emergency Service Zone, and enters this information into its 9-1-1 routing system. Thus there should be far less confusion about the location of the closest emergency dispatch center, as the ANI/ALI database has been standardized, and supplemented with vital information about the location of emergency response services. A 9-1-1 PERS platform has now been established in every province except Saskatchewan and Newfoundland. In Ontario, over 90 percent of households are now part of this system.[41]

These three stages of enhanced 9-1-1 service have progressed in Canada regardless of the problems associated with making 9-1-1 calls from a cellular phone. Yet, as is the United States, the E9-1-1 system quickly exposed the distinction between wireline calls with ANI/ALI capability, and wireless calls that had no enhanced capability at all. The policy problem in Canada is neatly expressed by this advice leaflet from the Canadian Wireless Telecommunications Association (CWTA):

> When you call 9-1-1 on your home telephone, your call is sent to the nearest emergency response centre. In many locations, the E9-1-1 network also passes along your telephone number and address so a 9-1-1 operator can call you back, if necessary, and help can be sent immediately to your exact location. Unlike your home telephone, mobile phones generally do not pass along the telephone number or any other customer information when you make a call. This means that, when you call 9-1-1, you have to give your complete number to the E9-1-1 operator—including your mobile phone area code. This is important because the operator may have to call you back if you are disconnected. In addition, you have to tell the operator, as best you can, exactly where you are. Remember, your mobile phone can be used anywhere service is available. Only you can provide your precise location or the location of the emergency.[42]

This problem with wireless calls was recognized from the genesis of the 9-1-1 PERS network. Only in the mid-1990s, however, did the Canadian PSAPS, who are less well organized than their American counterparts, voice strong concerns. Although the problem was initially debated within the CRTC's Canadian Industry Steering Committee (CISC), the CRTC was unwilling to take the kind of strong interventionist stance as did the FCC in the United States. So the issue migrated to the trade association of the wireless industry, the CWTA. Since 1997, an E9-1-1 Working Group has been working under the auspices of CWTA. This group includes membership from the four major cellular providers in Canada (Bell Mobility, Microcell, Telus Mobility, Rogers Wireless), from York and Toronto police departments, and from other representatives of the PSAP community. A member of the CRTC maintains a watching brief.

The main approach of this committee has been to organize and monitor trials. The first occurred in Calgary, Alberta from October 1999 to April 2000, largely as a result of a technical proposal submitted by Telus, the ILEC in the West. Four wireless service providers participated in the trial, which tested the interconnection between the provincial 9-1-1 platform, operated by Telus, the wireless carriers and the PSAPs. This enabled the delivery of information relating to the location of the wireless antenna receiving a 9-1-1 call, as well as the 10-digit telephone number of the mobile subscriber placing the call. This functionality is equivalent to what the FCC established as part of its Phase I requirements. The final report of this trial indicated that participants were generally satisfied that accurate information on call-back number and station location was provided to the E9-1-1 operators.[43] The results also demonstrated that the technical solutions could be applied in any other province with a similar E9-1-1 PERS delivery platform. However, the limited trial area and the involvement of only one PSAP meant that the delivery of call-back numbers to secondary PSAPs, and the associated routing issues, could not be properly examined. A second trial was therefore initiated in Toronto and North York in Ontario, a potentially more complex urban environment, involving more than one primary PSAP.[44] An examination of how this trial was conducted will demonstrate how wireless E9-1-1 is likely to be implemented in the rest of Ontario.

The trial participants were Toronto and York region PSAPs, Bell Canada, and the four wireless carriers (Bell Mobility, Rogers Wireless, Microcell and Telus Mobility). The goals were to interconnect wireless carriers to the existing 9-1-1 PERS platform, to display the 10-digit wireless Call Back number and cell site/sector identification, and to transfer information to secondary PSAPs. Technically, this process required the establishment of a separate, routable, and nondialable Emergency Service Routing Digit (ESRD), which assigns a ten-digit number to each cell-site/sector. The ESRD is then delivered to the ALI database where it is cross-referenced with street address. So when a customer dialed 9-1-1 on a wireless phone, the PSAP received both the wireless customer's 10-digit call-back number, as well as the 10-digit ESRD. The display of the ESRD, and the associated cell site sector address, is then used by the emergency personnel to identify the location of the originating cell site or sector where the 9-1-1 call entered the wireless network, using the existing ANI/ALI display terminal.[45]

During the trial period, each of the major participants was required to undertake certain responsibilities. The participating wireless carriers were expected to establish a contact point to investigate and respond to PSAP queries, and to define and establish a process for assigning an ESRD to cell site locations. PSAPS were expected to begin staff training and to set up a pretrial validation. And Bell Canada had to make upgrades to the hardware and software associated with its tandem switches. The trial enabled the assessment of the viability of the technology and architecture and allowed participants to test whether the routing of wireless 9-1-1 calls had improved, whether the default routing arrangements were appropriate in a wireless context, whether overflow calls were processed and routed properly, whether certain "call-management" features were compatible, whether the calls were successfully routed to the appropriate secondary PSAPs (for fire, police and ambulance), and whether existing PSAP display terminals could accurately display both call back number and the ESRD.

The trial exposed some technical complications relating to the existing voice, data and database technology and architecture and particularly the signaling arrangements between different switching technologies. The trial found that the routing of wireless 9-1-1 calls had improved, but not without some problems associated with the misrouting of calls to the wrong PSAP. False call-back numbers from unsubscribed handsets also continued to plague the system. But these are difficulties inherent in the mobile character of the technology. Callers may travel between several cell sites by the time the call is initially made, and the emergency dispatch is initiated. Moreover, the call sector may not necessarily be the nearest to where you are when the call is made; calls are rerouted when one sector is very busy. On the positive side, there was clear evidence that the combination of cellsite address, call-back number and other information allowed emergency response teams more accurately to pinpoint location and, in some instances, to save lives. The system also allowed PSAPs to identify abusive, frivolous or mistaken callers more effectively.[46]

As a result of the success of this trial, Wireless E9-1-1 is now available as a commercial tariff in Ontario, as well as in Alberta and BC. But implementation is dependent on the commercial decisions of the WSPs. So far all the wireless carriers, with the exception of Rogers AT&T Wireless have begun to offer enhanced 9-1-1 services, in conformity with a CRTC decision in 2001 that allowed Bell Canada to charge all WSPs a monthly rate of $0.02 for each of its wireless working telephone numbers equipped with outward calling, so that they might access Bell's 9-1-1 PERS network.[47] In March 2001, Microcell also asked the CRTC to order all phone companies (especially the ILECs in other provinces, such as Manitoba Telecom Services Inc. and Sasktel) to provide 9-1-1 network access services to wireless carriers. The CRTC is expected to rule on this issue in September 2002.[48]

No doubt implementation would be speedier if the PSAPs and the wireless carriers could have resolved a contentious dispute over the provision of subscriber records. Public safety agencies have sought wireless subscriber records, and want the ALI database to include the home or business address of all wireless customers. The wireless carriers believe that the ESRD solution is the only feasible locational information that can be transferred at this stage. They have also protested that such subscriber information is going to be very misleading in ascertaining the location of a wireless caller. Subscriber records are also notoriously unreliable given that many cellular phone customers buy prepaid packages, and have no incentive to provide accurate names and addresses for billing purposes. In one of its only interventions on E9-1-1 questions, the CRTC has ruled, in a decision regarding Microcell, that in an emergency, subscriber records could be of value to PSAPs.[49] But there has been no general ruling that these records should be provided as a matter of course. This position contrasts, therefore, with the requirements established under Section 5 of the U.S. WCPSA of 1999.

Discussions about Phase II implementation and the technologically, economically and politically controversial questions of how to transmit exact location coordinates to PSAPs have only just begun. The Toronto trial exposed the obvious limitations of Phase I implementation, and the complexities of integrating mobile networks with existing publicly switched networks. No doubt there will be extensive debates about the relative merits of most effective locational solution; through a handset-based GPS chip, or a triangulation method, or a combination of both.

Phase Two implementation will no doubt bring many benefits for the emergency response communities. On the other hand, the investment in infrastructure by the PSAPs is likely to be enormous. So far, Phase One implementation has had a minimal impact on the existing technical equipment of the PSAPs.

So, for now, the mood in Ontario, which seems the most advanced province, is that the cellular industry has yet to be convinced that comprehensive Phase One implementation is in their commercial interests. And no doubt there will be continued pressure from the emergency response community for a more assertive role to be played by the CRTC. Nevertheless, in contrast to the U.S. states discussed above, the common PERS platform, operated by a single ILEC in each province, reduces the burden and cost on municipal and provincial governments. It also renders unnecessary the kinds of functions performed by 9-1-1 call delivery companies. Moreover, with fewer WSPs, there is a greater likelihood that the industry can proceed in a more uniform way, and project a more united position through its trade association, the CWTA. In short, the number of actors involved, even in a complicated urban environment like Toronto, is considerably lower than in most U.S. states. In consequence, the chances of finding a cooperative and concerted solution to the challenging technical and regulatory problems are far higher than they are south of the 49th parallel.

Discussion and Conclusions

From a legal perspective, U.S. and Canadian policies on cellular phone privacy are somewhat different at this point, although in both countries privacy policies are still evolving. We will first briefly review those policies but are more concerned in our discussion here with examining how the "on the ground" implementation of wireless E9-1-1 affects the possibilities for privacy. Formal policies may point in one direction, while technological developments and organizational relationships may point in a different direction.

U.S. law places very tight restrictions on the information collected during a 9-1-1 call. Section 5 of the Wireless Communications and Public Safety Act (WCPSA) of 1999 amended Section 222 of the Communications Act of 1934 to require a customer to give prior authorization for any disclosure of call location information for a purpose other than the delivery of emergency services. In addition, however, several other laws including the Electronic Communications Privacy Act (ECPA), the Communications Assistance for Law Enforcement Act (CALEA), and most recently the USA PATRIOT Act all may speak to aspects of wireless location information.[50] The next several years promise to provide lawyers for the telecommunications carriers with interesting litigation as these issues are sorted out.

Wireless privacy questions have to this point been debated far more actively in the United States than in Canada. In other information privacy areas, Canadian policy is generally more protective of privacy than is that in the United States. A more comprehensive network of public and private sector legislation in Canada, overseen by federal and provincial privacy commissioners, normally means that privacy issues are articulated earlier and with more force than in the United States. But so far there has been very little public debate and virtually no involvement from the federal or provincial privacy commissioners.[51] Moreover, the CWTA has yet to articulate a public position on privacy, let alone produce a code of practice for its members, something associations in other sectors have done. A newly enacted federal law for the protection of personal information in the private sector certainly regulates how wireless service providers collect, use and disclose personal information. But to date, there has been little, if any, analysis of how this legislation applies to the locational data generated through mobile phone usage.[52]

Whether or not national legislation speaks to the privacy implications of E9-1-1, they have received little to no attention in the actual "on the ground" implementation of Phase I and Phase II in the United States and Canada. Instead state regulators, PSAP administrators, Wireless Service Providers, and Local Exchange Carriers are beleaguered with questions of funding, technology, and coordination. In the United States and Canada, the goal is to provide effective responses to wireless 9-1-1 calls and to enhance public safety. In the world of those administering these systems, privacy has been of only marginal and intermittent interest.

A comparative analysis of the three jurisdictions reveals several trends that are likely to push technologies or organizational configurations in ways that might not be hospitable for privacy protection. This is an area where "following the money" proves beneficial to teasing out the potential privacy implications.

Wireless E9-1-1 is pushing local governments to cooperate in larger, often regional, bodies. This is true for PSAP administration, funding, and mapping. Emergency response functions have traditionally been the responsibility of local governments—local police, fire, and ambulance services that often knew the geography and people in their jurisdictions through long day-to-day experience. Although there have been previous trends to centralization at the county or provincial levels, the local PSAPs were still granted autonomy to act in a way that best served their local communities. Such autonomy is less possible in a wireless environment. Cell phones and cell sites do not recognize the geographic boundaries and necessitate more cooperation among neighboring jurisdictions.

Additionally, the complexity of wireless mapping, locational database, and routing systems and the need to have these uniform for interfaces with ILECS and wireless service providers limits the real choices that local units are able to make. More and more of the PSAP operations have become standardized and, at least in the United States, are interconnected with more private sector telecom providers and third party vendors.

We saw that the two American jurisdictions have created organizations that are designed to provide more coordination and cooperation in the implementation of wireline and/or wireless E9-1-1. In Texas, the Commission on State Emergency Communications (CSEC) was established in 1985 to coordinate ERS throughout the state. In Virginia, the Wireless E9-1-1 Services Board was created within the Public Safety Communications Division of the State Department of Technology Planning to coordinate wireless E9-1-1. The public safety community is also increasingly developing concerted positions through the three 9-1-1 industry associations—National Emergency Number Association (NENA), Association of Public Safety Communications Officials (APCO), and National Association of State 9-1-1 Administrators (NASNA). The same need for coordination among local municipalities is, of course, observed in Canada. The difference, however, is that the system of ILECs, which enjoy province-wide monopolies, provide a natural arena to coordinate the mapping and call-routing functions. Moreover, the PERS network provided a standardized platform for the routing of 9-1-1 calls, before the wireless problem emerged.

These larger trends toward the coordination of 9-1-1 implementation, in both the U.S. and Canada, could have some conflicting implications for privacy protection. On the one hand, cooperation may lead to greater demands for the sharing of subscriber and locational information; the pressure from the Canadian PSAPs to integrate wireless subscriber records into the ANI/ALI system is evidence of such a trend. Conversely, however, institutional coordination can provide arenas for at least the consideration of common privacy problems, and perhaps for the articulation of a common position on privacy protection. An extensive fragmentation of E9-1-1 implementation is likely to be mirrored by a concomitant diversity of positions and practices about the treatment of personal information. In this respect, it is probable that the smaller number of stakeholders in the Canadian context, as well as a more developed statutory framework than in the United States, would militate in favor of the development of a concerted position on the protection of personal information.

At least in the United States, wireless E9-1-1 is creating a new revenue source at the state level. Funding of wireless E9-1-1 has become an important source of revenue for public and private actors. In both Texas and Virginia more funds are being generated by the fees and surcharges than are being currently used in implementation of Phase I and Phase II implementation. In Texas these funds go into the general revenue and are then appropriated by the legislature back to public safety or to other state needs. In Virginia, the funds go directly to the Wireless Board.

The new funding for wireless E9-1-1 is enabling a redistribution of funds to less wealthy parts of the state. In both states, some parts of the state subsidize other areas of the state. In Texas urban areas subsidize rural areas. In Virginia, areas with high cell phone penetration, which is likely to be urban and suburban areas, subsidize areas with less cell phone use.

To a certain extent, this redistributive effect is felt in the Canadian context, at least in large provinces like Ontario. But in the absence of a government mandate to develop wireless E9-1-1, market considerations are currently driving the behavior of both ILECs and WSPs. Thus, to date, the major costs of 9-1-1 service is passed from the ILEC, through the WSPs to the consumer, in accordance with CRTC tariff rulings. And Phase One implementation has so far involved quite limited infrastructure modification for the PSAPs. Phase Two implementation, as in the United States, is a very different story.

Wireless E9-1-1 has created new markets and new market actors. The PSAPs have become more prominent and influential market actors in the communications industry with their need to purchase maps, databases, and display formats to enable them to comply with wireless E9-1-1. In Texas centralization in the CSEC has created one actor. In Virginia, there has been somewhat less centralization except in the development of maps.

Market changes have not just been limited to the activities of state actors. Wireless carriers are under federal mandate to deliver locational data to PSAPs if the PSAP requests such a service. Wireless carriers have had to install both the means of locating calls and the means of transferring that location to the E9-1-1 network. They have usually outsourced both of these. While the integration of these location determining systems has been an economic and technical challenge, it is also fraught with profit making potential.

For example, Verizon Wireless and AT&T Wireless are, respectively, the 1st and 3d most dominant wireless carriers in the United States national market.[53] Both have chosen Allen Telecomm's Geometrix as their locational technology.[54] The system's centerpiece is the Geometrix Geolocation Control System. Located at a wireless operator's mobile switching center, it obtains data from Geometrix Wireless Location Sensors at base stations, performs location determination calculations, and reports location coordinates to Mobile Positioning Centers (MPCs). The E9-1-1 ANI/ALI databases are updated by querying these MPCs. Geometrix touts the system's ability to support and enhance value-added wireless location services. It also includes planning tools that can accurately predict the wireless service providers' network needs by using a combination of propagation models, terrain and morphology data, and location algorithms. Many other call location firms are competing to establish E9-1-1 services as the toe hold from which other location-based market offerings can be launched.[55]

Likewise, companies offering E9-1-1 call routing and delivery systems are positioned to expand their products and services. For example, as was discussed earlier, Intrado's management of the ANI/ALI database could be parlayed first into emergency notification systems then into commercial telephone notification services. Both of these services enable the client to quickly send identical messages to a large target population, which can be specified either as a geographic region or as a list of individuals. For public safety clients, the target database starts with an extract of 9-1-1- data. Private sector target notification uses the client's existing database, which will also be geo-coded. The notification systems are identical except in the source of their databases.[56]

The language by which telecommunications issues are framed is increasingly obscure. The debates about wireless E9-1-1 are replete with technological jargon and acronyms. All reports on the subject have lengthy glossaries, and all people working in the area comment on the need to know these acronyms. Although this may appear at first to be a somewhat trite observation, several key information privacy protection principles rely upon individuals being able to understand the environment in which their information is being collected, used, exchanged, and manipulated. One key obligation within Canada's private sector privacy law, for instance, is "transparency" both in terms of personal information practices, and organizational policies. But in the environment of wireless E9-1-1, the discourse is nontransparent to the lay person and, most likely, to political oversight bodies. The technological jargon and the acronyms camouflage the real-world meaning and implications.

We believe that these four trends contribute to two over-arching privacy implications. *First*, wireless E9-1-1 has produced an environment with evolving technologies, market actors, business models, and governance structures that is less sensitive to the privacy concerns of individuals and more attuned to surveillance of populations and centralization of functions. This is seen in the creation of more sophisticated and detailed maps, the invention of new location identifying devices, the development of new database managing companies, and the consolidation of local government activities. These market and governmental changes—all with the goal of realizing public safety—are not without other, perhaps unintended consequences. In general all of these systems make the "monitoring of everyday life" (Lyon 2001) more of a reality. In this environment "notice and consent," or the full package of data protection principles, does not begin to provide any meaningful policy response. These are systems, not just in a technological sense, but also in the broader political economic sense.

The *second* privacy implication is analogous to Lessig's and Reidenberg's[57] insights regarding the importance of "code" and system architecture in communications systems. The architecture of the ERS systems contains detailed maps, databases, and routing systems—as well as inter-operability among Local Exchange Carriers, Wireless Service Providers, PSAPS, and third party vendors. This architecture will dictate the privacy possibilities. And at this time privacy is not consciously being considered as a factor in the design of these systems. If it is not initially part of the planning, then privacy will be an afterthought and its protection will never be as secure as it needs to be. Furthermore, if the system architecture and the computer code itself regulates the potential for capturing personal information, then the differences we have seen between the implementation of wireless 9-1-1 in Canada and the United States are likely to be rendered less relevant over time.

Notes

1. This research was funded by National Science Foundation Grants no. SES–0083271, SES–0083348. The other participants in this project are Michael Curry (UCLA) and Charles Raab (University of Edinburgh).

2. Bennett, Colin and Priscilla M. Regan. 2002. "What Happens When You Make a 911 Call? Privacy and the Regulation of Cellular Technology in the United States and Canada." Presented at the 2002 Canadian Political Science Association, University of Toronto, May 30, 2002

3. Federal Communications Commission. "Report and Order and Further Notice of Proposed Rulemaking, In the Matter of Revision of the Commission's Rules to Ensure Compatibility with Enhanced 911 Emergency Calling systems (FCC 96–264)" July 26, 1996. Available at: http://www.fcc.gov/Bureaus/Wireless/Orders/1996/fcc96264.txt

4. Section 1 of the Communications Act of 1934 (47 U.S.C. 151). Available at : http://www.fcc.gov/Reports/1934new.pdf

5. Federal Communications Commission. "Report and Order. . ."

6. Federal Communications Commission. "Fact Sheet: FCC Wireless 911 Requirements" January 2001. Available at: http://www.fcc.gov/911/enhanced/released/factsheet_requirements.012001.pdf

7. Federal Communications Commission. "Fact Sheet: E–911 Phase II Decisions" October 2001. Available at: http://www.fcc.gov/Bureau/Wireless/News.Releases/2001/nw110127a.pdf.

8. *Wireless Communications and Public Safety Act of 1999*. PL 106–81.

9. Personal interview with Rob Huggins. Texas Commission on State Emergency Communications. July 2, 2002.

10. Texas Commission on State Emergency Communications. Rule 251.1. Available at: http://www.911.state.tx.us/files/pdfs/services/nurule251_1.pdf; Texas

Senate Committee on Economic Development, Subcommittee on State E9-1-1 Emergency Communications. 2000. "Interim Report to the 77th Texas Legislature." p. 5; Texas Association of Regional Councils. "Regional Councils in Texas." Available at: http://www.txregionalcouncil.org/what.htm.

11. Texas Senate Committee on Economic Development, "Interim Report . . ." p. 5.

12. Moreno, Mary. 2002. "City will drop from state E9-1-1 program." *Corpus Christi Caller-Times*. March 29, p. B1

13. Intrado, Inc. 2002. "2001 Annual Report: Intrado Informed Response; Making the Connection between Information and Location." Longmont, Colo.: Intrado, p. 4.

14. SCC Communications Corp. 2000. "9-1-1 SafetyNet: The Next-Generation Public Safety Communications Network." Boulder, CO: SCC Communications Corp.; Intrado, Inc. 2002. "Access the Facts: Investor Fact Sheet Q1—2002," Longmont, CO: Intrado, Inc.; Intrado, Inc., "2001 Annual Report. . . ."

15. "Emergency Petition of the Advisory Commission on State Emergency Communications and the Greater Harris County 9-1-1 Emergency Network for Declaratory Rulings Ordering Southwestern Bell Telephone Company to Unbundle its 9-1-1 Network and 9-1-1 Database Management System." Texas PUC Docket no. 20334, SOAH Docket no. 473-99-0155; "Draft Preliminary Order, Emergency Petition of the Advisory Commission on State Emergency Communications and the Greater Harris County 9-1-1 Emergency Network for Declaratory Rulings Ordering Southwestern Bell Telephone Company to Unbundle its 9-1-1 Network and 9-1-1 Database Management System." PUC Docket no. 20334, SOAH Docket no. 473-99-0155.

16. County Road and Bridge Act (as enacted in Texas Senate Bill 1091 in 1989).

17. Texas State Advisory Commission on State Emergency Communications. "9-1-1 Programs, Policies and Procedures." February 1999. § 2.02.

18. Personal interview with Patsy Greiner, Texas Commission on State Emergency Communications. July 2, 2002.

19. Personal interview with Rob Huggins.

20. Cornyn, John. 1999. Letter to James D. Goerke dated August 16, 1999. Open Records Decision no. 661. Available at: http://www.csec.state.tx.us/pdf/ord_661.pdf.

21. Personal interview with Patsy Greiner.

22. Texas Health and Safety Code. § 771.061

23. Texas State Advisory Commission on State Emergency Communications, "9-1-1 Programs. . ." § 1.06.

24. Cornyn, John. Letter to James D. Goerke.

25. Cornyn, John. Letter to James D. Goerke.

26. Personal interview with staff, Public Safety Communications Center, Fairfax County Police Department. July 29, 2002.

27. Spears-Dean, D. "Scope of Work for Tri-Cities wireless Workgroup." May 31,

2002. Available at: www.911.state.va.us/pdf/PM/TriCities%20-%20Scope%20of %20Work%20-%20Phase%20I%20and%20Phase%20II.pdf.

28. Virginia Department of Technology Planning, Division of Public Safety Communications. 2001. "Wireless E9-1-1 Services Board, FY2001 Annual Report." p. 8. Available online: http://www.va9-1-1.org.

29. Virginia Department of Technology Planning. "Wireless E9-1-1. . ." p. 1.

30. Commonwealth of Virginia, Division of Public Safety Communications. 2001. "Virginia Wireless E9-1-1 Implementation Guide (Revised 12/2001)." Available at: http://www.9-1-1.state.va.us/resource_info.htm

31. Virginia Department of Technology Planning. "Wireless E9-1-1. . ." p. 9.

32. AT&T Wireless Services (AWS) .2001.. "Memorandum of Understanding Re. E9-1-1 Phase I Service with Fairfax County, Virginia." Available at: http://www.9-1-1.state.va.us/sample_documents.htm

33. Personal interview with staff, Public Safety Communications, Commonwealth of Virginia.

34. Virginia Department of Technology Planning. "Wireless E9-1-1. . ." p. 7.

35. Virginia Wireless E9-1-1 Services Board. 2001. "Guidelines for the Submission of PSAP Cost Estimates for the Fiscal year beginning July 1, 2002 and ending June 30, 2003. (Adopted August 8, 2001 and Revised December 12, 2001)." p. 4. Available at: http://www.9-1-1.state.va.us/resource_info.htm.

36. In many municipalities, one department will play the role of both primary and secondary PSAP. Thus, in Toronto, the Toronto police department operates the primary PSAP for the entire city.

37. The main ILECs are Bell (operating in Ontario and Quebec), and Telus (for BC and Alberta). Until the late 1990s, all incumbent local exchange carriers were represented by a common trade association, known as Stentor.

38. Gow, Gordon. 2002. "Territorial Boundary Crossing and Regulatory Crossroads: The Case of Wireless Enhanced Emergency (9-1-1) Service in Canada." Paper delivered to the Third Wireless World Conference, University of Surrey, July 17–18, 2002.

39. Personal interview with Judy Tottman, Regional Manager 9-1-1 Service, Bell Canada. August 16, 2002. .

40. See for instance, the efforts by the City of Kingston to convince residents of the benefits of 9-1-1-PERS: "Therefore, in order to proceed with the application to improve emergency services for the Kingston community, 131 civic addresses have been identified to be changed to reflect numbering standards acceptable to Bell Canada. The City of Kingston is prepared to offer as much support and clerical assistance as possible to ensure this transition a smooth one. If your address is affected by the this you will receive a "change in address" package that includes a voucher towards the purchase of new civic address numbers for your home and other helpful information to get you started, such as most frequently asked questions, change in address forms and contact information for some of the common government agencies, organizations and household service providers." Available

at: http://www.city.kingston.on.ca/residents/emergency/9-1-1/index.asp.

41. Personal interview with Judy Tottman.

42. Canadian Wireless Telecommunications Association. "E9-1-1 Health and Safety Issues" Available at: http://www.cwta.ca/safety/E9-1-1/health–9-1-1.php3.

43. Alberta E9-1-1 Advisory Association. "Alberta Wireless 9-1-1 Trial Report" Available at: http://www.cwta.ca/safety/E9-1-1

44. There has also been a smaller and more limited trial in Nova Scotia.

45. Gow 2002, p. 9

46. Ontario 9-1-1 Advisory Board. "Wireless Enhanced 9-1-1 Trial Report" Available at: http://www.cwta.ca/safety/E9-1-1/

47. CRTC Order 2001–902. December21, 2001. Available at: http://www.crtc.gc.ca/archive/ENG/Orders/2001/o2001–902.htm

48. "CRTC to rule on 9-1-1 Wireless Ability," *Globe and Mail*, August 19, 2002

49. CRTC Order 2000–831. September 8, 2000. Available at: http://www.crtc.gc.ca/archive/ENG/Orders/2000/O2000–831.htm. This exhortation was repeated in Public Notice CRTC 2001–110 of October 31, 2001, which seeks comments on this and related issues. Available at: http://www.crtc.gc.ca/archive/ENG/Notices/2001/pt2001–110.htm.

50. Phillips, David J. 2003. "Beyond Privacy: Confronting Locational Surveillance in Wireless Communication." *Communication Law and Policy* 8:1–23.

51. One brief paper on mobile phone privacy issues was published by the Ontario Information and Privacy Commissioner: Ann Cavoukian and Mike Gurski, "Privacy in a Wireless World" Available at: http://www.ipc.on.ca/english/pubpres/ext-pub/wireless.htm.

52. The Protection of Personal Information and Electronic Documents Act (2000), applied in January 2001 to the "federally regulated" private sector, including telecommunications companies. Among other things, it establishes the conditions under which personal information might be used for "secondary purposes." In a nutshell, any "sensitive information" can be disclosed only with the express and positive consent of the person concerned. Nonsensitive information might be transferred using "implied consent" provisions (such as opt-out boxes). Whether locational information is sensitive or nonsensitive information will be a matter for the Federal Privacy Commissioner to decide.

53. "What Ails Wireless?" *Business Week*, April 1, 2002, p. 60.

54. "Allen Telecom's Grayson Wireless Division to Supply 9-1-1 Caller Location Systems to AT&T Wireless," January 16, 2002. Available at: http://www.geometrix9-1-1.com/newsrm/020116.html.

55. See, for example, www.cell-loc.com, www.triangulation.com, www.globallocate.com.

55. Intrado, Inc. "Intrado notification offerings." Available at: http://www.intrado.com/offerings/

55. Lessig, Lawrence. 1999. *Code and Other Laws of Cyberspace*. New York:

Basic Books. Reidenberg, J. 1998. "Lex Informatica: The Formulation of Information Policy Rules Through Technology." *Texas Law Review* 76:552–593.

References

Bennett, Colin, and Priscilla M. Regan. 2002. "What Happens When You Make a 911 Call? Privacy and the Regulation of Cellular Technology in the United States and Canada." Presented at the 2002 Canadian Political Science Association, University of Toronto, May 30, 2002.

"CRTC to rule on 9-1-1 Wireless Ability," *Globe and Mail*, August 19, 2002.

Gow, Gordon. 2002. "Territorial Boundary Crossing and Regulatory Crossroads: The Case of Wireless Enhanced Emergency (9-1-1) Service in Canada," Paper delivered to the Third Wireless World Conference, University of Surrey, July 17–18, 2002.

Lessig, Lawrence. 1999. *Code and Other Laws of Cyberspace*. New York: Basic Books.

Moreno, Mary. 2002. "City will drop from state E9-1-1 program," *Corpus Christi Caller-Times,* March 29, p. B1.

Phillips, David J. 2003. "Beyond Privacy: Confronting Locational Surveillance in Wireless Communication." *Communication Law and Policy* 8:1-23.

Reidenberg, J. 1998. "Lex Informatica: The Formulation of Information Policy Rules Through Technology." *Texas Law Review* 76:552–593.

"What Ails Wireless?" *Business Week*, April 1, 2002, p 60.

13

Creative Destruction in Emerging Markets: Privatizing Telecoms and the State

Lee W. McKnight, Paul M. Vaaler, Burkhard N. Schrage, and Raul L. Katz

I. Introduction

Joseph Schumpeter's (1934; 1939; 1943) term for describing cycles of innovation, "creative destruction," focuses on technological change and its simultaneously value-enhancing as well as value-destroying consequences.[1] Less attention has been paid to institutional innovation—and creative destruction—though it may have even more dramatic effects on firms and markets.[2] Telecommunications enterprise ("telecoms") privatization worldwide provides a telling example. Here, transfer of ownership and control of hundreds of enterprises with $ billions in asset value from state to private hands in the late twentieth century has undoubtedly induced substantial change in individual and organizational incentives and behaviors. But telecoms privatization is still relatively recent—less than 20 years old in industrialized countries and less than a decade old in many emerging markets. Where are we in terms of the institutional innovation it promises? What is the net effect in terms of value creation and destruction?

This chapter offers several counter-intuitive, but empirically derived, answers to these questions. Before offering you the reader our answers, however, we must set the stage by reviewing the relevant literature, and the conventional wisdom.

The short- to medium-term performance implications of that institutional change are currently in dispute with two theoretical models proposing different answers, but interestingly, relying on similar factors to reach those answers. What we call a "mainstream" theoretical model of enterprise privatization (e.g., Boycko et al. 1996) suggests that less state ownership and control as well as greater exposure to market forces will increase enterprise value by aligning more closely its incentives and behaviors with those of profit-maximizing shareholders. But what we also call an "alternative" theoretical model (e.g., Perotti 1995) suggests a different relationship between states, markets and privatizing enterprises: Performance may improve when the state retains a substantial equity stake and commits to substantial intervention in relevant markets of privatizing enterprises. Such state policies signal to shareholders the state commitment to ensure adequate returns on their risky investment, thus, also engendering investor interest in other state assets up for privatization.

These competing theoretical models imply different relationships between privatized enterprise performance, on the one hand, and state ownership and market experience—what we call temporal distance from the date of initial privatization—on the other. Less state ownership and greater temporal distance improve privatizing enterprise performance according to the mainstream view, but may detract from such performance according to the alternative view. Determining which of these research streams is supported empirically is important. From an academic research perspective, it gives us the chance to find out which theory is supported empirically in the case of telecoms. Such an investigation also offers insight on the value creating and destroying aspects of residual state ownership during the institutional transformation of telecoms from state agencies to private firms. Such insight would have important implications for telecoms policy-makers developing privatization programs, telecoms executives charged with the management of these enterprises in transition, and investors observing it all with an eye toward maximizing their returns.

At first glance, Schumpeter's creative destruction might seem out of place when talking about institutional change in telecoms. As originally articulated in the 1930s and 1940s or as revived and rearticulated by neo-Schumpeterians in the 1980s and 1990s (e.g., Nelson and Winter 1982; Heertje and Perlman 1990; Harris 1998), creative destruction means "carrying out of new combinations" of factors and products, and thereby challenging older modes of organizing and producing. Creative destruction implies competitive tension between the old and tried and the new and still largely untested. The tension may prove continued worth of the old ways. But if the untested "new combination" proves superior, the old gives way, the innovation diffuses, and in the long-run, there is value creation net of the costs of displacing the old. But that is the long-run outcome and a long-run outcome benefiting society generally. In the short- to medium-term, the added value may still be in the offing while the costs of displacing tried and tested combinations are at hand. And even in the long-run, competition between old and new combinations will not benefit all directly. In any contest there will be winners and losers, and those who choose poorly may find little solace in the promise of long-run benefits to society. For many, Schumpeter's dynamic may appear less like creative destruction and more like destruction, plain and simple.

Telecoms are illustrative. McKnight and his colleagues, for example (e.g., McKnight and Lehr 1998; Lehr and McKnight 2000; McKnight and Boroumand 2000; Vaaler and McKnight 2000), have analyzed recent Internet-based technologies for their potentially negative as well as positive implications for incumbent and new-entrant telecoms in the short- and medium-term. Internet-based telephony technology allows a new class of domestic and foreign computer equipment and software, as well as start-up telecoms, to invade incumbents' traditional voice communications market segments and destroy incumbent enterprise value. At the same time, however, growth of Internet-based data traffic along the existing backbone of incumbent enterprises provides new avenues for revenue growth.

Like transformation wrought by the emergence of Internet-based technologies, institutional transformation wrought by telecoms privatization worldwide has both value creating and destroying implications in the Schumpeterian sense. Our study assesses the impact of residual state commitments on that trade-off even as we test for empirical support of the two theoretical models of enterprise privatization. These competing theoretical models and the institutional creative destruction they imply are examined in the context of 15 privatizing telecoms from industrialized and emerging-market countries, and their shareholder reactions to 205 announcements of material merger and acquisition, joint venture, and alliance transactions taking place between 1986 and 2001. While recent reviews of the privatization literature note a well-developed empirical research on the operating performance of privatizing enterprises in industrialized and emerging-market countries (e.g., Megginson and Netter 2001), there is surprising little empirical research based on financial (shareholder) performance measures[3] and none to date examining shareholder returns linked to specific transactions taken by privatizing telecoms in particular. Using descriptive and regression analyses, we assess relationships between the abnormal returns associated with these announcements and the state ownership and temporal distance attributes of these privatizing telecoms.

In brief, we find rather weak support for mainstream model linking higher shareholder returns to lower levels of state ownership and greater temporal distance. By contrast, we find strong support for the alternative model linking higher shareholder returns to higher levels of state ownership and less temporal distance, particularly in case of shareholder returns following announcements by privatizing telecoms from emerging-market countries. Residual state ownership has a value-creating effect, at least in the short-term, during institutional transformation of telecoms through privatization.

To make these and other points in detail, the remainder of this study is organized into five additional sections. Section 2 immediately below summarizes the background on previous privatization practice and research, and provides more detailed exposition of mainstream and alternative models on enterprise privatization and performance. Section 3 articulates the alternative mainstream and alternative model hypotheses for empirical investigation. Section 4 details the methods used to implement the investigation including the equations, specific test statistics for assessing support for mainstream and alternative models, estimation approaches, data sources and sampling approach. Section 5 reports the results from descriptive and regression analyses of the sample. Section 6 concludes the study with discussion of the central results, implications, and future research directions.

II. Privatization Background

Overview of Privatization Policy and Research

The application of privatization policies during the last two decades has enjoyed global scope both in industrialized and emerging-market countries. Several researchers, including Guislain (1997) and Megginson and Netter (2001), have chronicled the progress of these policies on a country-by-country basis. In the industrialized world, for example, French governments in the mid-1980s and again in the mid-1990s privatized more than 30 companies including such state-controlled icons as auto-maker

Renault and France Telecom. The Japanese experience with privatization since the 1980s saw the largest enterprise sell-off in the world to date when NTT was sold to shareholders in 1987 and 1988. The subsequent spin-off of NTT's cellular division, NTT Do-Co-Mo, in late 1998 instantly created the third largest company in terms of market capitalization of the Nikkei index; NTT without Do-Co-Mo remained the largest. The U.S. experience with privatization in the 1980s also saw a substantial transfer of assets to private hands though many of these transfers involved state and local government-owned or controlled rather than federal government-owned or controlled assets (Vernon 1988). While not a formal privatization, the break-up of the regulated private telephone giant, ATT, in 1984 represented a fundamental change in U.S. telecoms industry structure, and spurred a wave of new entries in local and long-distance voice, data and cable media segments previously thought to be better served by a single dominant supplier.

Stanbury (1994) suggests that emerging-market countries should have led rather than followed the lead of industrialized countries in implementing privatization programs in the 1980s and 1990s. Fiscal concerns were more acute in emerging-market countries compared to industrialized countries and the burdening of maintaining state-owned or controlled enterprises more onerous. Ramamurti (1992) echoes this point by showing that countries running higher budget deficits, accruing more foreign debt, and experiencing greater productive inefficiency in the administration of state-owned enterprises—a description of many emerging-market countries in the 1980s and 1990s—are more likely to implement privatization policies. Despite their predisposition to embrace privatization policies, emerging-market countries may be stifled in the implementation of such policies because of the absence of key factors including professional management expertise, capital, or a stable legal and regulatory framework. Research by Galal et al. (1994) highlight the small absolute size of national economies and slower economic growth rates of many developing countries as potentially limiting factors in the successful implementation of state privatization programs. At a minimum, such country-level, industry- (regulatory) and enterprise-specific contingencies explain varying degrees of success in privatization programs across emerging-market countries in Latin America, Central and Eastern Europe, and elsewhere.

The Mainstream Model and Related Studies

Almost as soon as privatization policies were implemented, researchers sought to understand whether and why privatized enterprises performed differently. In these streams of research we can discern the development of mainstream and alternative views on performance in privatizing enterprises. After early research by Caves and Christenson (1981) in Canada, and Yarrow (1986) and Vickers and Yarrow (1988) in the UK suggested that privatized enterprises were no more productively efficient than their nationalized counterparts, a steady flow of empirical research led by Megginson and his collaborators (e.g., Megginson et al. 1994) established that, for a range of countries and industries, shifts from state to private ownership followed by decreasing state-owned equity were associated with superior operating returns, employee productivity and turnover in either top-management teams, directorial boards or both *over time*. The empirical research, summarized most recently and comprehensively in Megginson and Netter (2001), provides the main supporting evidence for the mainstream model implication that decreasing state ownership and increasing temporal distance are central to organizational change and value creation on privatizing enterprises.

Many of these observed changes in privatizing enterprise behavior and performance are justified in terms of the realignment of enterprise stakeholder incentives, particularly the incentives of enterprise owners (principals) and enterprise managers (agents) (Jensen and Meckling 1976). As Boycko, Shleifer and Vishny (1996) and others contend, private ownership immediately provides strong incentives for managers to innovate products and markets and create value for the firm and its shareholders. Where managers and the directorial boards overseeing them fail in this mission, wealth-maximizing shareholders can replace them. And where shareholders fail, the market for corporate control will lead to a transfer of shares to more vigilant holders willing to pay more. Timely, substantial post-privatization turnover in management and directors, as well as enhanced employee productivity and firm performance are consistent with this principal-agent perspective so central to the mainstream model.

Foreign investment by privatizing enterprises speeds the transformative process from state- to private shareholder-orientation. Kogut (1996) suggests that the positive contribution of foreign investment results from the greater access it provides privatizing enterprises to more sophisticated individuals and capabilities. Because foreign investment frequently involves a transfer of equity to foreign individuals and institutions, there is an added beneficial effect in the form of better monitoring of enterprise managers. These different factors raise the probability that the enterprise will be able to draw on a broader international menu of organizational practices associated with higher performance.

This may undermine the domestic state's role in guiding privatized enterprises; on the other hand, it also eventually endows the privatizing enterprise with a broader portfolio of competencies *outside* the control of the state. Indeed, foreign investment policies undertaken by privatizing enterprises may even have the principal purpose of simply raising the costs of state interference in enterprise affairs. States may become more hesitant to impose their political agendas on newly privatized enterprises if they anticipate a backlash from the foreign investment community (Guislain 1997).

The Alternative Model and Related Studies

Though contrasting in its key conclusions about the impact of state ownership and temporal distance on privatizing enterprise performance, the alternative model draws on many of the same theoretical perspectives. The concept of "credible privatization" espoused by Perotti and Guney (1993) and then more formally by Perotti (1995) is at the heart of the alternative model, which takes issue with the mainstream model's prescription of rapid and complete state divestment. Principal-agent assumptions in the alternative model limit the ability of shareholders (principals) to monitor and properly motivate managers (agents) in the privatizing enterprise and lead to two important insights. First, the sale by the state of equity in such enterprises might have to be discounted to reflect these principal-agent problems as well as broader problems in the enforcement of shareholder rights and in the development of corporate governance mechanisms. Directors, private shareholders and the market for corporate control back-stopping all of them may function quite inefficiently if at all in countries making the transition from planned to market economy. As Dyck (2001) points out, corporate governance problems may be particularly acute in many emerging-market countries. Without strong "private governance chains" to constrain top management opportunism, shareholders would demand a steep discount on the price of privatizing enterprise equity or might refuse to invest at all.

A second and related insight drawn from the alternative model suggests that state divestment of ownership should be gradual rather than immediate. The state would, therefore, remain as a substantial (though not controlling) shareholder in the privatizing enterprise in the short- to medium-term. With retention of substantial state ownership (but with effective control in the hands of enterprise managers), the state would communicate to anxious private shareholders an intent to share their economic fate and, thus, ensure minimal enterprise performance standards. This makes the privatization "credible." It may follow from state oversight of managerial agents complementing private shareholder oversight. It might also take the form of beneficial state intervention in the privatizing enterprise's various market relationships. Examples include state allocation of preferred landing rights to privatizing airlines, guarantees on long-term debt carried on privatizing electricity generators, or, as is often the case with telecoms, guarantees limiting competitive entry into lucrative market segments (Guislain 1997). Whether by providing additional oversight or by intervening in market relationships to ensure some minimal standard of performance, state investment and related commitments may assuage private shareholder concerns about privatizing enterprise performance in the near term.

For Ramamurti (2001), this process of state retreat from initial commitments represents a contemporary form of the obsolescing bargain phenomenon originally developed by Vernon (1971) to explain fluctuations in foreign direct investment by multinational corporations negotiating with host governments in the developing world. For Emmons (2000) the resulting tendency to renegotiate property rights is central to understanding enterprise privatization's "evolving bargain" between state and firm. Again, the state's tendency to pull back from initial commitments may be most acute in emerging-market countries where institutional development regarding the rule of law and respect for property rights and private enterprise are less well-developed (Murtha and Lenway 1994), where political business cycles make such a pull-back attractive to an elected incumbent government official seeking to retain office (Schipke 2001). In these and related contexts, privatization and post-privatization development policies are less likely to be sustained to the detriment of shareholder confidence and enterprise share value (Perotti and Laeven 2002).

Consistent with the alternative model's predictions, recent empirical evidence reported by Jones et al. (1999) leads them to argue that state enterprises should be partially privatized rather than sold off 100 percent when its initial demonstrable value based on recent operating experience is low relative to its intrinsic value. On the other hand, much more of the recent empirical research on shareholder returns for privatizing enterprises seems to support the mainstream view to date, with cross-country studies in Dewenter and Malatesta (2001), D'Souza and Megginson (1999) and Bortolotti et al. (2001) finding positive abnormal returns over time and a negative association between the abnormal returns and the percentage of state ownership.

This study seeks to complement this macro view of the privatizing enterprise's overall performance trend. It provides a more focused micro view of shareholder assessments around specific and material decisions taken by privatizing telecoms. By this approach, we gain important additional insight on the value creating and destroying effects of residual state ownership in privatizing telecoms as well as how such effects support either theoretical model of privatization summarized briefly above.

III. Hypotheses for Empirical Analysis

Our review of the privatization literature generally, and of the mainstream and alterative models specifically, lead to the two sets of competing hypotheses stated below. Consistent with the mainstream model we hypothesize that:

H1a: Shareholder returns are negatively related to the percentage of state ownership in a privatizing telecom taking material investment decisions.

H2a: Shareholder returns are positively related to the temporal distance of a privatizing telecom taking material investment decisions.

As we indicated above, the mainstream model anticipates the prospective benefits to enterprise decision-making of less state ownership and more temporal distance from state control. It anticipates the speedy development of enterprise incentives and corporate governance institutions to implement shareholder-wealth-maximizing strategies effectively. These mainstream model assumptions and hypotheses may seem best suited to, say, the United States, the UK, and other industrialized countries with well-developed share markets, corporate governance systems and property rights regimes.[4]

By contrast, state ownership and temporal distance in the alternative model are predicted to have opposite effects on shareholder returns associated with privatizing telecom investment decisions:

H1b: Shareholder returns are positively related (or show no relation at all) to the percentage of state ownership in a privatizing telecom taking material investment decisions.

H2b: Shareholder returns are negatively related to the temporal distance of a privatizing telecom taking material investment decisions.

The alternative model carries with it skepticism regarding the effectiveness of still-developing enterprise incentives and corporate governance structures. Indeed, there seems also to be concern in this view for the clarity, consistency, and enforceability of still-developing property rights. State participation in this context provides a partial and temporary palliative for privatizing enterprise managers and their shareholders. These alternative model assumptions and hypotheses seem best suited to, say, Brazil, Hungary, Thailand, and other emerging-market countries with still-developing share markets, corporate governance systems and property rights regimes.

IV. Methodology

Given the focus on financial performance associated with specific, material decisions taken by privatizing enterprise management, we chose an event study methodology, which uses share price or asset price changes to assess the performance implications of organizational decision-making. It is used primarily in the finance field, but has been increasingly applied to business strategy, accounting, law, organizational behavior, and marketing research questions (McWilliams and Siegel 1997).

Empirical Models

We use two empirical models to assess our four hypotheses. Consistent with standard event study methods, equation (1) below is used to estimate cumulative abnormal returns to shareholders related to privatizing telecom investment events.

$$\begin{aligned} CAR_{ijt} = {} & \beta_0 + \beta_1 percstate_{ijt} + \beta_2 \log(zeromon_{ijt}) + \beta_3 emgmkt_{ij} \\ & + \beta_4 percstate_{ijt}{}^* emgmkt_{ij} + \beta_5(\log(zeromon_{ijt})^* emgmkt_{ij}) \\ & + \beta_6(\log(zeromon_{ijt})^* percstate_{ijt}) + \beta_7 eventJV_{ijt} + \beta_8 eventMA_{ijt} \\ & + \beta_9 target_{ijt} + \beta_{10} \log(sales_{it}) + \beta_{11} roa_{it} + \beta_{12} pubexpgdp_{it} \\ & + \textstyle\sum \omega_{1-14} company_i + \sum \psi_{1-13} year_t + \mu_{ijt} \end{aligned} \tag{1}$$

In equation (1), the subscript i indicates the privatizing telecom, the subscript j is an investment event counter for each privatizing telecom i, and the subscript t indicates the year of the telecom investment event j announced by privatizing telecom i.

The dependent variable, *CAR*, designates the cumulative abnormal returns measured according to the methodology laid out above. We calculate CAR in equation (1) following Brown and Warner's (1985) standard event study methodology. We identify an investment event j, record its date as $T = 0$, and use daily data on the stock market returns for the privatizing telecom i from $T = -200$ to $T = -10$. These data permit estimates of expected shareholder returns over the investment event window of observation. The returns are expected to follow the equation:

$$E(r_{iT}) = a_i + r_{mT}$$

where $E(r_{iT})$ is the expected stock return of privatizing telecom i on day T, r_{mT} is the corresponding daily market return on the equal-weighted S&P 500 index and a_i is the intercept. For the privatizing telecom, its specific abnormal returns are calculated as:

$$AR_{iT} = r_{iT} - E(r_{iT})$$

which is the difference between the actual returns to privatizing telecom shareholders and the broader market returns over the same day in investment event window. Cumulative abnormal return ("CARs") simply add up these daily abnormal returns over the entire event window:

$$CAR_{ijt} = \textstyle\sum_T AR_{iT}.$$

We use two-, three- and five-day event windows to measure this market-based CAR.

The independent variables of central interest in equation (1) concern the privatizing telecom's percentage of state ownership and its temporal distance from the initial date of privatization. The variable, *percstate*, measures the percentage of equity held by the state at the end of the year of each investment event. The term, *zeromon*, is the number of months between the month of initial privatization and the month of the investment event. Since we have hypothesized that over time there is a convergence between formerly state-owned private enterprises over time, we take the natural log of *zeromon*, which has the effect of attributing greater weight to investment events closer to the date of initial privatization. We also interact *percstate* and *zeromon* with each other and with a dummy variable *emgmkt*, which assumes the value 1 if the firm is domiciled in an emerging market and 0 otherwise. Interaction with the emerging-market dummy permits us to assess differences in state ownership and temporal distance effects between privatizing telecoms from industrial versus emerging-market countries.

The right-hand side of equation (1) includes several controls for company-specific factors that may also explain shareholder returns associated with an investment event. Following previous event studies examining M&A or JV transactions (e.g., Grover 2001; Fuller et al. 2002; Park et al. 2002), equation (1) also controls for size (the natural log of *sales*), measured as the company revenues in US$, and profitability (*roa*) measured as company operating income divided by net assets in US$. We also control explicitly for one country-level variable thought to affect CARs, annual change in the percentage of GDP comprised by public (government) expenditure (*pubexpgpd*). Ramamurti (2000) argues that shifts in public policy favoring less state involvement in the economy and greater privatization encourages the speedy development of institutions favorable to private enterprise ownership, including corporate, labor, and broader regulatory law reform. The *pubexpgdp* term serves as a proxy for such shifts and is measured as the difference of the percentage in the year of an investment event less the percentage in the previous year. It is expected to have a negative sign.

Equation (1) also includes other controls, including dummies for privatizing telecoms (*company*), years (*year*), and investment event types in our sample. While the individual privatizing telecom and year dummies are straightforward, the investment event dummies merit brief explanation. For data on investment events, we used the Securities Data Corporation's Mergers & Acquisitions ("M&A") Database (SDC 2002), which provides comprehensive coverage of mergers, acquisitions (both as acquirer and target), seasoned equity offerings, joint ventures ("JVs") and strategic alliance ("Alliance") announcements. As additional controls, therefore, we include three different investment event dummies. The variable *eventJV* takes the value of 1 when the investment event is the announcement of a joint venture resulting in the creation of a third-party entity but where there is no equity transferred from one party directly to any of the others. The variable *eventMA* takes on the value of 1 when the investment event is the announcement of merger or acquisition transaction involving the direct transfer of equity from one party to another and where the privatizing telecom is deemed by SDC to be the acquiring company. The variable *target* takes on a value of 1 if the same MA investment occurs but the privatizing telecom is deemed by SDC to be the equity-giving company. Alliance investment events are the omitted category, and are the same type of event as a JV except that no third-party entity is created.

In order to check for robustness of results obtained from the multivariate regression analysis, we estimate a second model which has identical independent variables to equation (1), but a different dependent variable, *abpos*, a 0–1 indicator of whether an investment event resulted in a positive or negative cumulative abnormal return to the privatizing telecom shareholders over the observation window. We define this dummy variable as

$$abpos_{ijt} = \begin{cases} 0 \text{ if } \mathrm{CAR}_{\mathrm{ijt}} \leq 0 \\ 1 \text{ if } \mathrm{CAR}_{\mathrm{ijt}} > 0 \end{cases}$$

With *abpos*, equation (2) below permits assessment of the effects state ownership and temporal distance may have on shareholder returns independent of the magnitude of such returns. It considers instead the trends in the frequency of favorable (positive) investment event returns.

$$\begin{aligned} abpos_{ijt} = {} & \beta_0 + \beta_1 percstate_{ijt} + \beta_2 \log(zeromon_{ijt}) + \beta_3 emgmkt_{ij} \\ & + \beta_4(percstate_{ijt}{}^*emgmkt_{ij}) + \beta_5(\log(zeromon_{ijt}){}^*emgmkt_{ij}) \\ & + \beta_6(\log(zeromon_{ijt}){}^*percstate_{ijt}) + \beta_7 eventJV_{ijt} + \beta_8 eventMA_{ijt} \\ & + \beta_9 target_{ijt} + \beta_{10}\log(sales_{it}) + \beta_{11} roa_{it} + \beta_{12} pubexpgdp_{it} \\ & + \textstyle\sum \omega_{1-14} company_i + \sum \psi_{1-13} year_t + \mu_{ijt} \end{aligned} \tag{2}$$

Turning to our four hypotheses, equations (1) and (2) facilitate straightforward tests. Hypotheses 1a and 2a make mainstream model predictions that privatizing telecoms will take investment decisions resulting in higher (more frequently positive) CARs as the percentage of state ownership decreases and temporal distance increases. This implies a negative coefficient sign on *percstate* and a positive coefficient sign on *log(zeromon)*. In terms of the each equation, the hypothesis tests reduce to:

H1a: $\beta_1 < 0$ and *H2a:* $\beta_2 > 0$

This prediction is challenged by the alternative model of privatization, which predicts positive coefficient sign on *percstate* and a negative coefficient sign on *log(zeromon)*. The hypothesis tests reduce to:

H1b: $\beta_1 \geq 0$ and *H2b:* $\beta_2 < 0$

An interesting subsidiary analysis interacts *percstate* and *log(zeromon)* with the the emerging-market country indicator *emgmkt*. The impact of state ownership and temporal distance may be different for privatizing telecoms from emerging-market countries suited to alternative view assumptions versus those from industrialized countries suited to mainstream view assumptions. If so, then the coefficient sign on *percstate* emgmkt* interaction should be positive relative to the coefficient sign on *percstate* alone, which represents the state ownership impact for privatizing telecoms from industrialized countries. Similarly, the coefficient sign on *log(zeromon)*emgmkt* interaction should be negative relative to *log(zeromon)* alone. In terms of equations (1) and (2), these subsidiary propositions will be supported if: $\beta_4 > 0$ and $\beta_5 < 0$.

Estimation Strategy

As McKinlay (1997) recently pointed out, event study methods used today are remarkably similar to those developed by Brown (1968) and Fama et al. (1969) more than 30 years ago. Our own approach to estimating equation (1) follows the standard method closely, though instead of using OLS, we estimate equation (1) using a generalized least squares ("GLS") estimator, which for our sample of privatizing telecom events includes robust standard errors to correct for possible heteroskedasticity and clustering on privatizing telecoms. We also calculated Cook's Distance statistics to check for outliers and eliminated 19 observations with extreme D values (D > 0.02).

Estimating equation (2) differs from the estimation approach for equation (1). Owing to the 0–1 limitations on the dependent variable in equation (2), we employ a probit model. As with the estimators in equation (1), we use robust standard errors and adjust for clustering on privatizing telecoms.

Data Sources and Sampling

To obtain our sample of privatizing telecoms, we turned to the "Telecom/Data Networking" category of Bank of New York's Depository Directory (Bank of New York 2002). This directory lists all firms that have issued depository receipts ("DRs") in the United States, whether they are traded on regulated exchanges or on over-the-counter and whether they are sponsored or not. By limiting our data to privatizing telecoms with DRs in the United States, we were able to control for several factors, and assess investment event CARs for privatizing telecoms from different countries with a common currency ($) and against a single (U.S.) stock market index of returns.

From this data source we sampled firms operating in the fixed-line telecommunications business, with a history of state ownership or effective state control, and having experienced either the sale of former state-owned equity or the release from *de facto* control of such equity by the state since 1980. This resulted in 18 privatizing telecoms, 15 of which were previously wholly owned by the state, and three of which had *de jure* private owners but were under *de facto* state control (i.e., Telecom Italia, Telefónica de España and Philippine Long-Distance Telephone Company). We noted the date of initial sale of equity, either through private placement, public offering of shares, material asset sale, voucher distribution or related means as the date of initial privatization for the 15 previously state-owned telecoms. For the remaining three telecoms, we followed an approach taken by Vaaler (2001) and noted their date of initial privatization as the date of fixed-line telecom operation deregulation, which, in each case also shifted *de facto* control to private owners.

From this group of 18 privatizing telecoms, we eliminated nonoperating (corporate holding company) firms and those for which there was no data on DR prices from the Center for Research in Security Prices database (CRSP 2002). Our final sample reported in table 13.1 comprised 15 privatizing telecoms, 11 of which were domiciled in industrialized countries (i.e., British Telecom, Deutsche Telekom, France Telecom, Hellenic Telecom, KPN (Netherlands), New Zealand Telecom, Nippon Telephone & Telegraph, Portugal Telecom, TDK (Denmark), Telecom Italia, and Telefónica de España) and four of which came from emerging-market countries (Korea Telecom, Philippine Long Distance Telephone Company, Rostelecom (Russia), and Teléfonos de Mexico). Dates of initial

Table 13.1
Telecom Firms in the Sample

Firm	Date of privatization	Number of events
British Telecom	Nov 1984	69
Deutsche Telekom	Nov 1996	17
France Telecom	Oct 1997	17
Hellenic Telecom	Jan 1996	4
Korea Telecom	Nov 1993	5
KPN (Netherlands)	June 1994	13
New Zealand Telecom	July 1991	2
Nippon Telegraph & Telephone	Nov 1986	43
Philippine Long Distance Telephone Company	Dec 1993	3
Portugal Telecom	June 1995	11
Rostelecom	July 1997	1
TDK (Denmark)	May 1994	3
Telecom Italia	Nov 1985	20
Telefónica de España	Oct 1989	23
Teléfonos de Mexico	May 1991	1

privatization ranged from 1984 (British Telecom) to 1997 (France Telecom and Rostelecom), with the majority undergoing initial privatization in the early to mid-1990s.

Our two dependent variables, CARs (*CAR*) and the 0–1 positive CARs indicator (*abpos*), are both derived from DR and broader stock market price data associated with privatizing telecom investment events. Accordingly, we collected data on prices in US$ for DRs from CRSP and noted the daily percent returns for each of the 15 privatizing telecoms. To compare them with broader market returns over a comparable period, we also obtained from CRSP daily percent returns of the equally weighted Standard and Poor's ("S&P") 500 index.

For data on investment events (*eventJV*, *eventMA*, *target*), recall that we used SDC data and their investment event designations: M&A (acquirer or target), JV and Alliance. We included as M&A (target) investment events secondary equity offerings by privatizing telecoms.

We then screened these investment events for their materiality to shareholders. If announcement of the investment event appeared in SEC filings or was reported in the American editions of the *Wall Street Journal*, the

Table 13.2
Frequency of Events by Category

Event	Number	Percentage
Joint Venture	77	34.4%
Alliance	90	40.2%
M&A (Target)	36	16.1%
M&A (Acquirer)	21	9.4%
Total	224	100%

Financial Times, or the Reuters News Network, it was deemed material. Finally, we screened the remaining investment events to eliminate those occurring prior to the issuance of the privatizing telecom's DR, or if two investment events for the same privatizing telecom were reported within an interval of five business days. These screens resulted in a sample of 224 investment events occurring between 1986 and 2001 and are presented in table 13.2.

To estimate equations (1) and (2), we also required additional data on the privatizing telecoms, and their respective countries of domicile. 20-F filings from the U.S. Securities and Exchange Commission ("SEC")[5] provided information on year-to-year changes in the percentage of state ownership (*percstate*) and permitted confirmation of all initial privatization dates (*zeromon*). Using Compustat (2002) corporate-level data, we obtained information on annual sales (*sales*), net income and assets (*roa*), market capitalization, and shares outstanding. Using S&P's Emerging Market Database, we grouped the 15 privatizing telecoms into industrial and emerging-market countries (*emgmkt*). The World Bank's World Development Indicators database (World Bank 2002) provided data on aggregate yearly government spending as percentage of country GDP (*pubexpgdp*).

V. Results

Together, our results presented in table 13.3 indicate that both state ownership and temporal distance affect shareholder returns from privatizing telecom investment events, and in line with the alternative rather than mainstream model. We find only tepid support for the mainstream model's hypothesized negative effect of state ownership (H1a) and positive effect of temporal distance (H2a) on shareholder returns. By contrast, coefficient signs and significance show support for the alternative model, particularly in the case of privatizing telecoms from emerging-market countries: State ownership is positively related (H1b) and temporal distance negatively related (H2b) to shareholder assessments of investment events undertaken by privatizing telecoms from emerging-market countries. This, in turn, suggests that distinctions between industrialized versus emerging-market country status are quite important for understanding the impact of the privatization-related factors on telecom investment decision-making quality. These points are discussed in greater detail below.

Descriptive Analyses

Column 1 in table 13.3 reports descriptive statistics for our sample. Average net returns for the sample is a healthy 4.2 percent with surprisingly little difference between ROA for industrialized versus emerging-market telecoms. Substantial variation in other variables, however, suggests that inclusion of terms distinguishing industrialized from emerging-market country telecom effects is warranted. Average size in terms of sales is approximately $36 billion, but telecoms from industrialized countries exhibit substantially greater average sales ($37 billion) than telecoms from emerging-markets ($3.5 billion). Average state ownership over the period sampled is roughly 26 percent with more than 80 percent of telecom investment events taking place in the context of some residual state ownership. On the other hand, industrialized telecoms register lower average state ownership (26 percent) compared to their emerging-market counterparts (37 percent). Timing of investment events also illustrates this divide. The average investment event took place approximately 111 months after privatization for the total sample, but for emerging-market telecoms, the average is only 59 months. A summary interpretation of these statistics suggests that all of our privatizing telecoms were busy with investments in the later half of the 1990s. Industrialized telecoms were doing so with much less state involvement and more experience in the market place, while their emerging-market counterparts were doing so with substantial residual state ownership and little temporal distance from their days as a state ministry.

Regression Analyses

Column 2 of table 13.3 reports results from GLS estimation of equation (2). State ownership (*percstate*) captures CARs effects for privatizing telecoms from industrialized countries only, while state ownership effect differences for emerging-market telecoms are captured in the *percstate* emgmkt* interaction term. Consistent with the mainstream model, *percstate* exhibits a negative sign but is not significantly different from zero at commonly acceptable levels. Supporting the alternative model and associated Hypothesis 1b, however, we observe significant (at the 5 percent level) positive effects (0.060) on CARs related to the percentage state ownership of privatizing telecoms from emerging-market countries. Addition of *percstate* and *percstate*emgmkt* terms ($\beta_1 + \beta_4$) yield a significant (at the 5 percent level) positive coefficient (0.057), indicating a positive state ownership effect for privatizing emerging-market telecoms relative to zero (rather than merely relative to state ownership effects for privatizing telecoms from industrialized countries).

Practically speaking, the results suggest that a 1 percent increase in state ownership results in a 0.06 percent increase in CARs to shareholders. Positive state ownership effects in emerging-market contexts may be explained by the relatively underdeveloped nature of public (e.g., securities regulation) and private institutions (e.g., credit rating agencies). In such contexts, increased state ownership and incentives to monitor enterprise managers and investments more closely may have positive performance effects outweighing negative effects from possible state interference in enterprise decisions calculated to enhance shareholder wealth. Alternatively, increased state ownership could also raise state incentives to ensure some minimum level of privatizing enterprise performance through intervention in the privatizing firm's various market relationships. In either case, private shareholders are beneficiaries and view privatizing enterprise decisions more favorably.

The alternative model and its associated Hypothesis 2b predicts that this benefit to shareholders decreases as temporal distance increases while the mainstream model and its associated Hypothesis 2a predicts a positive relationship between temporal distance and shareholder returns. Again, results in column 2 generally fail to support the mainstream model. The coefficient on the temporal distance for privatizing telecoms from

Table 13.3
Regression Results[a,b]

		Dependent variable		
		3-Day CAR	Probability of positive 3-Day CAR	Probability of positive 3-day CAR
	Estimator/event window			
Variable	(1) Mean (std. dev.)	(2) GLS	(3) Probit	(4) dprobit[d]
percstate [β_1]	26.3	–0.002	–0.211*	–.084
	(23.8)	(0.002)	(0.114)	
log (zeromon) [β_2]	4.49	–0.043	–1.861	–.742
	(0.74)	(0.052)	(2.068)	
emgmkt [β_3]	0.04	dropped	dropped	
	(0.19)			
*percstate * emgmkt* [β_4]	1.49	0.060**	19.491***	7.770
	(8.28)	(0.025)	(0.000)	
*log (zeromon) * emgmkt* [β_5]	0.21	–0.415**	–128.090***	–51.066
	(0.91)	(0.157)	(0.000)	
*log (zeromon) * percstate* [β_6]	109.61	0.000	0.024	.009
	(99.79)	(0.000)	(0.026)	
eventJV [β_7]	0.34	0.023*	0.655***	.254
	(0.47)	(0.010)	(0.242)	
eventMA [β_8]	0.25	0.026	0.641***	.245
	(0.43)	(0.015)	(0.242)	

target [β_9]	0.16	–0.019	–0.377	–.148
	(0.36)	(0.015)	(0.428)	
log (sales) [β_{10}]	10.10	–0.028	–4.465***	–1.780
	(0.98)	(0.034)	(1.181)	
roa [β_{11}]	0.04	–0.361***	–25.175***	–10.036
	(0.03)	(0.094)	(6.602)	
pubexpgdp [β_{12}]	–0.03	0.158	13.097***	5.221
	(0.02)	(0.123)	(2.723)	
Constant [β_0]		0.520*	56.694***	
		(0.273)	(14.172)	
Observations	224	205	199	199
R–squared[e]		0.26	0.17	0.17

*significant at 10%; **significant at 5%; ***significant at 1%.

a. Generalized Least Square (column 2) and Probit-related estimators (columns 3–4) with observations clustered on companies. Robust standard errors were obtained using the Huber-White estimator of variance (Stata Corp., 2001) and are reported in parentheses.

b. Reported results (columns 2–4) include year and company dummy variables. The majority of coefficient estimates for these dummies are significant at $p < 0.05$ or higher levels. Joint significance of dummies are also significant at $p < 0.05$ or higher levels. These results are available from the authors on request.

c. The dependent variable is a dummy variable assuming the value of 1 if an event is associated with a positive CAR; otherwise it takes the value of 0. A 3–day window was used to estimate CARs. Results using 2- and 5-day windows are consistent with these results and available from the authors on request.

d. Reported results (column 4) are obtained using Stata's (2001) dprobit routine. The dprobit routine reports the percentage change in probability of a positive CAR given a unit increase in the independent variable.

e. Pseudo R^2 statistics are reported (columns 3–4).

industrialized countries (*log(zeromon)*) is negative rather than positive as the mainstream model and Hypothesis 2a predict. They are not significant at any commonly acceptable level. By contrast again, the coefficient for differences in emerging-market privatizing telecom temporal distance effects (*log(zeromon)*emgmkt*) is both negative (–0.415) and significant at the 5 percent level. This strong support for the alternative model and Hypothesis 2b is confirmed when we add the two effects ($\beta_1 + \beta_4$) to assess temporal distance effects relative to zero for privatizing telecoms from emerging-market countries. The results are negative (–0.459) and significant at the 5 percent level.

Practically speaking, an increase in temporal distance from 1 to 2 years decreases CARs to privatizing telecoms from emerging-market telecoms by approximately 32 percent, while an increase in temporal distance from 2 to 3 years decreases CARs by an additional 19 percent. Read together with the state ownership effects, we see that higher shareholder returns derived from signaling commitment to support the privatizing telecom may substantially decrease and, indeed, quickly turn negative, with the passage of time and inevitable pressures on the state to reverse such commitments for financial and or political gain.

Columns 3–4 of table 13.3 report results from probit regression using the positive CARs 0–1 indicator as the dependent variable. They confirm strong support for the alternative model of investment decision-making quality for privatizing telecoms from emerging-market countries. They also confirm the rather tepid support for the mainstream model we observed in the GLS results. In column 3's probit regression results, we observe for the state ownership term (*percstate*) a negative coefficient (–0.211) that is significant at the 10 percent level. While consistent with the mainstream model, the 10 percent level of significance suggests caution in concluding support for Hypothesis 1a. On the other hand, state ownership effect differences for telecoms from emerging markets (*percstate*emgmkt*) are positive (19.491), significant at the 1 percent level, and practically substantial. Holding other factors at their mean values, a 1 percent increase in state ownership of an emerging-market telecom results in an approximately 7 percent change in the probability that an investment event will generate positive CARs relative to similarly situated telecoms from industrialized countries. Temporal distance effects for emerging- market telecoms relative to industrialized telecoms (*log(zero-mon)**emgmkt*) echo strong support for the alternative model and Hypothesis 2b. The coefficient is negative (–128.090) and significant at the 1 percent level. Holding other factors at their mean values, an increase in temporal distance from 10 to 27 months decreases the probability of an investment event leading to positive CARs for privatizing telecoms from emerging-market telecoms by approximately 51 percent relative to similarly situated industrialized telecoms.

Illustration and Practical Implications of Results

Results from analysis of specific privatizing telecoms and their investment events provide helpful complementary insight to the regression results we summarized above. For one such case we turn to Russia's Rostelecom and its July 9, 1999 announcement of a joint venture with a UK-based partner, Sweet and Great Northern Telegraph Company, to invest in the Russian company RTC Page. RTC Page possessed a license to operate a national paging system based on the digital ERMES standard. During the three-day event window around the announcement date, Rostelecom's DR experience returns of cumulative shareholder returns of 1.70 percent. The (S&P 500) market-adjusted returns for the same period approximated 0 percent, thus, CARs were also approximately 1.70 percent. During this investment event window, the Russian government owned 45 percent of Rostelecom's equity. Rostelecom's temporal distance from initial privatization in July 1997 was approximately 24 months. Using coefficients from column 2 of table 13.3, we calculate that the increase in shareholder returns to Rostelecom would be approximately 0.22 percent higher had the state's equity share in Rostelecom been 49 percent rather than only 45 percent (0.057*.04*100%). On the other hand, if the joint venture decision had been announced 30 rather than only 24 months after initial privatization CARs would be expected to fall by approximately 10 percent ([–0.459*(ln(30)]—[–0.459*(ln(24)]).

Given Rostelecom's market capitalization of approximately $1.4 billion in July 1999, even small changes in CARs could have had substantial impact on the privatizing telecom's financial performance. A 0.22 percent increase in CARs translates into approximately $3.1 million in additional market capitalization. A 10 percent decrease in CARs results

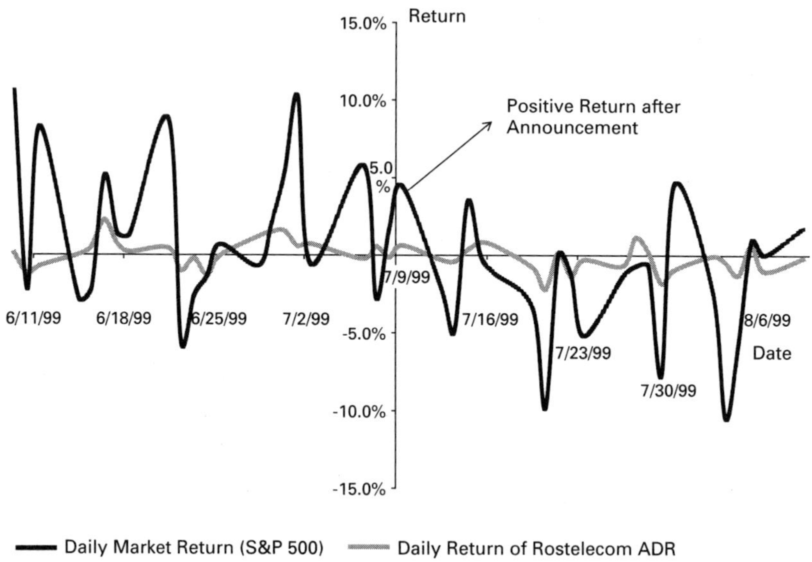

Figure 13.1
Daily ADR and Market Returns 30 Days Before and After Rostelecom's Joint Venture Announcement on July 9, 1999.

in \$140 million in lost market capitalization. Like Rostelecom, other privatizing telecoms from emerging markets may be pressed to make material investments immediately after initial privatization when the state's equity share is still substantial and its commitments to the enterprise and its private shareholders more credible.

VI. Discussion and Conclusion

We conclude, or perhaps we should say, hypothesize, that the next wave of creative destruction and institutional innovation in telecommunications will sweep away the notion that the state has no role to play in enabling market success. On reflection, we should not be surprised that the state retains an important role in supporting shareholder value in what are called, after all, "emerging-market" nations.

This study provided an empirical test of two competing theoretical models concerning privatization and the residual performance effects of state factors viewed from a shareholder perspective. Our hypotheses derived from the two models predicted quite different effects for state ownership and temporal distance on the financial performance of privatizing telecoms. The mainstream model indicated that residual state ownership is not desired by shareholders, and that governments were generally incapable of making valuable commitments to shareholders. The alternative model, however, suggested that shareholders do have incentives, at least in the short term, to keep the state involved as a non-controlling owner. Clearly there was, at best, only tepid support for the mainstream model, but quite substantial evidence supporting the alternative model describing "credible" privatization, particularly for telecoms from emerging-market countries.

These results raise interesting implications for our broader research interest: Understanding the value creating and destroying implications of institutional innovation and change. Our results shed light on that issue, even as they also adjudicate between mainstream and alternative models of privatization. Investment event returns for our privatizing telecoms are contingent. Higher quality investment decision-making, at least from a shareholder perspective, does not necessarily materialize the moment telecoms first transfer equity from public (state) to private hands. Nor do any positive indications from shareholders necessarily persist for long after the equity transfer from public to private hands begins. In the short- to medium-term, institutional transformation from state agency to private firm is dependent on many contingencies, including ironically, those related to residual state commitments. By managing those commitments, particularly in emerging-market contexts, public policy-makers and telecoms managers can together create value, or at least minimize value destruction in the short- to medium-term, the interim period when Schumpeter's dynamic can produce the most turbulence.

Our analysis of privatizing telecoms raises a host of issues for future study. For example, the regression results indicate that shareholders reward privatizing telecoms engaged in cooperative forms of foreign investment such as joint ventures and strategic alliances. Future study might examine links between these cooperative ventures and the sharing of risk and resources, including know-how, that they entail. Other future research might explore in greater detail differences in the decision-making calculus of privatizing telecoms from industrialized democracies versus emerging-market countries. Such follow-on work may provide additional insight on investment strategies helpful to managers seeking competitive advantages for their privatizing firms, and greater value-creation (or, at least, less value destruction) for their shareholders.

Notes

1. The authors gratefully acknowledge support for this research from the School of Information Studies, Syracuse University, the Program on Internet and Telecoms Convergence, Massachusetts Institute of Technology, Booz-Allen & Hamilton, Inc., Tufts University's Fletcher School of Law & Diplomacy, and in particular, Fletcher's Hitachi Center for Technology & International Affairs. This paper draws substantially from previous work by the co-authors including McKnight *et al.* (2001), Vaaler (2001) and especially, Schrage (2002).

2. Several contributions to McKnight *et al.* (2001) might represent exceptions to this trend.

3. Empirical studies on post privatization long-run shareholder returns for cross-country samples are provided in Dewenter and Malatesta (2001), Megginson *et al.* (1999) and Bortolotti *et al.* (2001) who examine privatizing telecoms shareholding returns but not with respect to specific events as in this study. There are numerous single country studies cited in Megginson and Netter's (2001) exhaustive review of the privatization literature.

4. Interestingly, however, the mainstream model's authors, Maxim Boycko, Andrei Shleifer and Robert Vishny, applied this model to analysis of privatization programs in emerging-market countries such as Russia.

5. 20-F filings are required annually for the registration of securities by foreign private issuers pursuant to section 12(b) or (g) of the US Securities Exchange Act of 1934.

References

Bank of New York. *Global DR Directory*. Available from Bank of New York. http://www.adrbny.com/dr_directory.jsp (18 June 2002).

Bortolotti, Bernardo, Juliet D'Souza, and Marcella Fantini, William L. Megginson. "Sources of Performance Improvement in Privatized Firms: A Clinical Study of the Global Telecommunications Industry." University of Oklahoma Working Paper, 2001.

Boycko, Maxim, Andrei Shleifer, and Robert Vishny. "A Theory of Privatization." *Economic Journal* 106 (1996): 309–319.

Brown, Stephen J., and Jerold B. Warner. "Measuring Security Price Performance." *Journal of Financial Economics* 8 (1980): 3–31.

Brown, Stephen J., and Jerold B. Warner. "Using Daily Stock Returns: The Case of Event Studies." *Journal of Financial Economics* 14 (1985): 3–31.

Compustat. *S&P Compustat Database*. New York: McGraw-Hill, 2002.

D'Souza, Juliet, and William L. Megginson. "The Financial and Operating Performance of Newly Privatized Firms During the 1990s." *Journal of Finance* 54 (1999): 1397–1438.

Dewenter, Kathryn, and Paul H. Malatesta. "State-Owned and Privately-Owned Firms: An Empirical Analysis of Profitability, Leverage, and Labour Intensity." *American Economic Review* 91 (2001): 320–334.

Dyck, Alexander. "Privatization and Corporate Governance: Principles, Evidence, and Future Challenges." *World Bank Research Observer* 16 (2001): 59–84.

Eckbo, B. Espen, Vojislav Maksimovic, and Joseph Williams. "Consistent Estimation of Cross-Sectional Models in Event Studies." *Review of Financial Studies* 3 (1990): 343–366.

Fama, Eugene F., Kenneth R. French, Michael C. Jensen, and Richard Roll. "The Adjustment of Stock Prices to New Information." *International Economic Review* 10 (1969): 1–21.

Fuller, Kathleen, Jeff Netter, and Mike Stegemoller. "What Do Returns to Acquiring Firms Tell Us? Evidence from Firms that Make Many Acquisitions." *Journal of Finance* 57 (2002): 1763–1793.

Grover, Varun, and Chang K. Wilcox. "Valuation of Mergers and Acquisitions in the Telecommunications Industry: A Study on Diversification and Firm Size." *Information & Management* 38 (2001): 459–471.

Guislan, Pierre. *The Privatization Challenge. A Strategic, Legal, and Institutional Analysis of International Experience.* World Bank Regional and Sectoral Studies Series. Washington, D.C.: The World Bank (1997).

Harris, Richard. "The Internet as a GPT: Factor Market Implications." In *General Purpose Technologies and Economic Growth*, edited by Elhannan Helpman. Cambridge, Mass.: MIT Press (1998): 145–166.

Heertje, A., and M. Perlman, eds. *Evolving Technology and Market Structure: Studies in Schumpeterian Economics.* Ann Arbor: University of Michigan Press, 1990.

Jensen, Michael, and William Meckling. "Theory of the Firm: Managerial Behavior, Agency Costs and the Ownership Structure." *Journal of Financial Economics* 3 (1976): 305–360.

Jones, Steven L., William L. Megginson, Robert C. Nash, and Jeffry M. Netter, "Share Issue Privatizations as Financial Means to Political and Economic Ends." *Journal of Financial Economics* 53 (1999): 217–253.

Kogut, Bruce. "Direct Investment, Experimentation, and Corporate Governance in Transition Economies." In *Corporate Governance in Central Europe and Russia: Volume 1: Banks, Funds and Foreign Investors,* edited by R. Frydman, C. Gray, and A. Rapaczynski. Budapest, Hungary: Central European University Press, 2001.

Lehr, William, and Lee McKnight. "Show Me the Money. Agents and Contracts in Service Level Agreement Markets," *Info. The Journal of Policy, Regulation and Strategy for Telecommunications* 1 (2002): 24–36; An earlier version was presented as "A Broadband Access Market Framework: Towards Consumer Service Level Agreements." Paper Presentation. Washington, D.C.: 28th Annual TPRC (2000) TPRC's webpage, August 25, 2002 http://tprc.org/ (August 25, 2002).

Lenway, Stefanie A., and Tom P. Murtha. "The State as Strategist in International Business Research." *Journal of International Business Studies* 25 (1994): 513–535.

McKinlay, A. Craig. "Event Studies in Economics and Finance." *Journal of Economic Literature* 35 (1997): 13–39.

McKnight, Lee W., and Jahaingir Boroumand. "Pricing Internet Services: After Flat Rate." *Telecommunications Policy* (Special Issue on Next Generation Internet) 24 (2000): 565–590.

McKnight, Lee W., Paul Vaaler, and Raul Katz, eds. *Creative Destruction: Business Survival Strategies in the Global Internet Economy.* Cambridge, Mass.: MIT Press, 2001, 2002, Toyo Kezai (Japanese translation) 2003.

McKnight, Lee, and William Lehr. "Next Generation Internet Bandwidth Markets." *Communications & Strategies* (Special Issue on Internet: Une nouvelle donné économique) 32 (1998): 91–106.

McWilliams, Abagail, and Donald Siegel. "Event studies in management research: theoretical and empirical issues." *Academy of Management Journal* 40 (1997): 626–657.

Megginson, William L., and Jeffrey M. Netter. "From State to Market: A Survey Of Empirical Studies On Privatization." *Journal of Economic Literature* 39 (2001): 321–389.

Megginson, William L., Robert C. Nash, and Matthias van Randerborgh. "The Financial and Operating Performance of Newly Privatized Firms: An International Empirical Analysis." *Journal of Finance* 49 (1994): 403–452.

Megginson, William L. "Appendix Detailing Share Issue Privatization Offerings, 1961–2000." William Megginson's webpage, <http://faculty-staff.ou.edu/M/William.L.Megginson–1> (August 25, 2002).

Nelson, Richard, and Sydney Winter. *An Evolutionary Theory of Economic Change.* Cambridge, Mass.: Harvard University Press, 1982.

Park, Myeong-Cheol, Dong-Hoon Yang, Changi Nam, and Young-Wook Ha. "Mergers and Acquisitions in the Telecommunications Industry: Myths and Reality." *Etri Journal* 24 (2002): 56–64.

Perotti, Enrico, and Luc Laeven. "Confidence Building in Emerging Stock Markets." Working Paper. Washington, D.C.: World Bank, 2002.

Perotti, Enrico, and Serhat E. Guney. "The Structure of Privatization Plans." *Financial Management* 22 (1993): 84–98.

Perotti, Enrico. "Credible Privatization." *American Economic Review* 85 (1995): 847–859.

Prabhala, Nagpurnanand R. "Conditional Methods in Event Studies and an Equilibrium Justification for Standard Event-Study Procedures." *Review of Financial Studies* 10 (1997): 1–38.

Ramamurti, Ravi. "A Multilevel Model of Privatization in Emerging Economies." *Academy of Management Review* 25 (2000): 525–550.

Ramamurti, Ravi. "The Obsolescing 'Bargaining Model'? MNC-Host Developing Country Relations Revisited." *Journal of International Business Studies* 32 (2001): 23–39.

Ramamurti, Ravi. "Why Are Developing Countries Privatizing?" *Journal of International Business Studies* 23 (1992): 225–249.

Schmidt, Klaus M. "The Costs and Benefits of Privatization: An Incomplete Contracts Approach." *Journal of Law, Economics, and Organization* 12 (1996): 1–24.

Schrage, Burkhard. *Three Essays on Investment and Risk Assessment in Emerging Market Countries*. Unpublished Doctoral Dissertation. The Fletcher School, Tufts University, Boston, Mass., 2002.

Schumpeter, Joseph. *Business Cycles* (2 Vols.). New York: McGraw-Hill, 1939.

Schumpeter, Joseph. *Capitalism, Socialism and Democracy*. London: Allen & Unwin, 1943.

Schumpeter, Joseph. *The Theory of Economic Development*. Cambridge, Mass.: Harvard University Press, 1934.

Securities Data Corporation. *Mergers & Acquisitions Database*. New York: Thomson Financial, 2002.

Shleifer, Andrei, and Robert W. Vishny. "Politicians and Firms." *Quarterly Journal of Economics* 109 (1994): 995–1025.

Standard and Poor's. *Emerging Markets Data Base*. New York: McGraw-Hill, 2002.

StataCorp. *Stata Statistical Software: Release 7.0*. College Station, Tex.: Stata Corporation, 2001.

Tandon, Pankaj. "Welfare Effects of Privatization: Some Evidence from Mexico." *Boston University International Law Journal* 13 (1995): 329–330.

Vaaler, Paul M. "Creating and Destroying Shareholder Value Across Borders." In *Creative Destruction. Business Survival Strategies in the Global Internet Economy*, edited by Lee W. McKnight, Paul M. Vaaler, and Raul L. Katz. Cambridge, Mass.: MIT Press, 2001, 145–166.

Vaaler, Paul M., and Lee W. McKnight. "Creative Destruction in the Internet Economy: The Internet's Impact on Enterprise Valuation." *Communications & Strategies* 40 (2000): 193–210.

Vernon, Raymond. *Sovereignty at Bay: The Multinational Spread of US Enterprises*. New York: Basic Books, 1971.

World Bank. *World Development Indicators*. Washington, D.C.: World Bank, 2002.

14

The Potential Relevance to the United States of the European Union's Newly Adopted Regulatory Framework for Telecommunications

J. Scott Marcus

I. Convergence: A Challenge to the Regulator

Not so long ago, specific services and the associated networks were closely intertwined. Telecommunications networks delivered voice telephony. Broadcast systems delivered radio and television. The introduction of cable television and satellite transmission resulted in only a marginal increase in complexity.

Today, one can no longer say that the service and the network are inextricably intertwined. Voice telephony is delivered over wireline telecommunications, wireless, cable and the Internet. Radio and television programming are delivered over radio, cable and, to a limited but growing degree, the Internet. Indeed, the Internet is fundamental to the challenges of convergence, insofar as it totally decouples the application from the underlying mechanisms of transmission.

Convergence poses vexing problems for the regulator. In the United States, the Communications Act of 1934[1] (the statute governing telecommunications regulation) provides for substantially different treatment for wireline, mobile wireless, and cable-based services. To the extent that the Act fails to account for present technical and market realities, notably including the rapid growth of the Internet, there may be the risk of irrational results, regulatory arbitrage, or distortions in the development of technology or coverage.

Convergence is by no means confined to the United States. It is a global phenomenon. Responses, however, have varied from region to region.

The European Union's telecommunications regulatory framework adopted in March 2002 represents a bold and innovative response to the challenges of convergence.[2] It recognizes that much of telecommunications regulation exists as a means of addressing potential and actual abuses of market power. With that in mind, the E.U. attempts a comprehensive, technology-neutral approach to regulation, which borrows concepts of market definition and of market power from competition law.

This chapter assesses potential strengths and weaknesses of the E.U. approach, and considers its possible relevance to the very different legal and regulatory framework in the United States. It addresses the following questions, among others. First, why is it that the two systems appear to frequently generate similar results? When might the two systems generate different results, and why?

Perhaps most intriguing of all: *Why* do we regulate the things that we regulate? What light does the new E.U. regulatory framework shed on this question?

We consider first the U.S. telecommunications regulatory system, and then that of the European Union. We consider each system in terms of its regulatory framework, its competition law framework, the ability of regulators to obtain the information they need and to protect sensitive third party data, the support for deregulation, and the balance struck between centralization and decentralization. We then evaluate specific outcomes of the U.S. regulatory system, and then pose the question in each case as to whether the new E.U. system could potentially generate similar outcomes. We proceed to review briefly certain implementation challenges to the new E.U. system, and close by considering the potential relevance of the new European framework to regulatory practice in the United States.

II. Convergence and the U.S. Legal and Regulatory Framework

As previously noted, convergence has been widely recognized as representing a regulatory challenge. Particularly vexing issues relate to the regulatory treatment of broadband services over cable and wireline media, and potentially of IP telephony.

For example, a recent report from the National Academy of Sciences noted:

> The Telecommunications Act of 1996, which for the most part assumes the continued existence of a number of distinct services that run over distinct communications technologies and separate infrastructure, does not fully reflect the

convergent nature of broadband (different communications infrastructures are able to deliver a similar set of services using a common platform, the Internet).[3]

In this section, we consider the legal framework for telecommunications regulation in the United States.[4] We then proceed to consider merger and competition law in the United States, in order to gain a comparative sense of how it relates to equivalent practice in Europe.

Legal Framework of Telecommunications Regulation in the United States

Telecommunications in the United States is primarily governed by the Communications Act of 1934,[5] which was substantially amended, most notably by the Telecommunications Act of 1996.[6]

Within the Act, Title I establishes the structure and jurisdiction of the FCC, and also provides definitions used throughout the Act. Title II addresses the regulation of Common Carriers, which represent the traditional world of telephony. Title III concerns wireless services and broadcast Radio and television, while Title VI addresses the regulation of Cable Communications.

Title II contains a wide range of obligations applicable to telecommunications common carriers. These provisions govern, for instance, the prices they may charge for services,[7] obligations to publish those prices in tariffs,[8] limitations on their ability to discriminate,[9] and obligations to interconnect with other carriers and to provide collocation and Unbundled Network Elements.[10] Notably, there is a prohibition against Bell Operating Companies (BOCs) offering interLATA (long distance) services within their historic service areas until they have demonstrated that they have sufficiently opened their local markets to telecommunications competition within the state in question.[11]

These obligations are not applicable to wireless broadcasters or cable operators (except to the extent that they offer telecommunications services over their facilities). Broadcasters and cable operators are, however, subject to a different set of rules, many of which relate to the content that they carry, or to the spectrum over which wireless services operate.[12]

Under the Act, organizations that provide telecommunications services are held to be common carriers and thus subject to Title II regulation. *Telecommunications service* is defined as "the offering of telecommunications for a fee directly to the public, or to such classes of users as to be effectively available directly to the public, regardless of the facilities used."[13] *Telecommunications*, in turn, is defined as "the transmission, between or among points specified by the user, of information of the user's choosing, without change in the form or content of the information as sent and received."[14] The definitional category turns on the *nature of the service that is offered*, not necessarily on the technology over which it is offered.[15]

The Computer Inquiries

The *Computer Inquiries* were a series of FCC regulatory proceedings that addressed the perceived convergence between telecommunications and computing.[16] The Computer Inquiries strongly influenced the Telecommunications Act of 1996; at the same time, certain of the orders remain in effect today.

In *Computer I,* the Commission made two decisions that laid the foundation for its regulatory approach to services provided by computer data processing service providers. First, the Commission concluded that the public interest would not be served by regulating such data processing services, since the provision of such services was deemed to be "essentially competitive."[17] Second, while the Commission determined that the participation of common carriers in the data processing market would benefit consumers, it expressed concern that common carriers might engage in unfair competition. The dangers of unfair competition, the Commission explained, relate "primarily to the alleged ability of common carriers to favor their own data processing activities by discriminatory services, cross-subsidization, improper pricing of common carrier services, and related anticompetitive practices and activities."[18] Accordingly, the Commission concluded that there was a need for competitive safeguards, and it required common carriers seeking to offer data services to do so through a structurally separate affiliate.[19] These safeguards were intended to ensure that carriers would not "give any preferential treatment to their data processing affiliates" and that competing data service providers would therefore have nondiscriminatory access to the underlying communications components used in providing their services.[20]

The Commission continued its examination of these issues in the *Computer II* proceeding, which it initiated in 1976.[21] In *Computer II*, the Commission reaffirmed its basic regulatory approach to the provision of computer data services, but refined its analysis. In particular, the Commission, attempting to define and distinguish regulated telecommunications services and unregulated data services, created the categories of *basic* services and *enhanced* services.[22] The Commission also specified in greater detail the extent of structural separation required between the incumbent telephone provider and its enhanced services affiliate.[23]

In 1986, the Commission further extended this line of reasoning with its *Computer III* decision.[24] *Computer III* offered an alternative set of competitive safeguards to protect competitive providers of enhanced services. Specifically, the Commission gave AT&T and the BOCs that sought to provide enhanced services the option of continuing to comply with *Computer II's* strict separate subsidiary requirements, or alternatively of complying with new "nonstructural safeguards."

Finally, in order to prevent any improper shifting of costs from unregulated to regulated activities, the Commission, in its *Joint Cost* proceeding,[25] adopted new, and more detailed, accounting rules that applied to all incumbent local exchange carriers and to dominant interexchange carriers.[26]

Thus, in the Computer Inquiries, the Commission reaffirmed its commitment to its essential policy of regulating only the common carrier *basic* transmission service, while exempting *enhanced* services (which represented a blending of computation and communications) from common carrier regulation. Enhanced services did not themselves *provide* bottleneck facilities, but they *depended* on bottleneck facilities controlled by the traditional carriers. The FCC therefore concluded that enhanced services *per se* did not need to be regulated as basic (telecommunications) services. The equipment necessary to implement enhanced services was available on the open market. Barriers to entry were potentially low. The FCC wisely chose to let market forces drive the evolution of enhanced services, without regulatory interference.

At the same time, the Commission continued to emphasize the need for competitive safeguards to ensure that common carriers did not use their bottleneck facilities to compete unfairly against unaffiliated enhanced service providers.

The Telecommunications Act of 1996 formalized and codified the distinction between basic services (renamed *telecommunication services*) and enhanced services (renamed *information services*). The Act defines an information service as "the offering of a capability for generating, acquiring, storing, transforming, processing, retrieving, utilizing, or making available information via telecommunications, and includes electronic publishing, but does not include any use of any such capability for the management, control, or operation of a telecommunications system or the management of a telecommunications service."[27]

The Regulatory Framework and the Internet

The *Computer I, II,* and *III* rulings and their embodiment in the Telecommunications Act of 1996 represent the underpinnings of U.S. deregulatory policy toward the Internet. On the one hand, they led to the view that the Internet should be viewed as an enhanced service, and that the Internet consequently should not itself be subject to significant regulation. On the other hand, they sought to ensure that the traditional carriers would not be permitted to withhold or to discriminate in the provision of the building blocks essential to the creation of the Internet.

In 1998, the FCC prepared a report to Congress on the likely impact of the Internet, and of Internet telephony, on contributions to the Universal Service Fund (USF).[28] The USF is a mechanism whereby the price of telecommunications service in areas of low teledensity (e.g., rural areas) is subsidized in order to ensure that it is affordable to all. A number of senators, notably including Senator Stevens of Alaska, were concerned that unregulated Internet services, which were not obliged to contribute to the USF, would ultimately undermine the financial viability of the USF.

The Stevens Report confirmed that Internet access services should continue to be viewed as information services, consistent with longstanding FCC practice. It also analyzed IP telephony at length. In doing so, it established many of the underpinnings of current regulatory practice in the United States as regards converged services in general and the Internet in particular.

It is noteworthy that a telecommunications bill enacted a scant six years ago explicitly references the Internet in only two places—in section 230 (the "Communications Decency Act"), and in referencing the support of advanced services to schools and libraries in section 254(h) of the Act.[29] This dramatically illustrates the pace at which the technology and the marketplace have progressed in the intervening years.

Antitrust Analysis in the United States

In the United States, the relationship between telecommunications regulation and antitrust is complex. The FCC, as the independent regulatory body for communications, has statutory responsibility in a number of instances for determining the permissible portion of a national or local market that a single entity may own. It also has responsibility for restricting certain forms of cross ownership (for instance, between broadcast television and newspaper publishing in the same local market).

In the United States, antitrust concerns sometimes arise as a result of the conduct of a single firm. The American attitude to large corporations has always been somewhat ambivalent—we worry about the power that large corporations wield, and yet at the same time we appreciate the potential benefits associated with the economies of scale and scope that they command. Consequently, it is not held to be a problem for a firm to possess market power; rather, what is problematic is the *abuse* of that market power.

Somewhat different antitrust issues may present themselves when two companies attempt to merge, particularly when the merger would dramatically expand their presence in a relevant market. One of two U.S. agencies will take the lead in investigating any merger—either the Federal Trade Commission (FTC), or the Department of Justice (DOJ).[30] In either case, the relevant agency determines whether the merger would constitute a violation of competition law.[31] In parallel with this evaluation, the FCC assesses the same merger using a very different standard: Does it serve the public interest?[32]

The DOJ/FTC Horizontal Merger Guidelines set forth the methodology that these enforcement agencies will apply in analyzing horizontal mergers (mergers between participants in the same industry).[33] The guidelines attempt to provide a rigorous economic methodology for evaluating the prospective impact of a merger.

Under the Guidelines, one begins by defining relevant markets. A relevant product market is defined as ". . . a product or group of products such that a hypothetical profit maximizing firm that was the only present and future seller of those products likely would impose at least a 'small but significant and nontransitory increase in price.'"[34] In applying this definition, the antitrust authorities employ a "smallest market principle." That is, they begin by identifying a narrow group of products that includes a product or products of the merging firms. They then consider the effect of a "small but significant and nontransitory" increase in price on a hypothetical monopolist that was the sole supplier of that product or products. If the price increase would result in such a large reduction in demand that the price increase would have been unprofitable, then the next best substitute or substitutes would be added to the relevant product group. The agency applies this procedure iteratively until it has identified the narrowest group of products where the price increase would be profitable. This group of products would then constitute the relevant product market.[35]

The agency then proceeds to identify participants in the relevant product market,[36] and to determine the market shares of the market participants (typically based on dollar sales or shipments). A shorthand tool that is often used to assess the impact of a prospective merger is the *Herfindal-Hirschman Index (HHI)*. "The HHI is calculated by summing the squares of the individual market shares of all the participants."[37] In a perfectly monopolized market, the HHI would be 10,000; in a market with a vast number of tiny competitors, it would approach zero. The HHI is thus a measure of *relative concentration*. In a highly concentrated market (HHI greater than 1,800 after a merger), a merger that results in an increase in the HHI of 100 or more is felt *ceteris paribus* to "potentially raise significant competitive concerns."[38]

With this information in hand, the agency proceeds to analyze the likely competitive effects of a proposed merger, considering all relevant factors, including the likelihood of subsequent competitive entry, and any beneficial efficiencies that might flow from the merger.

The DOJ or FTC will coordinate with the FCC insofar as possible (see below) during a merger review; however, there is no assurance that FTC/DOJ market definitions and competitive threats will be directly reflected in FCC regulatory policy.

Investigative Authority and Access to Information

In assessing a merger, one needs a great deal of information. Typically, much of the relevant information is in the hands of the merging parties, not initially in those of the competition authorities.

The Department of Justice is an investigative agency. When it needs information relevant to a merger, it generally issues a *Civil Investigative Demand (CID)*, which has legal force similar to that of a subpoena. Information received pursuant to a CID is maintained in strict confidence, much as would be the case in a criminal prosecution.

The FCC is not an investigative agency, but rather an administrative agency subject to the Administrative Procedures Act (APA).[39] Nonetheless, it has full statutory authority to use compulsory process to obtain information when necessary.[40] Furthermore, the parties to a merger will tend to be motivated to respond in order to gain permission to consummate the transaction.

In general, external documents received in connection with a "permit and disclose" proceeding must be placed in the public record; however, sensitive documents can be made subject to protective order.[41] Under the APA, all participating parties are in general entitled to see any material submitted by any other party to proceeding; consequently, third parties may be reluctant to provide information, especially where there is threat of retaliation from the merging parties.

Deregulation

A number of specific deregulatory initiatives are described later in this chapter. The primary statutory *mechanisms* for deregulation are the FCC's *forbearance authority*, and the *Biennial Review*.

The Telecommunications Act of 1996 directs the FCC to *forbear* (refrain) from applying any provision of the Act where analysis of the relevant market leads the FCC to conclude that associated charges are neither unreasonable nor discriminatory, and where forbearance does not harm the consumers and is generally in the public interest.[42] In doing so, the FCC must specifically consider whether forbearance will promote competitive market conditions.

The FCC is also required to conduct a Biennial Review of all of its regulations issued pursuant to the Act to determine whether any are "no longer necessary in the public interest as the result of meaningful economic competition."[43] The Biennial Review seeks to ensure that any deregulatory opportunities will be examined not less frequently than at two year intervals.

Centralization versus Decentralization

The United States is a federal system. The Federal government has responsibility for interstate communications, while the states have responsibility for activities within their state. In the case of the Internet, the FCC has taken the position that its traffic is interstate, and thus not subject to state or local jurisdiction.

In practice, the relationship is complex. States regulate many aspects of local telephone competition, including local interconnection agreements. Local or municipal governments generally establish franchise arrangements for cable operators. This division of authority is sometimes problematic, but it also is sometimes a source of strength and resiliency for the U.S. regulatory system, enabling support for local preferences, and also providing a more flexible vehicle in some cases for local experimentation with new and innovative regulatory models.

Convergence places special challenges on these complex national/state/municipal interrelationships. First, it impacts the players in somewhat different ways—and their interests are not fully aligned. Second, it slows the speed with which regulation can respond to changes in the marketplace, because regulation must adapt in different layers.

III. The New European Regulatory Framework

The European Union has been playing a progressively larger role in the regulation of telecommunications. In March 2002, the European Union adopted a new regulatory framework that effectively standardizes the regulatory framework for all E.U. member states.

An unusual confluence of factors appears to have motivated the E.U. to take a fresh and daring look at telecommunications regulation. First, E.U. regulations required a comprehensive regulatory review by the end of 1999. Second, the E.U. *per se* was not burdened with as long a history of preexisting regulation as is the United States. Moreover, most E.U. member states have migrated only in the last few years from government ownership of telecommunications, primarily on a monopoly basis, to private ownership and competition. They are, in consequence, acutely aware of the benefits of competitive free market mechanisms. They are technologically sophisticated, and recognize the impact of convergence. They also understand that, in the European context, even where there is consensus for change, it can be time-consuming or challenging to translate that consensus into legislation—therefore, when they make a change, it has to last for quite some time. Finally, there are ongoing tensions within the European Union between a strong internal-market role for the European Commission, the executive arm of the E.U., and freedom for member states to act as they wish. These tensions can be particularly acute when a sector, such as telecommunications, is still in the process of opening to competition for the first time. All of these factors contributed to the willingness of the E.U. to make so substantial a break with the past.

The Europeans recognized that the bulk of all telecommunications regulation deals, in one way or another, with responses to market power. In particular, they associate the possession of *Significant Market Power (SMP)* with obligations that could include transparency,[44] nondiscrimination,[45] accounting separation,[46] access to and use of specific network facilities (including Unbundled Network Elements [UNEs], wholesale obligations, collocation, and interconnection),[47] price controls and cost accounting,[48] making necessary leased lines available,[49] and carrier selection and preselection.[50]

The basic concept of the regulation is simple and straightforward. The European Commission will begin by defining a series of relevant telecommunications markets, and by providing a set of guidelines for determining the presence or absence of market power, all based on methodologies borrowed from competition law and economics. Within each market, the National Regulatory Authority (NRA) in each member state will determine whether one or more parties possess Significant Market Power (SMP). If SMP exists, the NRA will impose appropriate obligations from the set noted in the previous paragraph, taking into account the specifics of the particular marketplace in question.[51] These obligations are imposed *ex ante*, based on the presence of SMP—it is not necessary to demonstrate that market power has been abused. Conversely, if the NRA fails to find SMP, then any such obligations that may already be in place must be rolled back.

In doing so, the E.U. seeks to move completely away from technology-specific and service-specific legislation. This is a significant and dramatic innovation.

We now consider each element of the framework in greater detail.

Market Definition

In the new framework, it is the European Commission, the executive branch of the European Union, that provides a Recommendation on Relevant Product and Service Markets, "in accordance with the principles of competition law."[52] Annex I of the Framework Directive provides an initial list of such markets.

National Regulatory Authorities then take the European Commission's recommendation and define markets within their geographic territories. They are to take "the utmost account" of the recommendation, but the Framework Directive also envisions that NRA definitions might diverge from those of the European Commission in some instances.

The European Commission may also adopt a Decision identifying transnational markets, markets that span all or a substantial portion of the E.U.[53] In these markets, additional procedures are required to ensure that NRAs work in concert with one another.

The process for market definition is described in a document referred to as "the Guidelines."[54] The Guidelines adopt a common framework for National Regulatory Authorities (NRAs) and National Competition Authorities (NCAs), with the recognition that this should ideally lead to equivalent market definitions; however, the Guidelines recognize that the European Commission or national competition authorities may in some instances diverge from market definitions established by European Commission or national regulators for good and valid reasons. They are dealing with somewhat different issues.

European competition law is similar to that of the United States as regards market definition. The economic procedure employed is based on a hypothetical monopolist test, assuming a "small but significant, lasting increase" of 5 percent to 10 percent in price of a product or service.[55] The relevant market then includes all products and services that are readily substitutable for the services in question.[56]

This market definition immediately addresses a number of fundamental convergence issues, and technological neutrality is a direct consequence. As the Guidelines note:

> Although the aspect of the end use of a product or service is closely related to its physical characteristics, different kind of products or services may be used for the same end. For instance, consumers may use dissimilar services such as cable and satellite connections for the same purpose, namely to access the Internet. In such a case, both services (cable and satellite access services) may be included in the same product market. Conversely, paging services and mobile telephony services, which may appear to be capable of offering the same service, that is, dispatching of two-way short messages, may be found to belong to distinct product markets in view of their different perceptions by consumers as regards their functionality and end use.[57]

Significant Market Power (SMP)

Per the Framework Directive, "[a]n undertaking shall be deemed to have significant market power if, either individually or jointly with others, it enjoys a position equivalent to dominance, that is to say a position of economic strength affording it the power to behave to an appreciable extent independently of competitors, customers and ultimately consumers."[58]

The Guidelines distinguish between determining market power *ex post* and *ex ante*. In an *ex ante* world, the only meaningful measure of market power is the ability "of the undertaking concerned to raise prices by restricting output without incurring a significant loss of sales or revenues."[59]

As a proxy for market power, the Guidelines suggest computing market shares, typically based on sales volume or sales value. SMP is normally viewed as being a factor only where the market share exceeds 40 percent. Where the market share exceeds 50 percent, SMP is assumed to be present.[60]

This notion of concentration is roughly equivalent to that of a highly concentrated market, as described in the DOJ/FTC guidelines. A market share of 40–50 percent would imply an HHI of at least 1,600 to 2,500, assuming that all other market participants were extremely small. Note that an HHI of 1,800 or greater implies a highly concentrated market to the DOJ. Thus, the level of concentration at which the U.S. and E.U. would consider a market to be problematic are in the same general range.

The Guidelines also deal with market power in upstream or downstream vertically related markets,[61] and with collective dominance.[62]

Access Requirements

As previously noted, the E.U. framework requires NRAs to impose appropriate remedies *ex ante* from the list of possible options[63] where one or more firms are found to have SMP, but to eliminate restrictions absent SMP:

Where a national regulatory authority concludes that the market is effectively competitive, it shall not impose or maintain any of the specific regulatory obligations referred to in paragraph 2 of this Article. In cases where sector specific regulatory obligations already exist, it shall withdraw such obligations placed on undertakings in that relevant market. An appropriate period of notice shall be given to parties affected by such a withdrawal of obligations.

Where a national regulatory authority determines that a relevant market is not effectively competitive, it shall identify undertakings with significant market power on that market . . . and the national regulatory authority shall on such undertakings impose appropriate specific regulatory obligations referred to in paragraph 2 of this Article or maintain or amend such obligations where they already exist.[64]

Investigative Authority and Access to Information

When the European Commission assesses a merger, it has full authority to issue information requests with subpoena-like legal force, and it also has the obligation to protect confidential information that it receives pursuant to those requests. In these regards, its authority is similar to that of the U.S. DOJ or FTC.

The new framework recognizes the need for regulators to obtain data on which to base market definitions and determination of SMP. It accords NRAs rights and responsibilities equivalent to those of NCAs:

> Member States shall ensure that undertakings providing electronic communications networks and services provide all the information, including financial information, necessary for national regulatory authorities to ensure conformity with the provisions of, or decisions made in accordance with, this Directive and the Specific Directives. These undertakings shall provide such information promptly on request and to the timescales and level of detail required by the national regulatory authority. The information requested by the national regulatory shall be proportionate to the performance of that task. The national regulatory authority shall give the reasons justifying its request for information.[65]

The E.U. regulatory framework also establishes parameters whereby NRAs can exchange the data that they thus obtain with NCAs, the European Commission, and other NRAs, but only to the extent necessary and proportionate to enable implementation of the Framework.[66]

Deregulation

Under the new Framework, regulation and deregulation are handled symmetrically. Where SMP is present, appropriate remedies must be applied. Where SMP is absent, those remedies may not be applied, and if already present must be removed.

No specific timeframe is specified.

Centralization versus Decentralization

If the U.S. is a federal system, the E.U. might be said to be more akin to the U.S. under the Articles of Confederation, particularly in regard to areas such as foreign policy, defense and internal security.[67] The European historical experience has differed from that of the United States, and the European system is in consequence significantly less centralized than that of the U.S. today in many respects.

The tensions of centralization and decentralization that have been fought over in the United States for many decades are arguably even more intense in the European context. In most respects, E.U. member states are sovereign states. They work together in certain ways in order to achieve specific goals, such as uniform competition policy or a single currency.

In establishing a common regulatory framework, it was necessary to delicately balance the prerogatives of NRAs against the needs of the single market, and the prerogatives of the European Commission in maintaining that single market.

The balance that was struck preserves the ability, in general, of NRAs to operate unilaterally, but with notice to the European Commission and to other NRAs. The European Commission retains the ability to require that a market definition or a designation of SMP be withdrawn where it would create a barrier to the single European marketplace, or would be incompatible with the E.U. policy objectives embodied in Article 8 of the Framework Directive.[68]

A particularly knotty case relates to transnational markets, markets that span all or a substantial portion of the E.U. "In the case of transnational markets . . ., the national regulatory authorities concerned shall jointly conduct the market analysis taking the utmost account of the Guidelines and decide on any imposition, maintenance, amendment or withdrawal of regulatory obligations referred to in paragraph 2 of this Article in a concerted fashion."[69]

For there to be tension between centralization and decentralization in the implementation of the new telecommunications regulatory framework in the E.U. is perhaps not surprising—similar tensions have existed in many political systems, and in many eras.[70]

Benefits

There is much to be said for the new E.U. framework. It attempts to address convergence by using fluid market definitions instead of enshrining technology-based definitions within the law. It thus offers the potential of regulating at a velocity that approaches that of the changes in underlying technology and marketplace.The notion of regulating in a completely technology-neutral fashion is promising. If one service is substitutable for another, then it should be subject to roughly the same regulatory constraints, irrespective of the technologies used to deliver the services. This is a very elegant and appealing concept; however, it does not sit well with regulatory practice in the United States, as we shall see.At the same time, the proof of this pudding must lie in its eating—and significant questions remain. We take up this topic later in the chapter.

IV. Comparative Results

It is impossible to say exactly how the new European framework will be applied in practice, either by the European Commission or by the NRAs. It is nonetheless an interesting thought exercise to consider how it *might* be applied, and to compare the results to those of U.S. regulatory practice in a number of specific instances.

It is perhaps not meaningful to ask, "What would the Europeans do?" More meaningful is to ask, "Is this a plausible outcome in the context of the European framework?"

The examples that follow are drawn from well-established precedent, particularly in the area of traditional telecommunications services. We necessarily refrain from commenting on matters currently before the Commission.

Computer Inquiries

We noted earlier that, in the Computer Inquiries, the FCC ruled that enhanced services should not be regulated because they implicated no bottleneck facilities, and did not entail a significant risk of monopolization. This notion was carried forward in the Telecommunications Act of 1996, with its introduction of the concept of information services, and represents a key foundation block for deregulatory U.S. policies toward the Internet.

This result would appear to be entirely consistent with the E.U. regulatory framework. In the absence of SMP, none of the remedies for SMP should be applied.

Competitive Carrier Proceeding

In 1979, the FCC initiated the *Competitive Carrier* proceeding[71] to consider how its regulations should be modified for new firms entering formerly monopoly markets. In a series of orders, the Commission distinguished two kinds of carriers—those with individual market power (dominant carriers) and those without market power (nondominant carriers).[72] The Commission found AT&T's Long Lines Department, which provided interstate long-distance services, to be dominant in the interstate, long-distance market (including the long-distance private line market). It also found AT&T's 23 local telephone companies as well as independent, incumbent local telephone companies to be dominant, because they "possess control of essential facilities."[73] The Commission further found that specialized common carriers and resale carriers, both of which provided interstate, long-distance services in competition with AT&T, to be nondominant.

The Commission determined that nondominant carriers were unable to charge unreasonable rates or to engage in discriminatory practices that would contravene the requirements of the Communications Act, both because they lacked market power and because affected customers always had the option of taking service from an incumbent dominant carrier whose rates, terms, and conditions for interstate services remained subject to close scrutiny by the Commission.[74] Accordingly, the Commission gradually relaxed its regulations of nondominant carriers. Specifically, the Commission eliminated rate regulation for nondominant carriers and presumed that tariffs filed by nondominant carriers were reasonable and lawful. It also streamlined tariff filing requirements, which, *inter alia*, had required dominant carriers to file tariffs with notice periods of up to 120 days, and to submit cost support with their tariffs. For nondominant carriers, in contrast, the Commission required only that tariffs be filed on 14 days notice and did not require any cost support. Finally, the Commission reduced existing Section 214 requirements, which required dominant carriers to file a request for authorization before constructing new lines; under the Commission's streamlined rules, nondominant carriers only had to file a simple, semi-annual report on circuit additions, but did not have to obtain prior authorization.[75]

Again, these regulatory outcomes would appear to be entirely consistent with European thinking. Retail tariff regulations flow from the possession of SMP (which is roughly equivalent to U.S. concepts of market dominance); in the absence of SMP, there should be neither rate regulation nor the obligation to publish retail tariffs.[76]

Streamlining the Regulation of AT&T

As competition developed in the interstate, long-distance market, the Commission initiated two proceedings to determine whether it should streamline its regulation of AT&T, the sole dominant long-distance carrier. In 1990, the Commission initiated the *Interstate Interexchange Competition* proceeding to consider streamlining the regulation of certain AT&T services.[77] After analyzing the level of competition for particular classes of long-distance service, the Commission found that certain services provided by AT&T had become "substantially competitive," and accordingly, it streamlined the regulation of those services.[78] Specifically, for services that it found to be subject to substantial competition, the Commission removed those services from price cap regulation (i.e., eliminated rate regulation), reduced the notice period for tariff filings relating to those services; and eliminated the cost-support requirement for those tariffed services.[79] In addition, the Commission permitted AT&T and other interstate long-distance carriers to offer services pursuant to individually negotiated contracts (i.e., to offer contract tariffs).[80]

Subsequently, AT&T filed a petition to be reclassified as a nondominant carrier in the provision of interstate interexchange services. In 1995, the Commission granted AT&T's motion, based on its finding that "AT&T lacked individual market power in the interstate, domestic, interexchange market."[81] Thus, the Commission freed AT&T from price cap regulation for all of its domestic, interstate, interexchange services, subjected it to the same streamlined tariffing and Section 214 regulations that applied to its nondominant competitors, and eliminated certain accounting and reporting requirements applicable only to dominant carriers.[82] In 1986, the Commission reclassified AT&T as nondominant in the market for international services.[83]

Once again, this seems to be altogether consistent with European thinking. Once SMP has been alleviated, competitive safeguards are no longer necessary and should be eliminated.

Obligations for Interconnection, Resale of Retail Services, Unbundled Network Elements (UNEs), and Collocation

Section 251 of the Act provides for a very modest series of obligations for local exchange carriers in general[84] (including *competitive local exchange carriers [CLECs]*), but an extensive series of additional obligations for incumbent local exchange carriers (ILECs).[85] Notable among these are obligations to provide:

(2) INTERCONNECTION The duty to provide, for the facilities and equipment of any requesting telecommunications carrier, interconnection with the local exchange carrier's network (A) for the transmission and routing of telephone exchange service and exchange access; (B) at any technically feasible point within the carrier's network; (C) that is at least equal in quality to that provided by the local exchange carrier to itself or to any subsidiary, affiliate, or any other party to which the carrier provides interconnection; and (D) on rates, terms, and conditions that are just, reasonable, and nondiscriminatory, . . .

(3) UNBUNDLED ACCESS The duty to provide, to any requesting telecommunications carrier for the provision of a telecommunications service, nondiscriminatory access to network elements on an unbundled basis at any technically feasible point on rates, terms, and conditions that are just, reasonable, and nondiscriminatory . . .

(4) RESALE The duty (A) to offer for resale at wholesale rates any telecommunications service that the carrier provides at retail to subscribers who are not telecommunications carriers; . . .

(6) COLLOCATION The duty to provide, on rates, terms, and conditions that are just, reasonable, and nondiscriminatory, for physical collocation of equipment necessary for interconnection or access to unbundled network elements at the premises of the local exchange carrier, . . .

If we assume *arguendo* that ILECs possess SMP, then this regulatory outcome appears to be precisely analogous to that described in the Access Directive. Article 12, "Obligations of access to, and use of, specific network facilities," enumerates a number of obligations that NRAs may impose upon undertakings that possess SMP, including obligations:

(a) to give third parties access to specified network elements and/or facilities, including unbundled access to the local loop; . . .

(d) to provide specified services on a wholesale basis for resale by third parties; . . .

(f) to provide co-location or other forms of facility sharing, including duct, building, or mast sharing; . . .
(i) to interconnect networks or network facilities.

A significant difference between the two regulatory systems, however, entails the manner in which such constraints might be lifted if market conditions were to change and if effective competition were to emerge.

Under the European framework, the NRA should in theory automatically lift these obligations if market conditions were to change over time in such a way that the undertaking in question no longer possessed SMP.

The equivalent mechanism in the United States would be for the FCC to forbear from imposing portions of section 251(c). As previously noted, the Act provides the FCC with authority to forbear from imposing any regulation or any provision of the Act where the FCC determines that such forbearance is in the public interest, is not necessary to protect consumers, and is not needed to prevent discriminatory, unjust or unreasonable charges or terms and conditions.[86] In determining to forbear, the Act explicitly asks the Commission to weigh the competitive impact of forbearance.

As it happens, however, the Act specifically prohibits the FCC from forbearing from applying requirements under sections 251(c) or 271 until ". . . those requirements have been fully implemented."[87] This might in practice be somewhat circuitous, and perhaps less certain in its execution than the European solution, but the net effect could potentially be precisely analogous to that envisioned in the European framework—once SMP has been eliminated, the remedies to SMP must be rolled back.

Entry of Bell Operating Companies into Long Distance

One of the most significant sections of the Telecommunications Act of 1996 is Section 271. Section 271 prohibits Bell operating companies (BOCs) or their affiliates from offering interLATA (i.e., long distance) services in any in-region state[88] until and unless the BOC in question can demonstrate to the satisfaction of state and federal authorities that it is providing access and interconnection to competitors in that state. Section 271 includes a fourteen point checklist of conditions that the BOC must demonstrably meet in order to be granted authorization to provide interLATA services in that state.

This may not directly fit the European model, but it is consistent in spirit with it. The E.U. framework does not envision a prohibition on a carrier's ability to provide a vertically integrated service as one of the listed regulatory remedies to SMP; indeed, member states may only prevent a carrier from providing networks and services for overriding reasons of public policy, public security or public health.[89] One might view the BOCs as having possessed SMP in 1996 (which is not an unreasonable assumption, considering that they were formed through a consent decree). The notion, then, that a regulatory remedy to SMP should be lifted once effective competition has been established is entirely consistent with the European model.

Rates for Cable Service

Video services are subject to different rules, but many of the underlying principles are the same as those for common carriers. As one conspicuous example, "[i]f the Commission finds that a cable system is subject to effective competition, the rates for the provision of cable service by such system shall not be subject to regulation by the Commission or by a State or franchising authority. . . ."[90] This is entirely consistent with the new EC framework, in that regulatory rate setting is inappropriate in the absence of SMP.

V. Implementation Challenges

The new European regulatory framework appears to be both comprehensive and theoretically elegant. Implementation issues might nonetheless significantly impact its practical effectiveness.

Are there aspects of implementation that are particularly worrisome?

The Role of the European Commission versus that of the NRAs

As we have seen, the Framework represents delicate compromises between granting new powers to the European Commission and preserving the autonomy of the Member State NRAs. On balance, the new framework increases centralization of the European Union insofar as telecommunications regulation is concerned. One might reasonably expect that the new framework will drive an increase in regulatory consistency across the member states,[91] but possibly at some loss in the ability of the system as a whole to reflect diverse local needs or to enable innovative experiments at the Member State level.

This tension between centralization and decentralization would appear to represent a potentially significant "fault line" in the implementation of the new regulatory framework. The ability of European Commission and NRA regulators to apply the system in a sensitive and appropriate manner, and to find workable day-to-day compromises, may play a large role in determining the success of the new framework in practice.

The framework envisions possible differences in judgment among NRAs, and between NRAs and the European Commission, and it includes mechanisms for resolving those differences. It is difficult to predict how well those mechanisms will work in practice. This is an area that bears close watching.

Emerging or Nascent Services

The definition of SMP is, by default, based on market share. In many cases, emerging new services represent a challenge to the power of entrenched incumbents, and thus represent an enhancement to competition.

There is, however, a risk in regard to new services. A provider of a new service might initially—thanks, perhaps, to first mover advantages—possess a large market share of a tiny, emerging market. If this were to be interpreted as SMP, there is a risk that the regulatory apparatus of the state would be brought to bear in a way that impedes competitive entry instead of fostering it.

The Guidelines recognize this, and note that emerging markets "should not be subject to inappropriate ex-ante regulation. This is because premature imposition of ex-ante regulation may unduly influence the competitive conditions taking shape within a new and emerging market. At the same time, foreclosure of such emerging markets by the leading undertaking should be prevented."[92]

In principle, this would appear to represent appropriate guidance. In practice, it may be difficult for NRAs to determine whether the imposition of ex ante regulation is appropriate or not, and it is natural to wonder whether different NRAs will be able to apply this guidance in a consistent way across the E.U.

VI. Relevance to the United States

As we have just seen, in a great many cases the new European regulatory framework might well tend to reach conclusions similar to those which we reach in the United States. Given that the methodologies are radically different, why should the results be so similar?

Biologists speak of *convergent evolution*. Two unrelated species may evolve functionally equivalent organs in order to deal with similar environmental stresses. The human eye is not the same as that of a fruit fly, but they perform the same function.[93]

Analogously, the new E.U. framework and the U.S. regulatory environment tend to address similar issues in similar ways, not necessarily because of equivalent methodologies, but rather because our policy objectives, broadly stated, are similar. We are trying to solve roughly the same problems.

There are, however, important distinctions to be drawn. In the United States, our laws and regulations contain specific regulatory *outcomes*, while the E.U. framework defines a *process* for reaching similar results.

If both methodologies potentially lead to roughly equivalent regulatory outcomes, is there reason to prefer one methodology to the other?

The E.U. framework is extremely logical, and has as we have seen the potential to generate good results. In addition, it has certain advantages in comparison with the U.S. methodology:

- In many instances, the notion of SMP more accurately expresses the need for regulation than does the U.S. equivalent regulatory category.
- The notion that certain regulatory impositions should be imposed in the presence of SMP, and lifted in its absence, may express regulatory desiderata and the desired *timing* of regulation and deregulation more clearly and more simply than do equivalent U.S. statutes.
- In leaving the determination of SMP, and of suitable remedies, to regulation rather than to statute, the European system may be able to respond to change more nimbly than that in the United States.
- The European system arguably deals with technology convergence, which blurs regulatory categories, far more effectively than that of the United States.

Thus, there would seem to be much to recommend the European framework. Unfortunately, the European approach does not fit neatly into U.S. regulatory practice. It is important to bear in mind that the Europeans were able to initiate this monumental overhaul of their system because they had far less relevant regulatory history to contend with than do we in the United States. They were thus able, with the benefit of experience, to revisit and rewrite their regulation anew.

Our law and our history do not lend themselves to direct application of the E.U. framework. The law, as we have seen, is based on regulatory categories that imperfectly correspond to market power. More significantly, the law embodies a complex history that reflects innumerable social compacts. The Communications Act of 1934 was itself an agglomeration of earlier practice. Title III, dealing with radio, was added after the fact. The FCC subsequently established regulations for cable television, which subsequently led to the Cable Television Act of 1992 and then to Title VI of the Act.

In the U.S. system, the balances between regulation and deregulation, and between federal, state and local authority all entailed delicate compromises. The European framework is elegant in its simplicity and directness, but it does not capture those nuances.

There would also be certain practical difficulties in any direct application of the European framework in this country. The E.U. framework depends, as we have seen, on acquisition of sufficient data to enable NRAs to unambiguously determine relevant markets and the possession of SMP. In the United States, however, the FCC is the national regulatory authority. The FCC lacks the authority to get the information that it would need and may also lack the ability to protect that information from public disclosure.

Additional challenges exist. Europeans may tend to trust governments more than they trust corporations. In the United States, it is largely the reverse. It is not clear that Americans would be willing to give regulators such broad authority.

The E.U. telecommunications regulatory framework nonetheless provides a convenient and natural way to think about the public policy implications of many of the choices that confront the FCC. As we have seen, the E.U. framework often provides a very simple and direct way of visualizing regulatory outcomes. It could be a very useful exercise for the FCC to use the European methodology as a means of visualizing and understanding the public policy implications of the most challenging regulatory decisions that we confront.

Acknowledgments

The author is deeply indebted to Donald K. Stockdale, Jr. of the Office of Plans and Policy (OPP) of the FCC, and has borrowed extensively from his work. Donald Abelson (FCC International Bureau [IB]) provided invaluable guidance and support. Matthew Conway (recently of the UK Department of Trade and Industry [DTI]) provided an extensive and extraordinarily helpful review. The author is also grateful for the helpful comments and suggestions of many other colleagues on both sides of the Atlantic, including Patricia Cooper (FCC IB), Sherille Ismail (FCC OPP), Peter Scott (European Commission), Paul Verhoef (European Commission), Tracey Weisler (FCC IB), and Irene Wu (FCC IB).

This document is available on the FCC's World Wide Web site at http://www.fcc.gov/opp.

Notes

1. 48 Stat. 1064 (1934) (codified as amended at 47 U.S.C. §§151 *et seq.*), hereinafter *the Act*.

2. Indeed, the framework is in large part a response to convergence challenges raised in the "Green Paper" of 1997. "Regulation needs to be transparent, clear and proportional and distinguish between transport (transmission of signals) and content. This implies a more horizontal approach to regulation with a homogenous treatment of all transport network infrastructure and associated services, irrespective of the nature of the services carried." See http://europa.eu.int/ISPO/convergencegp/ip164en.html

3. *Broadband: Bringing Home the Bits,* Computer Science and Telecommunications Board, National Academy of Sciences, January, 2002, pg. 32.

4. In this section, we deal with telecommunications regulation in its present form. For a treatment of the history of telecommunications regulation in this country, as it relates to competition and deregulation, see Donald K. Stockdale, "The Regulation, Deregulation and Nonregulation of Telecommunications and the Internet in the United States" (unpublished manuscript, 2001). Portions of what follows appeared in that paper in a different form.

5. 48 Stat. 1064 (1934) (codified as amended at 47 U.S.C. §§151 *et seq.*), hereinafter *the Act.*

6. Telecommunications Act of 1996, Pub. L. no. 104–104, 110 Stat. 56 (1996 Act), codified at 47 U.S.C. §§ 151 *et seq.* (Hereinafter, all citations to the 1996 Act will be to the 1996 Act as it is codified in the United States Code.)

7. 47 U.S.C. §201.

8. 47 U.S.C. §203.

9. 47 U.S.C. §202.

10. 47 U.S.C. §251.

11. 47 U.S.C. § 271.

12. 47 U.S.C. §§ 301–396 and 601–653. Particularly noteworthy are "must carry" rules for cable (§ 612). Sherille Ismail suggests that differences in "must carry" regulatory treatment of cable compared to that of broadcast or DBS satellite may result, at least in part, from differences among these three in their degree of monopsony market power in the programming market. ("Achieving Regulatory Parity in Communications Policy," forthcoming)

13. 47 U.S.C. § 3.

14. Ibid.

15. See also the *Universal Service Report,* often referred to as the *Stevens Report,* at 59: "This functional approach is consistent with Congress's direction that the classification of a provider should not depend on the type of facilities used. A telecommunications service is a telecommunications service regardless of whether it is provided using wireline, wireless, cable, satellite, or some other infrastructure. Its classification depends rather on the nature of the service being offered to customers. . ."

16. *In the Matter of Regulatory and Policy Problems Presented by the Interdependence of Computer and Communications Services and Facilities, (hereinafter Computer I Inquiry),* 7 FCC 2d 11 (1966). See generally Huber, P., Kellogg, M., & Thorne, J. (1999), at 1086–1103; and Oxman, J. (1999).

17. The Commission specifically found "that there is ample evidence that data processing services of all kinds are becoming available . . . and that there are no natural or economic barriers to free entry into the market for these services." *Computer I,* Tentative Decision, 28 FCC 2d 291, at para. 20 (1970)

18. *Computer I,* Final Decision and Order, 28 FCC 2d 267, at para. 12 (1971).

19. *Id.,* at paras. 12 *et seq.*

20. *Id.,* at para. 21.

21. *In the Matter of Amendment of Section 64.702 of the Commission's Rules and Regulations (Second Computer Inquiry),* (hereinafter *Computer II*), Notice of Inquiry and Proposed Rulemaking, 61 FCC 2d 103 (1976).

22. The Commission defined the term "basic" service, which referred to traditional common carrier telecommunications offerings as "the offering of transmission capacity for the movement of information." *Computer II,* Final Decision, (*Computer II Final Decision*), 77 FCC 2d 584, at para. 93 (1980). The Commission defined "enhanced services" as:

. . . services, offered over common carrier transmission facilities used in interstate communications, which employ computer processing applications that act on the format, content, code, protocol, or similar aspects of the subscriber's transmitted information; provide the subscriber additional, different or restructured information; or involve subscriber interaction with stored information. (46 C.F.R. § 64.702(a)).

23. *Computer II Final Decision,* 77 FCC 2d at paras. 190–266.

24. *In the Matter of Amendment of Section 64.702 of the Commission's Rules and Regulations,* Report and Order, (*Computer III*), 104 FCC 2d 958 (1986), *vacated California v. FCC,* 905 F.2d 1217 (9th Cir. 1990).

25. *Separation of Costs of Regulated Telephone Service from Costs of Non-regulated Activities,* Report and Order, (*Joint Cost Order*), 2 FCC Rcd 1298 (1987), *on recons.,* 2 FCC Rcd 6283 (1984) (*Joint Cost Reconsideration Order*); *on further recons.,* 3 FCC Rcd 6701 (1988) (*Joint Cost Further Reconsideration Order*).

26. In *Computer III,* the Commission also imposed new rules governing disclosure of network changes and the handling of customer proprietary network information. *Computer III Order,* 104 FCC 2d, at paras. 241–65.

27. 47 U.S.C. §3(20).

28. *Federal-State Joint Board on Universal Service,* CC Docket no. 96–45, Report to Congress ("*Stevens Report*"), 13 FCC Rcd 11501, 11516–17 (1998), FCC document 98067.pdf.

29. This brief reference appears in 47 U.S.C. § 271. In addition, U.S.C. § 254 refers to "advanced services," while section 706 of the 1996 Act refers to broadband as "advanced telecommunications capability"—arguably, there are many *implicit* references to the Internet.

30. In recent years, for instance, the Department of Justice analyzed the WorldCom/MCI merger and the attempted WorldCom/Sprint merger, while the Federal Trade Commission analyzed the AOL/Time Warner merger. Note that the FTC has no jurisdiction over common carriers.

31. The competition law provisions applicable to mergers are contained in section 1 of the Sherman Act and section 7 of the Clayton Act.

32. 47 U.S.C. sections 214, 310 and 314.

33. U.S. Department of Justice and Federal Trade Commission, *Horizontal Merger Guidelines,* 57 Fed. Reg. 41557 (April 2, 1992, as revised April 8, 1997), http://www.ftc.gov/bc/docs/horizmer.htm.

34. DOJ/FTC Merger Guidelines §1.1, "Market Definition.".

35. Ibid.

36. Ibid., §1.3, "Identification of Firms that Participate in the Relevant Market." The Guidelines necessarily consider the possibility of supply response.

37. Ibid. §1.5, "Concentration and Market Shares."

38. Ibid.

39. 5 U.S.C. §551 *et seq.*

40. 47 U.S.C. 409(e): ". . . the Commission shall have the power to require by subpena (sic) the attendance and testimony of witnesses and the production of all books, papers, schedules of charges, contracts, agreements, and documents relating to any matter under investigation."

41. A more complex question relates to requests for sensitive information made pursuant to the *Freedom of Information Act (FOIA).*

42. 47 U.S.C. § 160.

43. 47 U.S.C. § 161.

44. *Directive 2002/19/EC of the European Parliament and of the Council of 7 March 2002 on access to, and interconnection of, electronic communications networks and associated facilities (Access Directive),* Official Journal of the European Communities, L 108, April 24, 2002, Article 9.

45. Ibid., article 10.

46. Ibid., article 11.

47. Ibid., article 12.

48. Ibid., article 13.

49. *Directive 2002/22/EC of the European Parliament and of the Council of 7 March 2002 on universal service and users' rights relating to electronic communications networks and services (Universal Service Directive),* Official Journal of the European Communities, L 108, April 24, 2002, Article 18.

50. Ibid., article 19.

51. There is no automatic presumption that any obligation will be appropriate. If a competition authority is about to act, for example, regulatory action may well be inappropriate.

52. *Directive 2002/20.EC of the European Parliament and of the Council of 7 March 2002 on a common regulatory framework for electronic communications networks and services (Framework Directive),* Official Journal of the European Communities, L 108, April 24, 2002, Article 15.

53. A preliminary draft Recommendation exists: "On relevant Product and Service Markets within the electronic communications sector susceptible to ex ante regulation in accordance with Directive 2002/.../EC of the European Parliament and of the Council on a common regulatory framework for electronic communication networks and services."

54. *Commission Working Document on Proposed New Regulatory Framework for Electronic Communications Networks and Services, Draft Guidelines on market analysis and the calculation of significant market power,* Brussels, March, 3, 2001.

55. Guidelines, at 31.

56. Ibid., at 35. "According to settled case-law, the relevant product/service market comprises all those products or services that are sufficiently interchangeable or substitutable, not only in terms of their objective characteristics, by virtue of which they are particularly suitable for satisfying the constant needs of consumers, but also in terms of the conditions of competition and/or the structure of supply and demand on the market in question. Products or services which are only to a small, or relative degree interchangeable with each other do not form part of the same market."

57. Ibid., at 36.

58. Framework Directive, Article 14, at 2.

59. Guidelines, at 65.

60. Ibid., at 67.

61. Ibid., at 74–76.

62. Ibid., at 77–79. The concept of collective dominance has become well established in European case law. By contrast, collective dominance is rarely raised as a concern in the U.S. unless there is actual evidence of collusion.

63. Framework Directive, Article 16, at 2.

64. Ibid., at 3–4.

65. Framework Directive, Article 5, at 1.

66. Ibid., at 2.

67. Cf. Guido Tabellini, "The Assignment of Tasks in an Evolving European Union," Centre for European Policy Studies, January 2002, pages 4–6.

68. Framework Directive, Article 7.

69. Framework Directive, Article 16, at 5.

70. Indeed, this is a classic problem in social sciences. Tabellini, op. cit., applies established theory to the EU environment, noting trade-offs between the ability to cope with heterogeneity of local preferences and to exploit local information, versus the impact of "spill-over effects" on specific public goods. He notes the need to "avoid excessive centralisation." He also draws a key distinction between the "bureaucratic accountability" that arguably characterizes Europe today, versus "democratic accountability."

71. *Policy and Rules Concerning Rates for Competitive Common Carrier Services and Facilities Authorizations Therefore,* CC Docket no. 79–252, Notice of Inquiry and Proposed Rulemaking, 77 FCC 2d 308 (1979); First Report and Order, 85 FCC 2d 1 (1980) (*First Report and Order*); Further Notice of Proposed Rulemaking, 84 FCC 2d 445 (1981) (*Further Notice of Proposed Rulemaking*); Second Further Notice of Proposed Rulemaking, FCC 82–187, 47 Fed. Reg. 17,308 (1982); Second Report and Order, 91 FCC 2d 59 (1982) (*Second Report and Order*); Order on Reconsideration, 93 FCC 2d 54 (1983); Third Report and Order, 48 Fed. Reg. 46,791 (1983) (*Third Report and Order*); Fourth Report and Order, 95 FCC 2d 554 (1983) (*Fourth Report and Order*); *vacated, AT&T v. FCC,* 978 F.2d 727 (D.C. Cir. 1992), *cert. denied, MCI Telecommunications Corp. v. AT&T,* 113 S.Ct. 3020 (1993); Fifth Report and Order, 98 FCC 2d 1191 (1984) (*Fifth Report and Order*); Sixth Report and Order, 99 FCC 2d 1020 (1985), *vacated MCI Telecommunications Corp. v. FCC,* 765 F.2d 1186 (D.C. Cir. 1985) (collectively referred to as the *Competitive Carrier* proceeding).

72. The Commission defined market power as "the ability to raise prices by restricting output" and as "the ability to raise and maintain prices above the competitive level without driving away so many customers as to make the increase unprofitable." See *Competitive Carrier Fourth Report and Order,* 95 FCC 2d at 558, para. 7.

73. *Competitive Carrier First Report and Order,* 85 FCC 2d at 22–24. The Commission specifically noted that it would "treat control of bottleneck facilities as prima facie evidence of market power requiring detailed regulatory scrutiny. 1at 21. The Commission also found Western Union, domestic satellite carriers, and miscellaneous common carriers that relay video signals to be dominant in various

relevant markets. *Id.*, at 24–28. It acknowledged, however, that market developments were likely to erode the market power of these carriers over time.

74. *Id.*, at 31.

75. *Competitive Carrier First Report and Order*, 85 FCC 2d at 31–37, 39–44. Subsequently, the Commission announced a policy of permissive "forbearance," under which it would forbear from applying the tariff filing requirements of Section 203 and the entry, exit, and construction authorization requirements of Section 214 to nondominant carriers. See *Competitive Carrier Second Report and Order*, 91 FCC 2d at 73; *Competitive Carrier Fourth Report and Order*, 95 FCC 2d at 557; *Competitive Carrier Fifth Report and Order*, 98 FCC 2d at 1193, 1209. In 1985, the Commission decided to shift from "permissive" to "mandatory" forbearance, thus requiring detariffing by all nondominant carriers. *Competitive Carrier, Sixth Report and Order*, 99 FCC 2d at 1030–32. Federal Court of Appeals reversed this finding, holding that the Commission lacked statutory authority to prohibit the filing of tariffs, and in a subsequent appeal, the court further found that the Commission lacked the authority to allow permissive detariffing. See *MCI v. FCC*, 765 F.2d 1186 (D.C. Cir. 1985); *AT&T v. FCC*, 1993 WL 260778 (D.C. Cir. 1993), *aff'd MCI v. AT&T*, 512 U.S. 218 (1994).

76. Universal Service Directive, article 17.

77. *Competition in the Interstate Interexchange Marketplace,* Notice of Proposed Rulemaking, 5 FCC Rcd 2627 (1990); Report and Order, 6 FCC Rcd 5880 (1991) (*First Interstate Interexchange Competition Order*); Memorandum Opinion and Order, 6 FCC Rcd 7569 (1991); Memorandum Opinion and Order, 7 FCC Rcd 2677 (1992); Memorandum Opinion and Order on Reconsideration, 8 FCC Rcd 2659 (1993); Second Report and Order, 8 FCC Rcd 3668 (1993) (*Second Interstate Interexchange Competition Order*); Memorandum Opinion and Order, 8 FCC Rcd 5046 (1993); Memorandum Opinion and Order on Reconsideration, 10 FCC Rcd 4562 (1995) (collectively referred to as the *Interstate Interexchange Competition* proceeding).

78. In the *First Interstate Interexchange Competition Order,* the Commission found that services provided to large- and medium-size business customers had become "substantially competitive, while in the *Second Interstate Interexchange Competition Order,* the Commission found that, with the introduction of 800 number portability, the market for 800 services (except for 800 directory assistance where AT&T had a monopoly) had become substantially competitive. See *First Interstate Interexchange Competition Order,* 6 FCC Rcd at 5911, para. 188; *Second Interstate Interexchange Competition Order,* 8 FCC Rcd at 3668, para. 1.

79. See *First Interstate Interexchange Competition Order,* 6 FCC Rcd at 5894, para. 74.

80. *Id.,* at 5897, at para. 91.

81. Motion of AT&T Corp. to Be Reclassified as a Nondominant Carrier, Order, 11 FCC Rcd 3271, para. 1, 3356, para. 164 (1995).

82. *Id.,* at 3281, para. 12.

83. *Motion of AT&T to Be Declared Non-dominant for International Service,*

11 FCC Rcd 17,963 (1996).

84. 47 U.S.C. 251(a) and 251(b).

85. 47 U.S.C. 251(c).

86. 47 U.S.C. § 160.

87. 47 U.S.C. § 160(d).

88. An in-region state is any of the states allocated to that Bell operating company under AT&T Consent Decree of August 24, 1982. 47 U.S.C. 271(i)(1).

89. Authorisation Directive, Article 3, at 1.

90. 47 U.S.C. § 543(a)(2).

91. Indeed, this is an explicit objective for the NRAs. Framework Directive, Article 8, at 3, especially (d).

92. "Draft Guidelines on market analysis and the assessment of significant market power under the Community regulatory framework for electronic communications networks and services," Brussels, 21.02.2002, para. 32.

93. See, for instance, Richard Dawkins, *Climbing Mount Improbable,* W. W. Norton & Company, New York, 1996, pages 19–22.

References

Telecommunications Act of 1996, Pub. L. No. 104–104, 110 Stat. 56 (1996 Act), codified at 47 U.S.C. §§151 et seq. The 1996 Act amended the Communications Act of 1934 (the Act).

U.S. Department of Justice and Federal Trade Commission, *Horizontal Merger Guidelines*, 57 Fed. Reg. 41557 (April 2, 1992, as revised April 8, 1997), http://www.ftc.gov/bc/docs/horizmer.htm.

Directive 2002/20.EC of the European Parliament and of the Council of 7 March 2002 on a common regulatory framework for electronic communications networks and services (Framework Directive), Official Journal of the European Communities, L 108, April 24, 2002.

Directive 2002/19/EC of the European Parliament and of the Council of 7 March 2002 on access to, and interconnection of, electronic communications networks and associated facilities (Access Directive), Official Journal of the European Communities, L 108, April 24, 2002.

Directive 2002/22/EC of the European Parliament and of the Council of 7 March 2002 on universal service and users' rights relating to electronic communications networks and services (Universal Service Directive), Official Journal of the European Communities, L 108, April 24, 2002.

Commission Working Document on Proposed New Regulatory Framework for Electronic Communications Networks and Services, Draft Guidelines on market analysis and the calculation of significant market power, Brussels, March, 3 2001.

Federal-State Joint Board on Universal Service, CC Docket No. 96–45, *Report to Congress* ("Universal Service Report"), 13 FCC Rcd 11501, 11516–17 33 (1998).

Computer I, Final Decision and Order, 28 FCC 2d 267.

In the Matter of Amendment of Section 64.702 of the Commission's Rules and Regulations (Second Computer Inquiry), Notice of Inquiry and Proposed Rulemaking, 61 FCC 2d 103 (1976).

In the Matter of Amendment of Section 64.702 of the Commission's Rules and Regulations, Report and Order, (*Computer III*), 104 FCC 2d 958 (1986), *vacated California v. FCC*, 905 F.2d 1217 (9th Cir. 1990).

Broadband: Bringing Home the Bits, Computer Science and Telecommunications Board, National Academy of Sciences, January 2002.

Index